DEVELOPMENTS IN SWEETENERS—3

CONTENTS OF VOLUMES 1 AND 2

Volume 1

Edited by C. A. M. HOUGH, K. J. PARKER and A. J. VLITOS

Volume 2

Edited by T. H. GRENBY, K. J. PARKER and M. G. LINDLEY

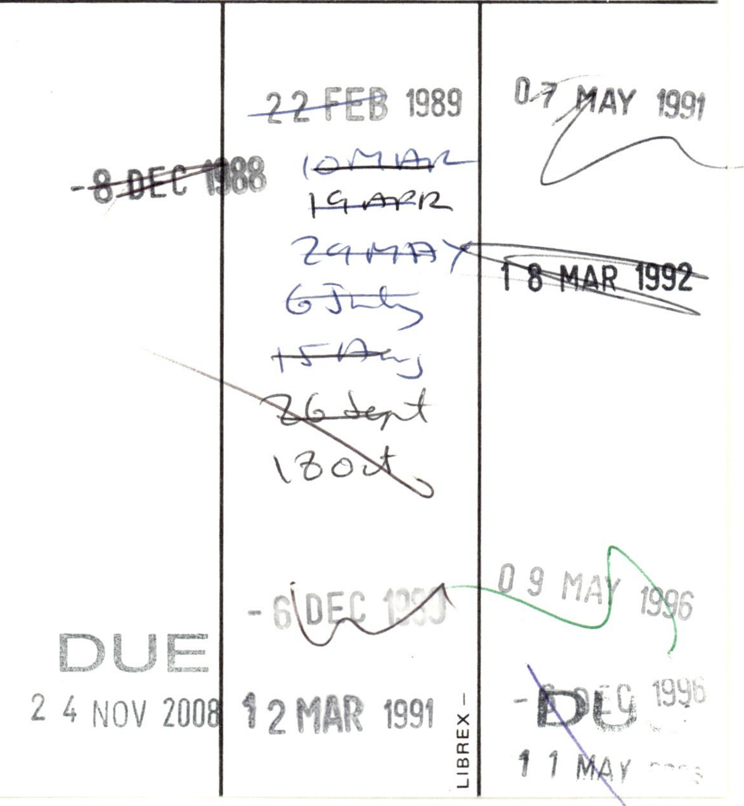

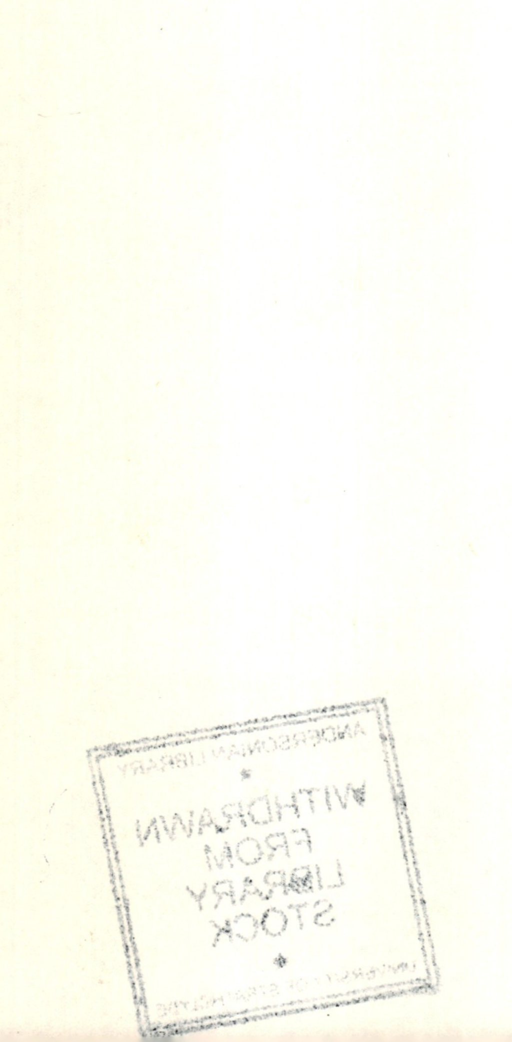
ANDERSONIAN LIBRARY
WITHDRAWN
FROM
LIBRARY
STOCK
UNIVERSITY OF STRATHCLYDE

DEVELOPMENTS IN SWEETENERS—3

Edited by

T. H. GRENBY

B.Sc., Ph.D., C.Chem, F.R.S.C.

Department of Oral Medicine and Pathology,
United Medical and Dental Schools,
Guy's Hospital, London, UK

ELSEVIER APPLIED SCIENCE
LONDON and NEW YORK

ELSEVIER APPLIED SCIENCE PUBLISHERS LTD
Crown House, Linton Road, Barking, Essex IG11 8JU, England

Sole Distributor in the USA and Canada
ELSEVIER SCIENCE PUBLISHING CO., INC.
52 Vanderbilt Avenue, New York, NY 10017, USA

WITH 54 TABLES AND 38 ILLUSTRATIONS

British Library Cataloguing in Publication Data

Developments in sweeteners.
3
1. Nonnutritive sweeteners
I. Grenby, T. H.
664'.5 TP422

The Library of Congress has cataloged this serial publication as follows:

Developments in sweeteners.—1- —London: Applied Science
Publishers, c1979–
v.: ill.; 23 cm.—(Developments series)

1. Sweeteners—Collected works. 2. Nonnutritive sweeteners—Collected works. I. Series.
TP421.D49 664.5—dc19 84-643161

ISBN 1-85166-104-2

Printed in Great Britain by Galliard (Printers) Ltd, Great Yarmouth

PREFACE

The first volume in this series on developments in sweeteners appeared in 1979 and the second in 1983. A comparison of their contents with this, the third volume, shows how the subject has expanded within just a few short years. Both the number of sweeteners available for use and the variety of food and drink products formulated with them have increased, and this sector of the market is still regarded as a major growth area.

It is the intention in this volume to provide an insight into the field as it is today, with a total of ten chapters covering some of the recent advances in intense sweeteners, bulk sweeteners, the body's metabolism and methods of dealing with these materials, their dental benefits from the standpoint of microbial and laboratory animal research, their use in speciality foods, drinks and table-top sweeteners, regulatory and ethical considerations, and finally a discussion of their usefulness in weight control.

Over the span of the last eight years since the publication of the first volume there have been a number of changes in emphasis in the ways of use and promotion of the sweeteners. For example, the 'multiple sweetener concept' has demonstrated the benefit of satisfying our requirements for sweetness by making use of a range of approved sweeteners rather than by excessive reliance on just one or two of them, so that the intake of each individual sweetener is kept low. This is one reason why there is a need for a variety of safe sweeteners with diverse properties suitable for specific purposes in different foods and drinks.

Another idea receiving attention in the media at present is that our food supply has become unnaturally 'artificial', and that 'synthetic' additives in food should be cut down. Some of the currently-available sweeteners fall into this category, as a result of which it seems that increasing attention is now being paid to the isolation and evaluation of natural sweeteners of plant origin, for example the stevia and rebaudiosides discussed in the first two chapters of this book.

Among the reasons put forward for the expanding need for non-sugar sweeteners are the health benefits they confer in weight control, acceptability for diabetics and improved dental health. Both their usefulness in weight control and some aspects of their dental advantages over sugar are discussed in the following pages. However, another function of the sweeteners, often overlooked, is the role they can play in food production and technology, and this is dealt with in the chapters on the properties and applications of the individual sweeteners, particularly in the contribution on sweeteners in the manufacture of special foods.

The continuing high level of interest in sweeteners is manifested not only by the amount of research in progress, but also by the effort being put into the evaluation and applications for regulatory approval of a variety of new materials, including peptides, following the success of aspartame, and chlorosugars. An exciting prospect is the preparation of sweet substances with a 'tailor-made' structure, now that the molecular configurations that confer a sweet taste on a compound are beginning to be known. These 'designer' sweeteners are moving us a long way from the first-generation materials that were discovered or synthesised and found to have a sweet taste just by chance. Other new developments, still in their infancy, are 'protein engineering' and the insertion of a gene for the formation of a protein sweetener into micro-organisms which can be cultured in the laboratory to yield a new breed of sweeteners. With a wider choice of sweeteners approved for use comes the prospect of targeting specific applications in foods and drinks with defined purposes and health benefits, aimed at the requirements of certain sectors of the population.

My thanks are due to all the authors for their efforts in preparing the individual chapters. I hope that this book will stimulate the interest both of workers already in the field, and also of those in other industries and disciplines who want to gain an insight into the present-day status of the expanding subject of sweeteners.

T. H. GRENBY

CONTENTS

LIST OF CONTRIBUTORS

J. Armstrong

The Boots Company PLC, Merchandise Technical Services, Nottingham NG2 3AA, UK

D. A. Booth

Customer Health Behaviour Research Programme, Department of Psychology, University of Birmingham, PO Box 363, Birmingham B15 2TT, UK

B. Crammer

Natural Products Laboratory, Department of Organic Chemistry, Hebrew University of Jerusalem, Jerusalem, Israel

C. H. den Uyl

CCA Biochem BV, PO Box 21, 4200 AA Gorinchem, The Netherlands

Ivan Fabry

Zentralfachschule der Deutschen Süsswarenwirtschaft, De-Leuw-Strasse 3–9, D-5650 Solingen-Gräfrath, Federal Republic of Germany

R. Havenaar

Department of General and Industrial Microbiology, CIVO/TNO Institutes, PO Box 360, 3700 AJ Zeist, The Netherlands

J. HOWELLS

The Boots Company PLC, Merchandise Technical Services, Nottingham NG2 3AA, UK

R. IKAN

Natural Products Laboratory, Department of Organic Chemistry, Hebrew University of Jerusalem, Jerusalem, Israel

K. G. JACKSON

The Boots Company PLC, Merchandise Technical Services, Nottingham NG2 3AA, UK

HARALD A. B. LINKE

Department of Microbiology, New York University Dental Center, 421 First Avenue, New York, NY 10010, USA

DONALD A. M. MACKAY

President, Applied Microbiology Inc., Building 5, Brooklyn Navy Yard, Brooklyn, New York 11205, USA

K. C. PHILLIPS

Keith Phillips & Associates, Highfields, Stockcross, Newbury, Berkshire RG16 8LL, UK

GÜNTHER SIEBERT

Division of Experimental Dentistry, University of Würzburg Medical School, Pleicherwall 2, D-8700 Würzburg, Federal Republic of Germany

SUSANNE C. ZIESENITZ

Division of Experimental Dentistry, University of Würzburg Medical School, Pleicherwall 2, D-8700 Würzburg, Federal Republic of Germany

Chapter 1

STEVIA: STEPS IN DEVELOPING A NEW SWEETENER

K. C. PHILLIPS

Keith Phillips & Associates, Newbury, Berkshire, UK

SUMMARY

Stevia is composed of several natural, heat-stable ent-*kaurene glycosides (steviol glycosides) whose intensities of sweetness and flavour profiles differ from each other and vary according to concentration and environment. Collectively they give stevia 100 to 300 times the sweetness of sucrose. Extracted from a variety of chrysanthemum,* Stevia rebaudiana, *the sweet principle is believed to have been used by the Paraguayan Indians for centuries. Since the 1950s the Japanese have developed its production and have overcome problems of refining to eliminate undesirable flavours. Following a slow start in the 1970s, stevia is now used in a wide range of food applications in Japan. Currently it is being evaluated for approval in the West. Experience of its use in man and data from animal feeding trials indicate that it is safe for human consumption. Results from studies* in vivo *are described here and compared with* in vitro *data.*

1. INTRODUCTION

Although human consumption of stevia began before the Spanish settlement of the country we now know as Paraguay, improved versions have been developed only recently. This chapter is an attempt to put into historical perspective the development of stevia from Paraguay where it

FIG. 1. *Stevia rebaudiana* Bertoni (Compositae).

was first discovered, through Japan where it was first refined and developed as a commercial product and into North America and Europe where it is now being evaluated.

2. ORIGIN IN PARAGUAY AND TRANSFER TO JAPAN

North-east Paraguay, in the region known as the Cordillera of Amambay, adjacent to the border with Brazil, is reported to be the original natural habitat of the plant *Stevia rebaudiana* Bertoni (Compositae)[1,2] (see Fig. 1). It is now cultivated extensively in the Far East (in particular Japan, Singapore, Taiwan, Malaysia, South Korea and China), as well as in parts of South America (including Brazil and Paraguay),[3] Israel[4] and California,[5] for the sweet diterpene glycosides which occur mainly in its leaves and which are known collectively as either *ent*-kaurene glycosides, diterpene glycosides, or more popularly as 'stevioside' or 'stevia' (see Fig. 2).

Known originally by the Paraguayan Guarani Indians as 'Caá-Hê-é' or 'Kaá-éhê' (meaning 'sweet herb'), it was first described in the scientific literature as *Eupatorium rebaudianum* by Moise Santiago Bertoni, Director of the College of Agriculture in Asuncion in 1899.[6] Following receipt of Bertoni's description together with a sample of the leaf from Cecil Gosling, the British Consul at Asuncion, Kew Gardens[1] concluded that it belonged to the genus *Stevia* rather than *Eupatorium*. Subsequently, in 1905, Bertoni redefined it as *Stevia rebaudiana*, a member of the Compositae.[7] The genus *Stevia* is named after Peter James Esteve, a Spanish Professor of Botany who died in 1566.

Gosling[1] described the plant: 'which has probably been known to the Indians since a hundred years or more and whose secret has as usual been so faithfully guarded by them, (it) grows in the Highlands of Amambai and near the source of the River Monday, not being, it is said, found further south than this. It is a modest shrub growing side by side with the weeds and luxuriant grasses of that district and only attains a height of a few inches. The leaves are small and the flowers still more diminutive.' He also notes that it is named after Professor Ovidio Rebaudi, of Asuncion. His description of the height as being only a few inches is misleading. Bertoni[7] describes it as typically 0·4–0·8 m high, which conforms more closely to our present knowledge of the wild forms of the plant.

It is reasonable to assume that the local Indians used stevia to sweeten their food and drink, especially mate, the traditional Paraguayan beverage prepared by steeping the crushed, dried leaves of *Ilex paraguayensis* in

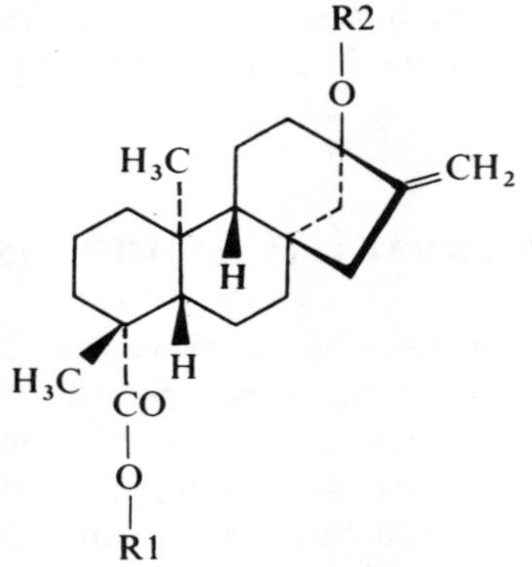

	Name	R1[a]	R2[a]
(A)	Most common steviol glycosides in *Stevia rebaudiana*:		
	Stevioside	—G	—OG(2,1)G
	Rebaudioside A	—G	—OG<(2,1)G; (3,1)G
	Rebaudioside C (dulcoside B)	—G	—OG<(2,1)Rh; (3,1)G
	Dulcoside A	—G	—OG(2,1)Rh
	Steviolbioside	—H	—OG(2,1)G
(B)	Other steviol glycosides occurring in only trace amounts:		
	Rebaudioside B	—H	—OG<(2,1)G; (3,1)G
	Rebaudioside D	—G(2,1)G	—OG<(2,1)G; (3,1)G
	Rebaudioside E	—G(2,1)G	—OG(2,1)G
(C)	The skeletal structure:		
	Steviol	—H	—OH

[a] G, beta-glucopyranosyl (glucose); Rh, alpha-rhamnopyranosyl (rhamnose).

FIG. 2. The steviol glycosides.

water.[8] However, Soejarto[2] reports that the sweetener is used only sporadically in this way today.

It is believed that the early Spanish settlers also used the leaf as a sweetener. It is perhaps significant that while the local Paraguayan name for the plant was Kaá-Hê-é, it is also recorded as Azuca-caá,[6] suggesting that the Spanish did use it as a substitute for sugar when they settled in the region in the sixteenth century.

However, two lines of evidence suggest that it was not of great significance to the natives. First, they valued honey as their principal source of sweetness; second, the herb was not developed commercially by the Spanish. This is in contrast to maté, in which the Jesuits established a significant trade by organising the Guaranis to collect the leaves of *Ilex paraguayensis* from the wild for export mainly to other parts of South America.[9]

Difficulty in propagating *Stevia rebaudiana* is probably the main reason for its limited use. Attempts to cultivate the plant in Paraguay soon after its 'discovery' by Bertoni were abandoned because it could not be grown easily from cuttings and the seeds were generally infertile.[10]

During the late 1940s and again in the 1960s interest was renewed, with attempts to develop a new sweetener industry in Paraguay based on stevia. They met with little success due to a combination of the difficulty in cultivation and limited financial resources. In 1945 Gattoni[11] prepared a report for the Medical Plants Division of the Instituto Agronomico Nacional of Paraguay. In recommending its development as an export crop he gave considerable details on extraction and economics.

Earlier, at the beginning of the Second World War, it had been proposed as a glass-house crop for production in the UK as a substitute for imported cane sugar.[12] The idea was not without precedent since a similar strategy adopted by Napoleon during the British blockade of France had led to the establishment of the beet sugar industry in Europe.

In the mid-1950s Japanese farmers required substitute crops for different reasons. Demand for rice as the traditional staple was beginning to decline in the post-war years, so rather than face the problems of surpluses with which we are so familiar in the West today, the Japanese set about finding alternatives. As net importers of sugar, the idea of creating an alternative indigenous source of sweetness, even if it did lack the bulking properties of sugar, appeared as attractive as it had done to the British and Napoleon previously. Thus in 1954 the Japanese Ministry of Agriculture began trials under the direction of Dr Tetsuya Sumita to develop the agriculture of *Stevia rebaudiana*, initially in Paraguay.

In 1956 the Japanese Government also provided resources for the toxicological evaluation of stevia under the direction of Professor Hiroshi Mitsuhashi at Hokkaido University.

3. JAPANESE DEVELOPMENT DURING THE 1970s

3.1. Japanese Agricultural Programme

A co-ordinated agricultural programme was established in approximately 50 growing areas in and around Japan as well as in Paraguay. New varieties had been developed which yielded a significantly higher concentration of the sweet diterpene glycosides than the 3–5% usually obtained from the leaves of the wild Paraguayan plants. The most favourable growing areas identified outside Paraguay were the warm temperate and subtropical regions of south Japan in Kyushu[13,14] and in the surrounding countries, notably Taiwan, Malaysia, Singapore and South Korea.

The possibility of obtaining stevia leaves from Paraguay was rejected for economic reasons which probably included the cost of continuing to subsidise Japanese farmers not to grow rice. Nonetheless, the Japanese-managed farms which had been established in Paraguay were still operating during the mid-1970s. Certainly A. J. Bertoni, a descendant of Moise Bertoni, worked in co-operation with the Japanese in Paraguay during this period. With another of Moise Bertoni's descendants, Ing. Hernando Bertoni, as the Minister for Agriculture and Husbandry in Asuncion, it is perhaps surprising that stevia is not better developed in Paraguay today.

Since the seeds of *Stevia rebaudiana* are of low fertility, propagation is from root stock.[15] Seedlings are normally transplanted during April and May, and being biennial, the plant can be harvested twice[14] or even three times[15] during the growing season.

The first seeds were introduced into Japan in 1971,[16] but it is not clear when or whether whole plants were introduced.

Judgement of commercial viability of crop yields was based largely on the support price for rice, cost of imported sucrose on a sweetness-equivalent basis and the market opportunity for a natural sweetener. Although there do not appear to have been any reports of serious outbreaks of disease or pest damage to *Stevia rebaudiana*, cultivation was developed in a number of separate growing areas largely to safeguard against such calamities. The Japanese food industry was therefore assured of a reliable source of the raw material.

3.2. Elucidation of the Structure of the Diterpene Glycosides

It is significant that during the early 1970s the precise identity of stevia as described in Fig. 2 was still unclear. In 1908 Rasenack[17] had extracted the leaves with hot alcohol to obtain a crystalline glycoside. However, the physical constants and analytical values reported leave some doubt as to the true identity of this material.

In 1909 Dieterich[18] extracted what he described as 'two sweet glucosidal components, Eupatorin and Rebaudin'. Eupatorin was soluble in alcohol, crystalline and had approximately 150 times the sweetness of sucrose, whereas rebaudin was insoluble in alcohol, amorphous and 180 times sweeter than sucrose. Dieterich concluded that because repeated purification of rebaudin by precipitation from methyl alcohol with ether still gave a product with 10–11% ash content it must be the potassium or sodium salt of eupatorin.

It was not until the work of Bridel and Lavieille which appeared in a series of reports in 1931[19–22] that significant progress was made. From aqueous alcoholic extraction of the leaves they obtained a crystalline glycoside with 300 times the sweetness of sucrose, which they called stevioside. They demonstrated that acid hydrolysis of stevioside yields an aglucone plus glucose as the only sugar component. Subsequently they reported that by enzymic hydrolysis using the hepato-pancreatic juice of the snail *Helix pomatia* they obtained another aglucone isomeric with the first. They called these aglucones isosteviol and steviol respectively. From determination of the molecular weights of steviol and stevioside they concluded that stevioside contained three glucose units.

During the 1950s a group based at Bethesda in the USA published a series of reports based on the use of ion-exchange for the first time to obtain 'nearly pure' stevioside which they analysed using a combination of hydrolytic and other chemical methods and paper chromatography. They confirmed the molecular structure of $C_{38}H_{60}O_{18}$,[23] arrived at by Bridel and Lavieille,[21] and also found that one glucose unit could be removed to produce what they called steviolbioside.[23] From this and further analyses they proposed that the three glucose units were linked to the aglucone as a pair and one single unit by beta-linkages.[24] They also proposed structures for steviol and isosteviol[25] which were later confirmed by the same group in 1963.[26]

Development of thin layer chromatography (TLC) and gas–liquid chromatography (GLC) enabled groups in Japan to separate the individual diterpene glycosides in the early 1970s. Much of this work was reported by Sakamoto and his co-workers at Hiroshima University. Using TLC they first reported separation of rebaudiosides A and B from stevioside.[27] Using ^{13}C nuclear magnetic resonance (NMR) they went on to determine the structure of rebaudioside A[28] followed by separation[29] and determination of the structures of rebaudiosides C,[29] D and E.[30] Simultaneously, Kobayashi *et al.* at Hokkaido University reported the isolation and structures of dulcosides A and B,[31] the latter of which was found to be identical with rebaudioside C[30] (Fig. 2).

The sequence of these major steps in the characterisation of stevia is summarised in Table 1.

TABLE 1
SIGNIFICANT STEPS IN THE ELUCIDATION OF THE SWEET DITERPENE GLYCOSIDES

Year	*Event*	*Reference*
1908	'Stevioside' crystallised	17
1909	'Eupatorin' and 'Rebaudin' identified	18
1931	'Stevioside' split into glucose and steviol	19–22
1955	Glucose beta-linkages to steviol proposed	24
1955	Steviolbioside produced from stevioside	23
1955	Steviol structure proposed	25
1963	Steviol and isosteviol structures confirmed	26
1970	First synthesis of stevioside	32
1975	Rebaudiosides A and B determined	27, 28
1977	Rebaudiosides C, D and E determined	29, 30
1977	Dulcoside A determined	31
1979	First synthesis of sweet analogues	33

Of the analytical methods developed for these determinations only ^{13}C NMR has proved to be entirely satisfactory.[34,35] TLC and GLC methods both suffer the disadvantage of relying on derivatisation, and TLC also suffers from having neither a suitable colour reagent nor adequate densitometric techniques for final determination.[27,36,37] Also, mass spectrometry had been attempted for the structural determinations but was found to be inadequate.[34,35] Subsequently, high-pressure liquid chromatography (HPLC) was developed for identification of the individual diterpene glycosides[38] and this has now become the standard technique[37,39–42] An enzymic method using crude hesperidinase has also been developed as a simple, rapid and specific technique for quantifying stevioside.[34,43]

3.3. Basis for Japanese Decision on Safety

By the early 1970s evidence for the safety and approval of what was then recognised as stevia, i.e. probably a mixture of stevioside and the other sweet diterpene glycosides, had been established to the satisfaction of the Japanese Ministry of Health and Welfare. Evidence was available from several sources, beginning with the apparent safe use for several centuries of the leaf and its crude extracts as a sweetener by the Paraguayan Indians. Confirmation of its safety in animals had begun in the 1920s and was expanded in Japan mainly by Mitsuhashi at Hokkaido University and Akashi at the Tama Research Laboratories.

3.3.1. Acute Toxicity

Bridel and Lavieille in 1931[44] reported the safety of their version of stevia in guinea pig, rabbit and rooster. In a report in the same year Pomaret and Lavieille[45] showed that 'stevioside' was excreted mostly without structural modification. They referred to the stevia in these tests as 'pure stevioside' and contrasted it with a 'yellowish impure form', an 'aqueous decoction' which previously they had demonstrated to be capable of haemolysing sheep red blood cells.

More detailed and controlled studies undertaken in Japan on rats and mice were reported by Akashi and Mitsuhashi separately in 1975. Akashi[46] measured the LD_{50} of three extracts:

'A' ('crude extract') obtained by treating leaves for 2 h with water at 80°C followed by filtration to obtain a dark-brown sample containing around 20% 'stevioside';

'B' ('purified extract') obtained by precipitating the crude extract with calcium hydroxide, filtering, refining with cationic and anionic ion exchange resins and concentrating to a yellow-brown sample containing 40–55% 'stevioside';

'C' ('stevioside crude crystal') obtained by treating 'B' with methanol and separating the insoluble white sample containing between 93 and 95% 'stevioside'.

The LD_{50}s for the 'stevioside' as measured by oral administration to mice were:

'A' 17 g/kg
'B' >42 g/kg
'C' >15 g/kg

Akashi reported that in the case of sample 'C' deaths were due to coagulation of the sample in the stomach as a result of the low solubility of the 'pure stevioside crystal'.

In his separate set of experiments, Mitsuhashi[47,48] obtained LD_{50}s which ranged from 1·59 g/kg for subcutaneous administration in male Wistar rats to more than 8·2 g/kg for both oral and subcutaneous administration in ICR strain mice. The stevia used is described as 'stevioside (powder form)'.

To relate these data to potential human consumption, the lowest LD_{50} measured for oral consumption (>8·2 g/kg by Mitsuhashi in rats and mice) represents more than 1968 times the likely human intake based on the following assumptions:

—the sweetness of stevia is 200 times that of sucrose;
—the average human body-weight is 60 kg;
—average daily intake of sucrose is 100 g/person per day;
—stevia can substitute for up to 50% of the daily average intake of sucrose on a sweetness-equivalent basis.

The highest LD_{50} measured in these experiments (for oral consumption by mice) is equivalent to more than 10 000 times the likely human consumption on this basis.

3.3.2. Sub-acute Toxicity
Akashi and Mitsuhashi, again reporting separately in 1975,[46–48] found no adverse effects from feeding respectively, 2·5 g/kg per day 'stevioside' to rats for one month and 5 g/kg per day 'purified extract' (Akashi's sample 'B' above) to rats for five months. Mitsuhashi describes his sample as being pure according to physical criteria following extraction and refining. The results, which are again quoted on a 'pure stevioside' basis, represent doses equivalent to 600 times the potential human intake on the assumptions stated in Section 3.3.1 above.

3.3.3. Mutagenicity
During the 1970s a substantial amount of research was undertaken in Japan to identify possible mutagenic or carcinogenic effects from stevia. In their 1977 review of cancer research the Japanese Ministry of Health and Welfare record studies on stevia in 11 separate locations.[49] None gives any cause for concern.

3.3.4. Effects on Reproduction
Planas and Kuc had reported from Montevideo in 1968[50] that Paraguayan Matto Grosso Indian women used a daily concoction of leaves of *Stevia rebaudiana*, including stalks as well as leaves and maybe roots, as a form of contraceptive. Planas went on to demonstrate a reduction in fertility of up to 79% in rats fed for 12 days a daily dose of a concoction of dry powdered leaf and stems boiled in water for 10 min and then filtered.

In 1975 Akashi and Yokoyama published the first report of attempts to repeat this work.[46] Using relatively crude extracts which nonetheless contained only leaf material rather than stalk or root they found no reduction in fertility. In the same publication they reported personal communications from several researchers outside Japan who had also tried to repeat Planas' work but without success. These included a report from Kuc who was then at Purdue University, Professor R. Whistler (Purdue University), Professor N. R. Farnsworth (University of Illinois) and Schering AG (Federal Republic of Germany). The conclusion reached by these researchers is that the contraceptive effects noted by Planas and Kuc were due to components of the plant which did not form part of the sweet material. Also, according to Akashi and Yokoyama,[46] Farnsworth was unable to reproduce the contraceptive effect even from the whole plant. Earlier, Farnsworth had also reported[51] a communication from Persinos that 'workers in other laboratories have not been able to confirm this activity'.

An investigation into a possible oestrogen-like effect in female rats and progesterone-like effect in female rabbits was also investigated.[48] No adverse effects were found.

3.3.5. Hypoglycaemic Effect

Claims had been made that extracts from *Stevia rebaudiana* were effective in the treatment of diabetes, though the inspiration for the investigations which generated these claims is unclear. Farnsworth[51] suggested that they could have been based on the 'Doctrine of Signatures' which proposes that sweet substances are effective in treating diseases which result in excessive outputs of sugar. All but one of the reports in support of a hypoglycaemic effect originate from Paraguay and Brazil.[52–55]

In 1966 Miguel[52] had somewhat inconsequentially reported from Brazil that 800 mg per day of 'stevia extract' was administered to patients suffering diabetes, but no side-effects were observed and the patients reported feeling healthy, although their diabetes continued to progress.

Oviedo reported from Paraguay in 1970[53] that both stevia extract containing water and dried stevia extract lowered the blood-sugar level in 25 healthy subjects not suffering from diabetes. No conclusions can be drawn from this report however, since insufficient details were given and the nature of the stevia extract was ill-defined.

The only reports to come from Japan on possible hypoglycaemic effects during the mid-1970s were carried out on rats, and they differed in their results. Akashi and Yokoyama[46] could detect no change in blood-sugar levels after feeding up to 7·0% stevioside in the diets of rats for 56 days. In contrast, Suzuki *et al.* reported in 1977[54] that they did achieve a significant reduction in blood-glucose levels after four weeks feeding 10% dried *Stevia rebaudiana* leaves, equivalent to approximately 0·5% stevioside in the diet.

They also found a reduction in liver glycogen at two weeks although the blood glucose was unaffected at that time. Later, in 1979, Lee *et al.*[55] reported that they could detect no effect after 56 days of feeding the equivalent of 0·5–1·0 g of stevia extract per day.

3.3.6. Metabolites

No evidence was available in the 1970s to suggest that steviol or isosteviol are produced from stevia in humans. The only enzyme systems known to degrade stevioside to steviol and glucose had been demonstrated in:

—hepatic pancreatic juice of the vineyard snail *Helix pomatia*;[22]
—gastric juice of the marine snail *Megalobalimus paranaguensis*;[56]
—pectinase;[57]
—crude hesperidinase.[27]

No enzyme in the human digestive system was known to have a similar action. The report by Vignais *et al.*[58] that steviol and isosteviol inhibit oxidative phosphorylation *in vitro* would not therefore have been considered significant, especially with the total lack of evidence for any toxic effect from the acute and sub-acute feeding data.

3.4. Commercial Development

Commercial development went ahead in Japan on the basis that stevia was safe and that the agriculture had been sufficiently well developed to provide a secure supply which could be expanded to meet demand. It was then up to the food industry to develop extraction processes and applications for stevia.

They were motivated by:

—a new supply of relatively cheap raw material compared with imported sugar;
—a limited availability of intense sweeteners due to cyclamate having been banned in 1969 and the safety of saccharin being in question;
—a product of natural origin.

However, the magnitude of the problem facing potential refiners was only beginning to emerge from a variety of features, namely:

—The leaves of *Stevia rebaudiana* contain around 42% of their dry weight as water-soluble material,[12] of which the diterpene glycosides constitute no more than 20% (Table 2) and typically, in Japan, constitute only 12–15%. In addition, many of the other components present affect the flavour profile, and therefore need to be removed.

TABLE 2
COMPOSITION OF THE MOST COMMON DITERPENE GLYCOSIDES IN LEAVES OF *Stevia rebaudiana*

Country of origin	*Content (% w/w in dry leaves)*[a]						*Reference*
	Stevioside	*Rebaudioside A*	*Rebaudioside C*	*Dulcoside A*	*Steviolbioside*	*Total*	
Japan	7·8	3·9	1·1	0·7	0·2	13·7	60
Japan	8·2	4·3	0·6	0·9	0·5	14·5	60
Japan	6·5	3·9	0·7	0·5	0·3	11·9	60
Japan	6·7	3·3	0·4	0·5	0·5	11·4	60
China	5·3	3·1	0·2	0·1	0·1	9·1	60
China	5·3	3·1	0·8	0·2	0·1	9·5	60
China	6·6	3·7	2·1	0·5	n.d.	12·9	42
S. Korea	7·6	2·3	1·8	1·0	0·2	12·9	60
S. Korea	5·5	2·5	1·4	0·7	n.d.	10·1	42
Taiwan	8·1	3·5	1·4	0·5	n.d.	13·5	42
Paraguay	4·6	1·9	0·9	0·4	n.d.	7·8	42
Paraguay	5·5	3·4	1·5	0·5	n.d.	12·9	42

[a] Determined by high-pressure liquid chromatography.
n.d. = none detected.

TABLE 3
PRINCIPAL PHYSICAL PROPERTIES OF THE MAJOR DITERPENE GLYCOSIDES IN *Stevia rebaudiana*

Diterpene glycoside	*Molecular weight*	*Melting point (°C)*
Stevioside	804	196–198
Rebaudioside A	966	242–244
Rebaudioside C	950	215–217
Dulcoside A	788	193–195
Steviolbioside	642	188–192

(The Ministry of Agriculture and Forestry, Food General Institute, Japan.)

—The profile, as well as the total concentration of the diterpene glycosides, vary according to location, season[36,59,60] and conditions of agriculture (Table 2).
—Properties of the individual diterpene glycosides vary especially with respect to sweetness, flavour, and solubility (Tables 3, 4 and 5).

3.4.1. Physical and Organoleptic Properties
Variation in the profile and total concentration of the diterpene glycosides is well illustrated by Kinghorn and Soejarto,[65] who summarised the results of analyses on a total of 67 samples of leaves of *Stevia rebaudiana* from Japan, Korea, Paraguay and Brazil which appeared in 17 separate reports in Japan between 1975 and 1982. In their examples, concentrations of stevioside range from 2·26 to 22·6% (w/w).

Of the other leaf components, Kinghorn and Soejarto[65] list triterpene, labdane diterpene, sterol, flavonoid glycoside, tannins and 31 volatile oils.

Stevia Company[66] suggest that volatile aromatic or essential oils, tannins and flavonoids contribute to the unpleasant flavours associated with stevia. They refer to these as 'taste qualifiers'. Soejarto *et al.*[2] suggest that the bitter taste common to many *Stevia* species is probably due to sesquiterpene lactones.

However, it should not be forgotten that stevioside itself contributes to the unpleasant flavour with a bitter note. The same is true of rebaudioside A although this is less marked. The ratios of sweet flavour to bitter note plus other flavours, for these compounds at flavour intensities equivalent to approximately 10% sucrose, have been assessed as 62:38 and 85:15 for stevioside and rebaudioside A respectively.[64] This clearly demonstrates the

TABLE 4

RELATIVE SWEETNESS OF DITERPENE GLYCOSIDES IN *Stevia rebaudiana* (SUCROSE = 1)

Diterpene glycoside	*Approximate equivalent concentration of sucrose (%)*													*Reference*
	0·7	*2·0*	*3·0*	*3·5*	*4·0*	*4·5*	*5·5*	*6·0*	*7·5*	*8·0*	*8·5*	*9·0*	*10·0*	
Stevioside (110–270)[a]														61
		240										128		62
				143					149			89		35
					150								100	63
													124	64
Rebaudioside A (150–320)[a]														61
		288						166						62
								242	149		85			35
													133	64
Rebaudioside D (200–250)[a]														61
							221			163		89		35
Rebaudioside E (150–200)[a]														61
						174		125			85			35
Dulcoside A and B (30);[a] (40–60)[a]														31, 61
	54													62
Rebaudioside B (10–15)[a]														61
Steviolbioside (0);[a] (10–15)[a]														23, 61
Steviol (0)[a]														61

[a] Range of equivalent sucrose concentration unspecified.

TABLE 5

TYPICAL FLAVOUR DESCRIPTIONS OF THE MAJOR DITERPENE GLYCOSIDES IN *Stevia rebaudiana*

Diterpene glycoside	*Comment*	*Reference*
Stevioside	'Slight bitterness, some astringency, aftertaste and fairly low general acceptability among the four diterpene glycosides evaluated'.	62
	'At purity 93–95% exhibits a persistent aftertaste, with bitterness and astringency and when only 50% pure a less desirable aftertaste is experienced'.	63
Rebaudioside A	'Least astringency, bitterness and most acceptable of stevioside, rebaudioside B, dulcosides A and B and compared favourably to sucrose in sensory attributes'.	62
Dulcoside A (also for dulcoside B)	'Taste characteristics similar to stevioside'.	62

advantage of increasing the proportion of rebaudioside A in stevia. From Table 2 it can be calculated that rebaudioside A varies between 25 and 33% of the total diterpene glycoside fraction and between 33 and 38% of the stevioside plus rebaudioside A fraction. The range for stevioside in these fractions is 43–60% and 62–70% respectively.

Dubois and Stephenson[64] suggest that the bitter flavour associated with these compounds may be eliminated by an increase in molecular hydrophilic character (see also Section 4.1.4).

Caryophyllene oxide and spathulenol have been found to account for approximately 43% of the essential oil obtained by steam distillation of a Brazilian leaf sample.[67] Fifty-two of more than 100 compounds detected in the oil were identified. Most were in trace amounts. No indication is given as to their potential flavour contribution.

The value of the heat-stability of the diterpene glycosides was perhaps not so well appreciated in the 1970s as it is today. At that time the only high-intensity sweeteners which had been used extensively by the food industry were saccharin and cyclamate, both of which exhibit moderate-to-good heat-stability. This feature has found greater significance in food applications (Section 4.3.2) than in extraction and refining.

3.4.2. Extraction Methods

Extraction procedures were developed which sought to overcome the limitations described above and exploit the more favourable characteristics of stevia.

Methods were developed using the basic procedures already outlined for preparation of material for toxicity studies (Section 3.3.1), namely aqueous extraction followed by refining and concentration. These refining methods have been classified by Kinghorn and Soejarto[65] into several types:

—solvent partition (mainly methanol/water);
—solvent partition incorporating a decolourising agent (mainly calcium hydroxide precipitation) to remove impurities;
—adsorption column chromatography;
—ion-exchange;
—electrolytic techniques;
—isolation procedures for rebaudioside A, rebaudioside C and dulcoside A.

From patent claims it would appear that ion-exchange for purification was the technique which received the greatest attention. Least well developed were electrolytic methods. Whilst many of these techniques were patented, it is suspected that many more have been operating and maintained in secrecy.

However, industry failed to provide a product which was free from unpleasant aftertaste for use in foods with delicate flavours. The only application in which stevia made any impact was in strongly-spiced products such as pickles, in which these undesirable flavours could be masked.

Consequently the volume of sales of stevia in Japan at the end of the 1970s was no more than 40 tonnes per year, which is roughly equivalent to only 12 000 tonnes of sucrose.

4. EXPANSION IN THE 1980s

Clearly the need to mask or remove the unpleasant aftertaste has been the main driving-force behind recent developments in the production of stevia. Success has resulted in a significant increase in sales for a wider range of applications than was possible with the early versions.

4.1. Improvements in Productivity

As the success of Japanese agricultural development produced higher overall yields of the diterpene glycosides from *Stevia rebaudiana*, emphasis turned towards increasing the proportion of the better-tasting rebaudioside A either by selecting new varieties or by discovering alternative sources by screening other species of *Stevia*.

Determining the botanical role of the diterpene glycosides has not apparently facilitated methods for raising their yields. Their role in the plant is similar to that of the gibberellins, to which they are structurally related by the steviol moiety which itself is an intermediate in their biosynthesis. Komai *et al.*[68] found that the diterpene glycosides behave like gibberellins in promoting the activity of alpha-amylase in embryoless barley seeds. They conclude that the activity of the six steviol glucosides used in their experiments is caused by the aglycone steviol because the difference in activity among them could not be regarded as significant. However, they could not determine whether the activity induced by these compounds is attributable to gibberellins converted metabolically in the tissues or to activity inherent to the compounds themselves. But Kinghorn and Soejarto[65] point out that plant-growth regulating activity is characteristic of a large number of naturally occurring compounds of diverse chemical structure.

4.1.1. Search for New Sources

In view of the long tradition of the use of stevia by the South American Indians it would appear unlikely that other species exist which would give better yields or better-tasting compounds. The genus *Stevia* is believed to consist of between 150 and 300 species.[65] Of 110 species of *Stevia* screened by Kinghorn *et al.*[69] only two were found to contain steviol glycosides. One was an herbarium specimen of *Stevia rebaudiana* collected in Paraguay in 1919 in which stevioside, rebaudioside A and rebaudioside B were detected. The other was an herbarium specimen of *S. phlebophylla* collected in Mexico in 1889, in which the only diterpene glycoside detected was stevioside, and then only in trace quantities. It is believed that this species might now be extinct.[69]

4.1.2. Selection of New Strains

Development of strains producing high concentrations of rebaudioside A would appear to be a more fruitful line of development. Several patent claims have been made for such plants, notably from Morita Kagaku

Kogyo,[70,71] which describe a type with a high rebaudioside A concentration and low stevioside and rebaudioside C concentrations. Another claim from the same company[72] describes a plant with rebaudioside content higher than stevioside at least six months after planting.

They describe the selection of this new variety 'AF3' developed since 1979, embracing the claims of the earlier patents.[70,71]

In October to December 1979 'ordinary' strains of *Stevia rebaudiana*, in which the ratio of stevioside:rebaudioside A (ST:RA) is of the order of 1:0·6, were cross-bred. The following spring the resulting seeds were grown under glass and transplanted to outside seed beds in May. During August young plants showing good growth, normal branching, large leaves and ST:RA of 1: > 1 were selected for grafting and crossbreeding in October to December.

Repetition yielded strains (AF2) with ST:RA = 1:1·5 in June 1981 and ST:RA = 1:2·56 to 9·10 in 1983, although the example given quotes the following specific values as percentage weight of the glycosides in dried leaves: stevioside, 7·3 (7·5 in 'ordinary' strain); rebaudioside A, 8·7 (2.3) and rebaudioside C, 0·7 (0·8).

Similar, if not identical, varieties have been developed in California by the Stevia Company who claim proprietorial rights through registration with the United States Department of Agriculture.[73] In addition, whereas propagation is usually from rootstock because of the limited fertility of the seeds, Stevia Company claim that their California strain has seeds of higher viability and also is more resistant to frost than normal varieties.

Although frost is not a hazard in South-east Asia and labour costs are less than in the West, these properties would facilitate production further north, thereby extending the sources of supply.

In their European Patent for 'rebaudioside A', Stevia Company[66] describe leaf material containing:

Glycoside	*Percentage w/w in leaves*
Stevioside	6·66
Rebaudioside A	5·62
Other diterpene glycosides	1·60
Total	13·88

In addition to being cultivated in California and the Far East, *Stevia rebaudiana* is also being cultivated in the semi-arid region on the border of the Negev Desert[74] which does not provide the warm temperate to subtropical climate ideal for it (Section 3.1).

4.1.3. Tissue Culture

Another potentially profitable approach to increasing productivity is tissue culture, initially to supply genetically homogeneous material for propagation, and ultimately to facilitate production in plant-free systems.

Although reports of this approach began to appear in Japan during the 1970s, according to Kinghorn and Soejarto,[65] only two resulted in the production of stevioside and then only on a laboratory scale.[75,76]

More recently, interest in this area has developed in France[77] and Taiwan.[78] Whereas Czechowiak *et al.*[77] developed a simple and rapid method of propagation from callus produced from specimens from the Belgian National Botanic Gardens, Hsing *et al.*[78] claim the first report of stevioside and rebaudioside A production from callus tissue. They recorded stevioside and rebaudioside A concentrations of 32·4% and 17·2% respectively of the dry weight of the callus after 31 days' culture, declining to 18·2% and 8·2% at 74 days.

Tamura *et al.*[79,80] suggest that although tissue culture in general should provide more genetically homogeneous plants and more consistent yields of the sweet glycosides than sexually or vegetatively propagated plants, stem-tip culture is significantly more rapid and provides consistently more homogeneous plants than callus culture. They describe a method for production of homogeneous plants which they advocate for commercial propagation, but they do not go on to propose the technique for plant-free production of the diterpene glycosides.

The latter approach is claimed by Koda *et al.* from Hiroshima University[81] for production of sweet glycosides from shoot tips propagated to primordia *in vitro* which are then developed into seedlings.

Another approach, which has not been attempted, is the development of genetically engineered micro-organisms such as have been developed for thaumatin production.[82–85] In view of the complexity of the synthetic pathway for diterpene glycosides it is unlikely that this approach would be successful.

4.1.4. Synthesis

Whilst synthesis of individual diterpene glycosides is unlikely to become economic,[4] analogues with improved taste characteristics have been developed.[86]

Much of this work has been undertaken by the Dynapol Laboratories in California[87–89] and is reviewed in more detail in the following chapter.[4] Recent patent claims which they have made for some of these products include: an alkali metal salt of a sulphopropyl ester of steviol or steviolbioside;[90] rebaudioside C non-glycoside polar ester which is biologically stable;[91] a range of analogues[92] and rebaudioside C, steviolbioside and steviolmonoside, in each of which each glucose ring has a polar organic substitute.[93]

Another approach has been to increase the yield of rebaudioside A by enzymic and chemical conversion of stevioside via rubusoside to rebaudioside A. Initiated by Kaneda *et al.*,[94–96] this work has been further developed at the Hebrew University in Jerusalem.[4] In a Japanese patent, Dick Fine Chemical and Dainippon Ink claim the use of a *Streptomyces* species in an aqueous solution or suspension of stevioside and beta-1,3-glucosyl sugar to effect this process.[97]

Although these methods should result in products with improved flavour, the resulting material cannot be regarded as natural. In strictly marketing terms this detracts significantly from their value.

4.2. Improved Extraction and Refining

Details of commercial processes are only likely to be revealed under the protection of patents. From patents published since 1980 the most favoured refining processes appear to involve aqueous or alcohol extraction, followed by precipitation or coagulation with filtration and a final clean-up on exchange resins before crystallisation and drying.

4.2.1. Aqueous Extraction

The benefits of a continuous rather than intermittent processing dictate that dried rather than fresh leaves are the normal raw material for initial extraction. However, Stevia Company[66] claim that heating leaves between 90 and 110°C, preferably in a stream of air, inert gases or steam, destroys and/or removes some of the undesirable flavours or 'taste qualifiers' (Section 3.4.1). According to Crammer,[4] at least one of the flavonoid taste qualifiers, apigenin, could be responsible for the contraceptive effect observed in rats by Planas and Kuc (Section 3.3.4). This procedure might therefore provide additional benefit in removing unwanted material before extraction. However, Planas and Kuc refer to their material as dried stevia leaves. No further details of their history since harvesting are given.

Other methods Stevia Company claim for removal of the taste qualifiers include extraction with water in the temperature range 25–59°C or extraction with solvents such as ethyl acetate, dioxane, methylene chloride, chloroform, tetrahydrofuran or ethylene dichloride either separately or in combination at no more than 10°C. This apparently does not reduce the concentration of the required diterpene glycosides in the remaining leaf material.

Although Stevia Company state that the use of water above approximately 100°C would also remove diterpene glycosides with the taste qualifiers, they do not then say how the remaining diterpene glycosides are extracted from the residual leaf material.

Hot water appears to be the preferred medium for extraction,[98,99] especially since the better-tasting rebaudioside A is more soluble than stevioside in water. However, some patents claim advantages from the use of solvents, such as ethanol,[100,101] methanol/chloroform[102] or glycerine, sorbitol or propylene glycol.[103] Liquid hydrogenated hydrolysed starch has also been suggested as an extraction medium.[104]

4.2.2. Use of Solvents

Many processes use solvents[105] for selective separation later in the refining process: to remove impurities using either butanol,[106] or ether, esters or organo-chlorides,[107] or aliphatic alcohols;[98] to crystallise stevioside preferentially using methanol[93,103,106,108,109] or ethanol;[107,109,110] to crystallise rebaudioside A preferentially using more than 70% ethanol;[111] for chromatographic separation using porous gel and organic solvents;[102] or to separate the individual diterpene glycosides chromatographically using alkyl alcohols.[112] However, chromatographic procedures are hardly likely to become commercially economic.

4.2.3. Precipitation and Coagulation

Use of precipitation, often in combination with solvent partition, has been well developed for removal of colour and other impurities. The most frequently quoted precipitant is calcium hydroxide.[98,100,110,113–115] Calcium orthophosphate has also been suggested[116] as well as barium hydroxide or barium acetate added to leaf and stem stock for subsequent filtration, and barium sulphate.[117]

Coagulants are also used to remove colloidal impurities.[114,118]

4.2.4. Adsorption and Ion-Exchange

A variety of materials, such as magnesium silicate aluminate,[119] synthetic macroreticular resin[120] and other adsorbent resins,[118,121] have been suggested for adsorption of impurities from aqueous solutions.

Cationic and ionic exchange resins are commonly used in the later stages of refining.[101,105,122,123]

Another refining procedure for which a patent has been claimed is the used of dialysis and ultrafiltration with gel filtration.[124]

Final separation is usually effected by crystallisation from aqueous solution by addition of solvent.

In processes which do not result in a stevia product with an acceptable taste, either due to failure to remove the taste qualifiers or an unfavourable ratio of stevioside to rebaudioside A, further techniques have been employed for improvement, such as blending with other, often sweet, materials. However, enzymic rearrangement, chemical modification or even treatment with ultrasonic waves[125] have all been proposed.

4.2.5. Enzymic Modification

In addition to enzymic rearrangement to convert stevioside to rebaudioside A (Section 4.1.4), other enzymic processes suggested for improving the flavour generally involve the transfer of glucose units from sugars to stevioside alone or with the other steviol glycosides using the appropriate transferase. Glucose sources proposed include: beta-1,4-glucosyl sugar;[126,127] alpha-glucose;[128] beta-1,3-glucosyl sugar;[129–131] cyclodextrin;[132] starch and/or its partial hydrolysate;[133] sucrose in the presence of yeast[134] and mannose.[135]

4.2.6. Blending

The most common materials claimed to be of value for blending appear to be cyclodextrins, for example beta-cyclodextrin specifically to mask the bitter taste[116] and/or for moulding,[136] gamma-cyclodextrin for masking bitterness and astringency,[137] and cyclodextrin with salt for general taste improvement.[138] Alternative blending materials include pullulan, dextran, locust bean gum[139] and L-histidine hydrochloride.[140]

Blending with lactose, starch or glucose can also be employed to reduce the hygroscopicity of stevia. It has been claimed that addition of crystals of stevioside to concentrated sucrose or glucose solutions has the additional advantage of accelerating their crystallisation.[141]

In addition to acting as a flavour-masking agent or controlling the humectant properties of stevioside, blending with bulking agents also provides products which are convenient to handle for both industrial and table-top applications. Materials for this are usually of either starch or sucrose origin.

Many patents have appeared since 1980 claiming new sweeteners for industrial and domestic use by blending: stevioside with sucrose;[141,142] stevia with glucose;[141] stevioside or stevia extract with D-xylose and sormatin;[143] stevioside with powdery sucrose to give a granular product which, with the addition of further sugar and molasses, can be moulded;[144]

TABLE 6
MEMBERSHIP OF THE STEVIA ASSOCIATION (STEVIA KONWAKAI)

1978	*1985*[226]
	Dainippon Ink Kagaku Kogyo Co. Ltd
Fuji Chemical Industries Co. Ltd	Fuji Kagaku Co. Ltd
Ikeda Saccharising Industry Co. Ltd	Ikeda Toka Kogyo Co. Ltd
Joban Plan Chemical Laboratory Co. Ltd	
Maruzen Chemical Industry Co. Ltd	Maruzen Kasei Co. Ltd
Mitsui Norin Co. Ltd	
Morita Chemical Industry Co. Ltd	Morita Kagaku Kogyo Co. Ltd
Nikken Chemicals Co. Ltd	Nikken Kagaku Co. Ltd
Sanyo–Kokusaku Pulp Co. Ltd	Sanyo–Kokusaku Pulp Co. Ltd
	Sekisui Kagaku Co. Ltd
Tama Biochemistry Laboratory Co. Ltd	Tama Seikagaku Co. Ltd
Takasago Perfumery Co. Ltd	
	Tokiwa Shokubutsu Kagaku Labs Co. Ltd
Toyo Ink Mfg Co. Ltd	Toyo Sugar Refining Co. Ltd

aqueous stevia extract with hydrogenated hydrolysed starch using sucrose fatty esters as surfactant and then spray-drying;[145] stevia dispersed in melted hydrogenated oil, solidified by cooling, crushing, dispersing in a solution containing protein and coagulating;[146] stevioside with sucrose by coating the sucrose;[147] stevia extract and/or alpha-glucosyl stevia in palatinose;[148] stevia extract by absorption onto monosaccharide, disaccharide and/or sugar alcohol for subsequent moulding and drying;[149] stevioside with D-sorbitol and lactic acid or sodium lactate;[150] stevia with natural sweeteners in aqueous pullulan to produce a sweetener in the form of a film;[151] stevioside with a jelly of hydrogenated malt sugar;[152] stevia with a solution of dextrose to form dextrose anhydride in a uniform mix with stevia;[153] stevioside with glycyrrhizin and potassium chloride;[154] stevia with calcium lactate, sodium tartrate and potassium chloride and/or sodium chloride blended into maltitol;[155] stevia with sucrose and maltitol for a low-calorie sweetener similar in sweetness to sucrose;[156] stevioside with aspartame to provide an acid-stable sweetener;[157] and stevioside with glucose in solution and converting to fructose in 'steviol glycoside'.[158]

4.3. Market Development

With the development of these new, improved versions of stevia, sales in

TABLE 7

SALES OF SATO STEVIA IN JAPAN BY APPLICATION, 1984

Application	*Sales (tonnes)*
Beverages	5
Ice-cream	20
Table-top sweetener	3
Bread (raisin type)	5
Candies	2
Pickle	5

Japan have risen from 40 tonnes per year at the end of the 1970s, through 80 tonnes in 1983 to over 300 tonnes in 1985, which is equivalent in sweetness to 30 000–60 000 tonnes of sucrose.

4.3.1. Companies

Several Japanese companies engaged in the production and refining of stevia formed an association in the 1970s which has operated to promote the use of stevia (Table 6). However, not all producers belong, so care must be taken in interpreting sales data published in Japan as these usually represent only sales by members of the Association. The estimates given above are for the total industry.

4.3.2. Applications

The increase in the range of applications for stevia brought about by introduction of these new versions is illustrated by the 1984 sales profile of Sato Stevia, a stevia refined without the use of solvents or enzymes launched by Sato Science Laboratories in 1983 (Table 7).

Although a detailed breakdown of all the sales of stevia in Japan is not available, a reasonable indication of the range of applications for which stevia is being used can be gained from the patent claims made in recent years. In addition to the table-top sweeteners described above a wide range of applications have been described.

Flavour enhancement

—of liquid isomerised sucrose by addition of stevioside and rebaudioside A for use in ice-cream and soft drinks;[159]

—of sweetness of chlorodeoxysugar by stevioside;[160]

—of sweetness of aspartame and cyclamate;[161]

—of fruits and vegetables by application of steviolbioside or its salt during growth;[162]

—(and ripening) of fruit and vegetables by application of alpha-glycosylstevioside;[163]
—of strawberries by addition of stevioside during growth;[164]
—(and aroma enhancement) of foods, drinks and medicinal products by addition of stevioside or other diterpene glycosides;[66]
—of low-grade rice by stevia;[165]
—of salt-free seasoning by stevioside;[166–168]
—of guava tea[169] and hisbiscus tea;[170]
—of a range of fruit and 'nutty' flavours by rebaudioside A in which, it is claimed, the proportions are critical;[66]
—of vitamin C (L-ascorbic acid) in powder form.[171]

Stevia can also be used to mask the taste of fatty acid glycerides and fatty acid esters.[172]

Confectionery
—with lactose, malt syrup and/or dextrin and glucose, fructose, sorbitol, maltitol, lactitol, isomerised sugar and/or honey and stevia to make hard confectionery,[173] such as popular fruit-flavoured and lemon-flavoured candy;
—for chewing-gum using steviolbioside or stevioside[174–176] and bubble-gum;[177]
—for jelly candy incorporating flavours such as kiwi, papaya, pineapple, guava, apple, orange, grape or strawberry.

Stevia has also been used in cake mixes, by combination with maltoligosyl-sucrose, sorbitol and glycine and/or DL-alanine.[178] Its heat-stability makes it especially suitable for these applications.

Soft drinks such as low- or reduced-calorie versions of the cola brand leaders are sweetened by stevia in combination with high fructose syrup.[174] It is also used in mixer drinks,[179] health drinks,[180,181] coffee and even sweet wine. Chang and Cook[182] found the stability of stevioside and rebaudioside A at room temperature in carbonated phosphoric- and citric-acidified beverages to be excellent over five- and three-month storage periods respectively. Although stevioside was also stable to sunlight, rebaudioside A degraded by approximately 20% when exposed to sunlight at 38 kilojoules per square metre.

Sauces and pickles continue to be a popular application for stevia.[183,184] The stability of stevia in these media is excellent. A substantial amount of data has been collected for this application.[65]

Stevia is also claimed to be useful as both an ingredient and processing aid in a variety of processes such as preparing pickles in which leaves of

Stevia rebaudiana are roasted and powdered for addition to the pickle[185] or in fermenting bananas in which the stevia is added with the yeast.[186] Other more typically Japanese applications include seasoning cuttle fish[187] and other fish products.[174]

Very many claims have been made of a more general nature for the use of stevia, stevioside or rebaudioside A in a variety of foods. Stevia Company in particular cover a wide range of claims for rebaudioside A as a flavour enhancer and ingredient in many foods, beverages, medicinal products, oral hygiene products and tobacco.[66]

Other non-food applications for stevia include: oral hygiene products,[188] such as toothpaste in combination with aspartame and stabilised with lecithin;[189] cosmetics,[190] one of which is specifically for rebaudioside A;[191] rodenticide in which stevioside is a direct replacement for sucrose;[192] in stimulating production of alpha-amylase;[193] and finally, as a plant growth regulator or promoter in which presumably its original biological function is exploited.[194–196] In a further extension of this application it is claimed that the culture liquor produced by a steviol glycoside-metabolising *Fusarium* species has plant growth regulating properties.[197]

4.4. Products

Clearly, many versions of stevia are now available in Japan. Each company producing stevia offers specifications which vary with respect to the ratio of stevioside to rebaudioside A and with respect to the degree of dilution with bulking agents such as lactose or starch.

In Brazil and Paraguay however, use of stevia is less developed, and it is sold largely for its 'health-giving' properties, especially for diabetics.[2,65]

4.5. Additional Safety Data

Investigations into the safety of stevia have been continuing in Japan, South Korea, Brazil and the USA since it was reviewed in Japan in the 1970s (Section 3.3).

Nevertheless, it has been stated that little, if any, toxicology has been undertaken on stevioside[174] and also that stevioside is 'not toxicologically acceptable',[198] which simply meant that insufficient data were available for the Scientific Committee for Food of the Commission of the European Communities, at the time of their review in September 1984. At that time stevia had already been in commercial use for 11 years in Japan.

Following Kinghorn and Soejarto's detailed review of the data available in 1985,[65] a significant report on chronic toxicity has been published[199] providing further support for the safety of stevia.

Safety data are available from three distinct areas:

—diterpene glycosides tested in laboratory animals;
—diterpene glycosides and their metabolites tested *in vitro*;
—experience in humans.

4.5.1. Diterpene Glycosides Tested in Laboratory Animals

4.5.1.1. Acute toxicity. Medon *et al.*[200] fed individual, refined diterpene glycosides to mice. This completed the studies begun in the 1970s on crude or 'refined stevioside' and the results are in accordance with the earlier findings (Table 8). They reported that no toxicity could be shown for the individual glycosides: stevioside, rebaudiosides A to C, steviolbioside and dulcoside A fed separately in 2·0 g/kg doses. On the assumptions described earlier (Section 3.3) this is equivalent to 480 times a reasonable human intake.

4.5.1.2. Sub-acute toxicity. Again, no data have been reported which give any cause for concern (see Table 8) apart from a South Korean study[55] which reported a significant decrease in serum lactose dehydrogenase activity from feeding an aqueous extract of *Stevia rebaudiana* leaves containing 50% stevioside to rats so that each rat received up to 0·5 g stevioside per day for two months. No other abnormalities were found from a detailed analysis of these rats sacrificed at 56 days. A further study by Tabarelli and Chagas[201] reported no untoward effects from feeding an aqueous extract to rats for seven weeks.

4.5.1.3. Chronic toxicity. In a two-year rat feeding study[199] 95·2% pure stevia extract was fed to 480 F344 rats at concentrations of 0%, 0·1%, 0·3% and 1% for 22 months in the case of males and 24 months in the case of females. At 6 and 12 months, 10 animals of each sex from each group were sacrificed for clinical and pathological tests. The 0·3% dose induced a slight growth retardation in both sexes, but the 1% dose reduced growth only transiently. General appearance and behaviour were the same in all groups, including the control. Mortality in the stevia groups at the end of the study was not significantly different from that in the controls. At six months a variety of changes were found in the results of urinary, haematological and blood biochemical examinations and in organ weights, but there were no such differences at 12 months nor at the end of the experiment. The incidence and severity of non-neoplastic and neoplastic changes were unrelated to the level of stevia extracts in the diet. The highest level of stevia extracts that caused no effects in rats was 550 mg/kg, under the conditions of the experiment. On the assumptions stated in Section 3.3, this is equivalent to more than 130 times a reasonable human intake.

These data are summarised with the other toxicity data in Table 8.

4.5.1.4. Effects on reproduction. Although none of the data produced during the 1970s supported Planas and Kuc's claim for contraceptive activity (Section 3.3.4), efforts were continued to ensure that no doubts remained. In particular, Mori *et al.*,[202] feeding 95·58% pure stevioside to rats at the sweetness-equivalent of 480 times a reasonable level of human intake (Section 3.3), provided further evidence for lack of contraceptive effect and for the absence of any effect on the development of the foetus (Table 9). The rats were fed for 60 days before mating in the case of six-week-old males, and for 14 days before mating for 11-week-old females, followed by seven days feeding after mating.

Also, Soejarto *et al.*[2] conducted a field study which included interviews with Paraguayan Indians in several regions where *Stevia rebaudiana* is cultivated and where it grows naturally, including Pedro Juan Caballero which is in the centre of the Amambay region, where the plant originated (Section 2). No evidence for its use as a contraceptive could be found from amongst either the long-established Indians or those who had recently settled in the region.

Several reports and early texts list herbs and plants which were used as medicines by the South American Indians, particularly those of the region we now know as Paraguay, but no reference to the use of *Stevia rebaudiana* or 'Caá-Hê-é' for this purpose has been found.[203–210] Reference is made, however, to *Stenosum variegatum*, which is reported to have been used by the Indians as a contraceptive.[211]

4.5.1.5. Mutagenicity. As noted in the chronic toxicity study described in Section 4.5.1.3, no indication of mutagenicity was found. The result of histopathological examination of 38 organs of both the male and female rats revealed no dose-related effects.[199]

4.5.2. Diterpene Glycosides and their Metabolites Tested in vitro

Although no mammalian enzyme system has been shown to be capable of degrading the diterpene glycosides (Section 3.3.6), Wingard *et al.*[212] and subsequently Dubois *et al.*[87] demonstrated *in vitro* that rat caecal flora degrades stevioside to steviol. By introducing carbon-14-labelled steviol into rat caeca and recovering it from the ligated bile ducts and urine, Wingard *et al.* also showed that steviol is absorbed through the lower gut.[212]

This has now been confirmed *in vivo* by Nakayama *et al.*,[213] who demonstrated that although uniformly tritiated stevioside passes unabsorbed through the rat intestine, microbial activity in the caecum degrades the

TABLE 8
SUMMARY OF *in vivo* TOXICITY REPORTS

Test	Material	Target	Dose[a]	Conclusion	Year	Reference
Acute toxicity (*single dose*)						
Subcutaneous	Crude extract	Guinea pig	0·1	Non-toxic	1931	45
Subcutaneous	Crude extract	Rabbit	0·2			
Intravenous	Crude extract	Rabbit	0·1			
Oral	Crude extract	Rooster	1·2	Non-toxic	1931	44
Oral (forced)	Crude extract	Mice	17	$LD_{50} = 17$ g/kg	1975	46
Oral (forced)	Pure extract (53% 'stevioside')	Mice	42	$LD_{50} > 42$ g/kg		
Oral (forced)	'Stevioside'	Mice	15	Insoluble		
Oral (forced)	'Stevioside'	Mice	8	$LD_{50} > 8$ g/kg	1975	47, 48
Subcutaneous	'Stevioside'	Mice	8	$LD_{50} > 8$ g/kg		
Intraperitoneal	'Stevioside'	Mice	2	$LD_{50} > 2$ g/kg		
Oral (forced)	'Stevioside'	Rats	8	$LD_{50} > 8$ g/kg		
Subcutaneous	'Stevioside'	Rats	8	$LD_{50} > 8$ g/kg		
Intraperitoneal	'Stevioside'	Rats	1	$LD_{50} > 1·5$ g/kg		

Gastric-intubation	'Stevioside' Rebaudiosides A–D Dulcoside A Steviolbioside	Mice	2	Non-toxic	1982	199
Sub-acute toxicity						
1 month	'Stevioside'	Rats	2·5	Non-toxic	1975	47, 48
3 months	Pure extract (53% 'stevioside')	Rats	5	Non-toxic	1975	46
2 months	50% 'stevioside'	Rats	0·5 g/rat	Increase in LDH	1979	55
7 weeks	4% 24-hour cold infusion of leaf	Rats	*ad lib*	Safe	1984	200
Chronic toxicity						
22 months	'Stevioside'[b]	Male rats	0·55	No dose-related effects	1985	199
24 months	'Stevioside'[b]	Female rats				

[a] As g/kg bodyweight per day unless otherwise stated (factor for equivalent human consumption = Dose × 240).

[b] Stevioside: 74·54%
Rebaudioside: 16·25%
Other: 4·46%
Total 95·25%

TABLE 9

SUMMARY OF REPRODUCTIVE DATA

Test	*Material*	*Target*	*Dose*[a]	*Conclusion*	*Year*	*Reference*
Pregnancy	Leaf and stem extract	Paraguayan Indians	0·25	Contraceptive	1968	50
Mating[b]						
−18 days	Whole plant extract	Rats (f)[c]	2	Reduction in fertility, 70%	1968	50
−60 days	Crude extract	Rats (m)				
−14 days	Crude extract	Rats (f)	0·4	Not contraceptive	1975	46
+7 days	Crude extract	Rats (f)				
−60 days	Pure extract	Rats (m)				
−14 days	Pure extract	Rats (f)	0·2	Not contraceptive		
+7 days	Pure extract	Rats (f)				
−60 days	'Stevioside'	Rats (m)			1975	46
−14 days	'Stevioside'	Rats (f)	0·1	Not contraceptive		
+7 days	'Stevioside'	Rats (f)				
−60 days	'Stevioside'	Rats (m)				
−14 days	(95·98% pure)	Rats (f)	2·0	Not contraceptive	1981	202
+7 days		Rats (f)		No effect on foetus		
Subcutaneous injections	'Stevioside'	Rats (f)	50 mg in 3 days	No oestrogen effect	1978	48
Subcutaneous injections	'Stevioside'	Rabbits (f)	250 mg in 5 days	No progesterone effect	1978	48

[a] As g/kg bodyweight per day (factor for equivalent human consumption = Dose × 240).
[b] A minus sign (−) indicates number of days before mating and (+) indicates number of days after mating.
[c] (f), female; (m), male.

stevioside to steviol and glucose. The stevioside was introduced into the stomach by tube in a single dose equivalent to approximately 30 times the likely human daily intake (Section 3.3.1), i.e. equivalent to 1·5 kg of sucrose in one dose.

From measurement of the timed sequence of distribution of tritiated hydrogen around the body during the subsequent 48 h, recovery of biliary unidentified steviol conjugates, recovery of steviol as the major metabolite in the faeces together with only small amounts of stevioside and steviolbioside and a very low excretion of radioactivity in the urine they inferred that under the conditions of this experiment:

—stevioside is not absorbed in the small intestine;
—stevioside is degraded by the bacterial flora in the caecum to steviol and glucose;
—the glucose is metabolised to water and carbon dioxide;
—the steviol is distributed throughout the body tissues and excreted via the bile duct as a conjugate and removed in the faeces.

4.5.2.1. Metabolic effects in vitro. Following the report of Vignais *et al.* of the inhibition of oxidative phosphorylation by steviol in rat liver mitochondria,[58] Kelmer-Bracht and others[214–219] reported that in intact cells steviol and isosteviol affect mitochondrial functions. Stevioside and steviolbioside were found to cross the cell membrane very slowly, and showed none of the activity of steviol on mitochondria in intact cells. However, it was found that the hexose carrier in the plasma membrane is inhibited by both the aglycones and the glycones, suggesting that carbohydrate metabolism could be influenced by all stevioside derivatives.

There is no correlation between these findings and the results of any animal studies or observations in man.

4.5.2.2. Mutagenic effects in vitro. Although extensive research, including the two-year rat feeding study (Section 4.5.1.5), into the mutagenic potential of the diterpene glycosides has revealed no mutagenic potential, it has been found that metabolically activated steviol produces a mutagen. Pezzuto *et al.*[218,219] used a forward mutation assay technique employing the S-9 fraction from homogenised livers of rats which had been treated with Aroclor 1254 or phenobarbital together with NADPH as the activating system and *Salmonella typhimurium* TM677 carrying the 'R-factor' plasmid pKM101 as the indicator. They conclude that a cytochrome P-450 mediated metabolic activation dependent on the double bond between steviol carbons 16 and 17 produces a mutagen as yet unidentified.

TABLE 10
SUMMARY OF MUTAGENICITY REPORTS

Test	*Material*	*Target*	*Conclusion*	*Year*	*Reference*
In vitro					
Ames	Crude extract Crude product 'Stevioside'[a]	Several strains	Non-mutagenic	1984	221
Forward mutation	Stevioside Steviolbioside Dulcoside A Rebaudiosides A, B	*Salmonella typhimurium* TM677	Non-mutagenic	1982	200
Forward mutation	As above, metabolically activated with Araclor-pretreated rat liver	*Salmonella typhimurium* TM677	Non-mutagenic	1982	200
Forward mutation	Steviol + treated rat liver fraction (S-9) + NADPH	*Salmonella typhimurium* TM677	Mutagenic	1985 1986	218 219
Forward mutation	Steviol (unmetabolised)	*Salmonella typhimurium* TM677	Non-mutagenic		
In vivo					
24 months	'Stevioside'[a] (0·55 g/kg per day)	Female rats	Non-carcinogenic	1985	199
22 months	'Stevioside'[a] (0·55 g/kg per day)	Male rats			

[a] Stevioside: 74·54%
Rebaudioside: 16·25%
Other: 4·46%
Total 95·25%

They also conclude that the 13-hydroxy group of steviol is required for the expression of mutagenicity since *ent*-kaurenoic acid is non-mutagenic and acetylation of steviol at this position prevents mutagenicity.[219]

They also showed[218] that metabolically activated steviol was extremely bactericidal at the concentrations at which mutagenicity was observed.

Since all stevia feeding studies *in vivo*, including the two-year rat study, have demonstrated a total lack of toxicity, the implication of these results should be treated with caution. This mutagenic effect *in vitro* is dependent on steviol being activated by what is normally an intracellular fraction at concentrations of steviol at least ten times that which would be used for stevioside in food products. In addition, steviol has been shown to be produced *in vivo* only in the rat, whose microbial flora is different from that of man,[65,220] at a concentration of stevioside of more than 100 times that which would be normally used in human food. Subsequent dilution in the digestive tract would increase this margin even further.

Non-metabolically activated steviol is not mutagenic.

A summary of the mutagenicity studies reported to date is given in Table 10.

4.5.2.3. Cariogenic study. Apart from a report of unidentified components of water and alcohol extracts of *Stevia rebaudiana* inhibiting several species of bacteria, including *Pseudomonas aeruginosa* and *Proteus vulgaris*,[65] studies on the antimicrobial activity of stevia have all been directed at determining its potential role in the reduction of dental caries.

Streptococcus mutans has been found to produce less acid when grown on stevioside as compared with sucrose, glucose or fructose.[222] Results from earlier, more detailed studies are summarised in Table 11.

Of the other intense sweeteners so far examined, only aspartame and saccharin have shown inhibitory action against *S. mutans*.[224] In the same experiments neither acesulfam-K nor cyclamate showed inhibitory activity. Thaumatin could not be evaluated because it precipitated from solution.

4.5.3. Experience in Humans

That stevia has been in use for centuries does not necessarily make it safe. However, material which has been consumed over long periods by primitive people has usually been selected partly as a result of its lack of apparent harm. This long record of use and the significant increase in consumption in Japan during the last 13 years without any suggestion of harmful effect must be considered as a factor in favour of its continued use. From consideration of the factors discussed in Section 4.5.2.2 concerning

TABLE 11
COMPARATIVE EFFECT OF STEVIA ON DENTAL CARIES BACTERIA[223]

Target	*Material*	*Inhibition*
Streptococcus mutans	Stevia, 0·5%	Marked
	Glucose, 0·5%	Marked
	Stevia + sorbitol	Moderate
	Glucose + xylitol	Marked
S. mutans HS-1	Stevia + sucrose	None
	Sucrose + xylitol	None
	Glucose + xylitol	None
S. mutans IB	Stevia + sucrose	None
	Sucrose + xylitol	Slight
S. mutans GS-5	Stevia + sucrose	None
	Sucrose + xylitol	Marked
Dextran sucrase	Stevia, 10 mM	18%
(*S. mutans*)	Glucose, 10 mM	10%
	Xylitol, 10 mM	5%
Invertase	Stevia	20%
	Sorbitol, xylitol	None
Lactobacillus plantarum	Xylitol	Total
	Sorbitol	Slight
	Stevia + sorbitol	Moderate
	Glucose + xylitol	Moderate
L. casei	Stevia	Total
	Xylitol	Total
	Sorbitol	Slight
	Stevia + sorbitol	Moderate
	Glucose + xylitol	Moderate

the concentration and location of steviol required to bring about the mutagenic effect *in vitro* and the positive data from the studies *in vivo* combined with this human experience, it is reasonable to conclude that stevia does not constitute a threat to health.

4.6. Future Expansion

Interest in the potential use of stevia in North America and Europe is increasing.

An application for approval is currently under consideration by the UK Ministry of Agriculture, Fisheries and Food (MAFF). An application has also been submitted to the Food Science Committee of the European Economic Committee in Brussels. Although it has been reported that

Atomergic of Long Island had petitioned the Food and Drug Administration for approval in the USA,[174] there is no record in the Federal Register of a stevia petition since 1977.

Countries in which stevia is currently being used include Japan, Paraguay, Brazil,[225] South Korea and The People's Republic of China.

As knowledge and availability of stevia increases it is anticipated that new opportunities will result in the development of synergistic blends of sweeteners and other flavours.[66,160,162]

At a time when consumers are expressing concern at the use of synthetic or artificial food ingredients, stevia presents a remarkable commercial opportunity as an additional natural sweetener; stevia, sucrose and thaumatin are the only sweeteners extracted and refined from plants without chemical or enzymic modification.

REFERENCES

1. Gosling, C. (1901). *Kew Bulletin*, (July–Sept.), 173.
2. Soejarto, D. D., Compadre, C. M., Medon, P. J., Kamath, S. K. and Kinghorn, A. D. (1983). *Econ. Bot.*, **37**, 71.
3. Soejarto, D. D., Kinghorn, A. D. and Farnsworth, N. R. (1982). *J. Nat. Prod. Lloydia*, **45**, 590.
4. Crammer, B. (1987). *Developments in Sweeteners—3*, T. H. Grenby (Ed.), Elsevier Applied Science Publishers, London.
5. Kinghorn, A. D., Nanayakkara, N. P. D., Soejarto, D. D., Medon, P. J. and Kamath, S. (1982). *J. Chromatog.*, **237**(3), 478.
6. Bertoni, M. S. (1899). *Revista de Agronomia de l'Assomption*, **1**, 35.
7. Bertoni, M. S. (1905). *Anales Cientificos Paraguayos*, **5**, 1–14.
8. Porter, R. H. (1950). *Econ. Botany*, **4**, 37.
9. Graham, H. N. (1984). *The Methylxanthine Beverages and Foods: Chemistry, Consumption, and Health Effects*, G. A. Spiller (Ed.), Alan R. Liss, New York.
10. Anon. (1920). *Bull. Imperial Inst.* **18**, 123.
11. Gattoni, L. A. (1945). *Report for Inter-America Tech. Service for Agric. Co-operation*, (*STICA*), Asuncion.
12. Bell, F. (1950). *Chem. Ind.*, (17 July), 897.
13. Kato, I. (1975). *Shokuhin Kogyo*, **18**(20), 44.
14. Akashi, H. (1977). *Shokuhin Kogyo*, **20**(24), 20.
15. Anon. (1984). *Chem. Week*, (14 November), 42.
16. Mitsuhashi, H. (1985). Personal communication to Stevia Corporation.
17. Rasenack, P. (1908). *Arbeiten aus dem Kaiserlichen Gesundheitsamte*, **28**, 420.
18. Dieterich, K. (1909). *Pharm. Zentralbl.*, **50**, 435.
19. Bridel, M. and Lavieille, R. (1931). *Bull Soc. Chim. Biol.*, **13**, 636.
20. Bridel, M. and Lavieille, R. (1931). *J. Pharm. Chim.*, **14**, 99, 154.
21. Bridel, M. and Lavieille, R. (1931). *Bull. Soc. Chim. Biol.*, **13**, 781.

22. Bridel, M. and Lavieille, R. (1931). *J. Pharm. Chim.*, **14**, 321.
23. Wood, H. B., Jr, Allerton, R., Diehl, H. W. and Fletcher, H. G. (1955). *J. Org. Chem.*, **20**, 875–83.
24. Vis, E. and Fletcher, H. G., Jr. (1956). *J. Amer. Chem. Soc.*, **78**, 4709.
25. Mosettig, E. and Nes, W. R. (1955). *J. Org. Chem.*, **20**, 884.
26. Mosettig, E., Beglinger, U., Dolder, F., Lichti, H., Quitt, P. and Waters, J. A. (1963). *J. Amer. Chem. Soc.*, **85**, 2305.
27. Sakamoto, I., Kohda, H., Murakami, K. and Tanaka, O. (1975). *J. Pharm. Soc. Japan*, **95**, 1507.
28. Kohda, H., Kasai, R., Yamasaki, K., Murakami, K. and Tanaka, O. (1976). *Phytochem.*, **15**, 981.
29. Sakamoto, I., Yamasaki, K. and Tanaka, O. (1977). *Chem. Pharm. Bull.*, **25**, 844.
30. Sakamoto, I., Yamasaki, K. and Tanaka, O. (1975). *Chem. Pharm. Bull.*, **25**, 3437.
31. Kobayashi, M., Horikawa, S., Degrandi, I. H., Ueno, J. and Mitsuhashi, H. (1977). *Phytochem.*, **16**(9), 1405.
32. Mori, K., Nakahara, Y. and Matsui, M. (1970). *Tetrahedron Lett.*, 2411.
33. Kamaya, S., Konishi, F. and Esaki, S. (1979). *Agric. Biol. Chem.*, **43**(9), 1863.
34. Tanaka, O. (1980). *Saengyak Hakhoe*, **11**, 219.
35. Tanaka, O. (1982). *Trends Anal. Chem.*, **1**, 246.
36. Mitsuhashi, M., Ueno, S. and Sumita, T. (1975). *J. Pharm. Soc. Japan*, **95**, 127.
37. Gagliardi, L., Amato, A., Basili, A., Cavazzuti, G. and Galeffi, C. (1986). *Anal. Chimica*, **76**, 39.
38. Hashimoto, Y., Moriyasu, M., Nakamura, S., Ishiguro, S. and Komuro, M. (1978). *J. Chromatog.*, **161**, 403.
39. Ahmed, M. S., Dobberstein, R. H. and Farnsworth, N. R. (1980). *J. Chromatog.*, **192**, 387.
40. Ahmed, M. S. and Dobberstein, R. H. (1982). *J. Chromatog.*, **236**, 523.
41. Chang, S. S. and Cook, J. M. (1983). *J. Agric. Food. Chem.*, **31**, 409.
42. Makapugay, H. C., Nanayakkara, N. P. D. and Kinghorn, A. D. (1984). *J. Chromatog.*, **283**, 390.
43. Mizukami, H., Shiba, K. and Hiromu, O. (1982). *Phytochem.*, **21**, 1927.
44. Bridel, M. and Lavieille, R. (1931). *J. Pharm. Chim.*, **14**, 99.
45. Pomaret, M. and Lavieille, R. (1931). *Bull. Soc. Chim. Biol.*, **13**, 1248.
46. Akashi, H. and Yokoyama, Y. (1975). *Shokuhin Kogyo*, **18**, 34.
47. Mitsuhashi, H. (1975). *Extraction of Stevioside, a New Sweetener. Study on the Investigation of Purification Methods and the Properties of Stevioside*, Hokkaido University, Hokkaido.
48. Mitsuhashi, H. (1978). *The Practical Use of Stevia and Research and Development Data*, O. Katayama, T. Sumita, K. Hayashi and H. Mitsuhashi (Eds), ISU Co. Ltd, Japan.
49. Ministry of Health. (1977). *Annual Report on Cancer Research*, Tokyo.
50. Planas, G. M. and Kuc, J. (1968). *Science*, **162**, 1007.
51. Farnsworth, N. R. (1973). *Cosmet. Perfum.*, **88**, 27.
52. Miguel, O. (1966). *Rev. Med. Parag.*, **7**, 200.
53. Oviedo, C. A., Fronciani, G., Moreno, R. and Maas, L. C. (1970). *Excerpta Med.*, **208**, 92.

54. SUZUKI, H., KASAI, T., SUMIHARA, M. and SUGISAWA, H. (1972). *J. Agric. Chem. Soc. Japan*, **51**(3), 171.
55. LEE, S. J., LEE, K. R., PARK, J. R., KIM, K. S. and TCHAI, B. S. (1979). *Hanguk Sikp'um Kwahakhoe Chi*, **11**, 224.
56. KELMER-BRACHT, A. M., KEMMELMEIER, F. S. and ISHII, E. L. (1985). *Arq. Biol. Technol.*, **28**, 431.
57. RUDDAT, M., HEFTMANN, E. and LANG, A. (1965). *Arch. Biochem. Biophys.*, **110**, 496.
58. VIGNAIS, P. V., DUEE, E. D., VIGNAIS, P. M. and HUET, J. (1966). *Biochem. Biophys. Acta*, **118**, 465.
59. MITSUHASHI, H., UENO, J. and SUMITA, T. (1975). *J. Pharm. Soc. Japan*, **95**, 1501.
60. STEVIA CORPORATION LTD (1986). *Sato Stevia Technical Information*, page 9.
61. KATAYAMA, O. (1979). *Food Chem. News Japan*, (Dec.), 6.
62. YOSHIKAWA, S., ISHIMA, T. and KATAYAMA, O. (1979). *Ab. Amer. Chem. Soc.*, (Apr.), 74.
63. ISIMA, N. and KAKAYAMA, O. (1976). *Shokuhin Sogo Kenkyusho Kenkyu Hokoku.*, **31**, 80.
64. DUBOIS, G. E. and STEPHENSON, R. A. (1985). *J. Med. Chem.*, **28**, 93.
65. KINGHORN, A. D. and SOEJARTO, D. D. (1985). *Economic and Medicinal Plant Research*, H. Wagner, H. Hikino and N. R. Farnsworth (Eds), Academic Press, London.
66. STEVIA CO. INC. (1985). European Patent 0154235 (unexamined).
67. MARTELLI, A. and FRATTINI, C. (1985). *Flavour and Fragrance J.*, **1**, 3.
68. KOMAI, K., IWAMURA, J., MORITA, T. and HAMADA, M. (1985). *J. Pest. Sci.*, **10**(1), 113.
69. KINGHORN, A. D., SOEJARTO, D. D., NANAYAKKARA, N. P. D., COMPADRE, C. M., MAKAPUGAY, H. C., HOVANEC-BROWN, J. M., MEDON, P. J. and KAMATH, S. K. (1984). *J. Nat. Prods—Lloydia*, **47**(3), 439.
70. MORITA KAGAKU KOGYO. (1984). Japanese Patent 59034826 (unexamined).
71. MORITA KAGAKU KOGYO. (1984). Japanese Patent 59045848 (unexamined).
72. MORITA KAGAKU KOGYO. (1985). Japanese Patent 60160823 (unexamined).
73. UNITED STATES DEPARTMENT OF AGRICULTURE. (1982). Plant Variety Protection Certificate No. 8200065.
74. CRAMMER, B. (1986). Personal communication.
75. KOMATSU, K., NOZAKI, W., TAKAMURA, M. and NAKAMINAMI, M. (1976). Jpn Kokai Tokkyo Koho, 76 19 169.
76. KOTANI, C. (1980). Jpn Kokai Tokkyo Koho, 80 19 009.
77. CZECHOWIAK, C., DUBOIS, J. and VASSEUR, J. (1984). *Comptes Rendus Acad. Sci.*, **298**(6), 173.
78. HSING, Y. I., SU, W. F. and CHANG, W. C. (1983). *Bot. Bull. Acad. Sinica*, **24**(2), 115.
79. TAMURA, Y., NAKAMURA, S., FUKUI, H. and TABATA, M. (1984). *Plant Cell Rep.*, **3**(5), 183.
80. TAMURA, Y., NAKAMURA, S., FUKUI, H. and TABATA, M. (1984). *Plant Cell Rep.*, **3**(5), 180.
81. HIROSHIMA UNIVERSITY. (1986). West German Patent 3520727 (unexamined).
82. EDENS, L. and VANDERWEL, H. (1985). *Trends Biotechnol.*, **3**, 61.
83. LEDEBOER, A. M., VERRIPS, C. T. and DEKKER, B. M. M. (1984). *Gene*, **30**, 23.

84. EDENS, L., BOM, I., LEDEBOER, A. M., MAAT, J., TOONEN, M. Y., VISSER, C. and VERRIPS, C. T. (1984). *Cell*, **37**, 629.
85. INTERNATIONAL GENETIC ENGINEERING INC. (1986). *Prospectus*, Bear, Stearns & Co Inc., New York.
86. WINGARD, R. E., JR, BROWN, J. P., ENDERLIN, F. E., DALE, J. A., HALE, R. L. and SEITZ, C. T. (1980). *Experientia*, **36**(5), 519–20.
87. DUBOIS, G. E., DIETRICH, P. S., LEE, J. F., MCGARRAUGH, G. V. and STEPHENSON, R. A. (1981). *J. Med. Chem.*, **24**(11), 1269–71.
88. DUBOIS, G. E., BUNES, L. A., DIETRICH, P. S. and STEPHENSON, R. A. (1984). *J. Agric. Food Chem.*, **32**(6), 1321–5.
89. DUBOIS, G. E. and STEPHENSON, R. A. (1985). *J. Med. Chem.*, **28**(1), 93–8.
90. DYNAPOL. (1982.) US Patent 4332830 (unexamined).
91. DYNAPOL. (1982). US Patent 4353889 (unexamined).
92. DYNAPOL. (1983). US Patent 4402990 (unexamined).
93. DYNAPOL. (1983). US Patent 4404367 (unexamined).
94. KANEDA, N., KASAI, K., YAMASAKI, K. and TANAK, O. (1977). *Chem. Pharm. Bull.* (*Tokyo*), **25**, 2466.
95. KASAI, R., KANEDA, N., TANAKA, O., YAMASAKI, K., SAKAMOTO, I., MORIMOTO, K., OKADA, K., KITAHATA, S. and FURUKAWA, K. (1981). *J. Chem. Soc. Japan*, **5**, 726.
96. TANAKA, O. (1980). *Saengyak Hakhoe Chi*, **11**, 219–27.
97. DAINIPPON INK CHEMICAL KK. (1984). Japanese Patent 59017996 (unexamined).
98. YAMADA, S. (1980). Japanese Patent 55162953 (unexamined).
99. DICK FINE CHEMICAL KK. (1981). Japanese Patent 56160962 (unexamined).
100. SEKISUI CHEMICAL IND KK. (1983). Japanese Patent 58212760 (unexamined).
101. SEKISUI CHEMICAL IND KK. (1984). JAPANESE PATENT 59042862 (unexamined).
102. BANSHU, CHOMIRYO. (1979). Japanese Patent 79040560 (examined).
103. SUN STAR. (1980). Japanese Patent 80046695 (examined).
104. SUN STAR. (1980). Japanese Patent 80026820 (examined).
105. MITSUBISHI. (1980). Japanese Patent 80023060 (examined).
106. FUJI FOOD KK. (1982). Japanese Patent 57046998 (unexamined).
107. MARUZEN KASEI. (1981). Japanese Patent 56109568 (unexamined).
108. AJINOMOTO KK. (1981). Japanese Patent 56121453 (unexamined).
109. AJINOMOTO KK. (1981). Japanese Patent 56121455 (unexamined).
110. MITSUBISHI ACETATE. (1983). Japanese Patent 58028246 (unexamined).
111. AJINOMOTO KK. (1981). Japanese Patent 56121454 (unexamined).
112. SUZUKI, F. K. INTERNATIONAL INC. (1982). US Patent 4361697 (unexamined).
113. YAMADA, S. (1980). Japanese Patent 7969719.
114. MITSUBISHI ACETATE. (1983). Japanese Patent 58028247 (unexamined).
115. SEKISUI CHEMICAL INDUSTRY KK. (1983). Japanese Patent 58212759 (unexamined).
116. DICK FINE CHEMICAL. (1982). Japanese Patent 57150359 (unexamined).
117. SHIN-NAKAMURA KAGAK. (1982). Japanese Patent 82042300 (examined).
118. TAMA SEIKAGAKU KK. (1982). Japanese Patent 57075992 (unexamined).
119. HAYASHIBARA BIOCHEMICAL. (1981). Japanese Patent 8204229 (examined).
120. TOYO SEITO KK. (1981). Japanese Patent 83011986 (examined).
121. TOYO SEITO KK. (1982). Japanese Patent 84033339 (examined).

122. AJINOMOTO. (1982). Japanese Patent 82033024 (examined).
123. SEISAN KAIHATSU KAGAKU. (1982). Japanese Patent 57007263 (unexamined).
124. NISSHIN FLOURMILL KK. (1982). Japanese Patent 82026738 (examined).
125. KONDO, K. (1985). Japanese Patent 60027360 (unexamined).
126. DAINIPPON INK CHEMICAL KK. (1983). Japanese Patent 58078562 (unexamined).
127. DAINIPPON INK CHEMICAL KK. (1983). Japanese Patent 58094367 (unexamined).
128. SANYO KOKUSAKU PULP. (1983). Japanese Patent 84051265 (examined).
129. DICK FINE CHEMICAL. (1983). Japanese Patent 58149697 (unexamined).
130. SANYO KOKUSAKU PULP. (1985). Japanese Patent 60037950 (unexamined).
131. DAINIPPON INK CHEMICAL KK. (1984). Japanese Patent 59048059 (unexamined).
132. SANYO KOKUSAKU PULP. (1984). Japanese Patent 59039268 (unexamined).
133. IKEDA TOKA KOGYO KK. (1984). Japanese Patent 59071662 (unexamined).
134. KONDO, K. (1984). Japanese Patent 59102372 (unexamined).
135. OGONTO KK. (1986). Japanese Patent 61005759 (unexamined).
136. MARUZEN KASEI KK. (1981). Japanese Patent 56042560 (unexamined).
137. SANRAKU OCEAN. (1985). Japanese Patent 60098957 (unexamined).
138. RIKEN VITAMIN CO. KK. (1985). Japanese Patent 60188035 (unexamined).
139. NIKKEN CHEMICAL KK. (1982). Japanese Patent 57150358 (unexamined).
140. AJINOMOTO KK. (1981). Japanese Patent 86012664 (unexamined).
141. DOWA KK. (1982). Japanese Patent 82016782 (examined).
142. FUJI FOODS KK. (1983). Japanese Patent 58040066 (unexamined).
143. UENO PHARMACEUTICAL KK. (1980). Japanese Patent 86012663 (examined).
144. MARUTOKU SEITO KK. (1982). Japanese Patent 57125699 (unexamined).
145. OGONTO KK. (1982). Japanese Patent 57129663 (unexamined).
146. KANEBO KK. (1983). Japanese Patent 58000871 (unexamined).
147. FUKUSHIMA, T. (1983). Japanese Patent 58020170 (unexamined).
148. MITSUI SEITO. (1983). Japanese Patent 831026 (examined).
149. MARUZEN KASEI KK. (1983). Japanese Patent 58036368 (unexamined).
150. TAKAGAMI, M. (1983). Japanese Patent 58040064 (unexamined).
151. LOTTE KK. (1985). Japanese Patent 85016217 (examined).
152. AKIBA, K. (1983). Japanese Patent 58141759 (unexamined).
153. SHOWA SANGYO KK. (1983). Japanese Patent 58198266 (unexamined).
154. KYOKUTO KAGAKU SANG. (1983). Japanese Patent 58216663 (unexamined).
155. MARUZEN KASEI KK. (1984). Japanese Patent 59183670 (unexamined).
156. SEKISUI CHEMICAL INDUSTRY KK. (1985). Japanese Patent 60075252 (unexamined).
157. AJINOMOTO KK. (1985). Japanese Patent 60221056 (unexamined).
158. OGONTO KK. (1986). Japanese Patent 61005760 (unexamined).
159. SANYO KOKUSAKU PULP. (1982). Japanese Patent 82058906 (examined).
160. TATE AND LYLE PLC. (1982). European Patent 64361 (unexamined).
161. BAKAL, A. (1985). 'Sweetener synergy in new product development', *Int. Sweetener Ass. Ann. Mtg., Rome.*
162. MIYAMOTO, H. (1983). Japanese Patent 58041808 (unexamined).
163. MIYAMOTO, H. (1983). Japanese Patent 58046010 (unexamined).
164. TAGI KAGAKU KK. (1984). Japanese Patent 84016732 (examined).

165. SANYO KOKUSAKU PULP. (1985). Japanese Patent 60214851 (unexamined).
166. TANPEI SEIYAKU KK. (1982). Japanese Patent 57186460 (unexamined).
167. NISSHIN OIL MILLS KK. (1984). Japanese Patent 84007429 (examined).
168. TANPEI SEIYAKU KK. (1982). Japanese Patent 57189663 (unexamined).
169. TAMADA, J. (1982). Japanese Patent 57083265 (unexamined).
170. HONAN SHOKUHIN KOGY. (1981). Japanese Patent 56029522 (unexamined).
171. TAKASHIMA, H. (1983). Japanese Patent 58140015 (unexamined).
172. YUKI GOSEI YAKUHIN. (1983). Japanese Patent 58047480 (unexamined).
173. NOBEL SEIKA KK. (1982). Japanese Patent 57071366 (unexamined).
174. HIGGINBOTHAM, J. D. (1983). *Developments in Sweeteners—2*, T. H. Grenby, K. J. Parker and M. G. Lindley (Eds), Applied Science Publishers, London.
175. MORITA, E. (1983). Japanese Patent 58101649 (unexamined).
176. KANEBO SHOKUHIN KK. (1985). Japanese Patent 85046939 (examined).
177. NABISCO BRANDS INC. (1984). European Patent 102081 (unexamined).
178. NISSHIN FLOUR MILL KK. (1981). Japanese Patent 56144038 (unexamined).
179. ASAHI BREWERIES KK. (1985). Japanese Patent 60098968 (unexamined).
180. SANYO KOKUSAKU PULP. (1984). Japanese Patent 84004114 (examined).
181. TOYO RIKEN KK. (1986). Japanese Patent 61001377 (unexamined).
182. CHANG, S. S. and COOK, J. M. (1983). *J. Agric. Food Chem.*, **31**(2), 409.
183. SANWA FOOD KK. (1984). Japanese Patent 59074965 (unexamined).
184. TOMOOKA KK. (1986). Japanese Patent 61040783 (unexamined).
185. HOHNEN OIL KK. (1983). Japanese Patent 83009653 (examined).
186. JO, S. (1984). Japanese Patent 59169461 (unexamined).
187. OSHIO, M. (1984). Japanese Patent 59156269 (unexamined).
188. LION CORPORATION. (1985). Japanese Patent 60130509 (unexamined).
189. AJINOMOTO KK. (1985). Japanese Patent 60048920 (unexamined).
190. NONOGAWA SHOJI. (1983). Japanese Patent 58219105 (unexamined).
191. NONOGAWA SHOJI. (1984). Japanese Patent 59139309 (unexamined).
192. VELSICOL CHEMICAL CORPORATION. (1985). British Patent 211443 (examined).
193. MORITA KAGAKU KOGYO. (1982). Japanese Patent 57206389 (unexamined).
194. MORITA KAGAKU KOGYO. (1982). Japanese Patent 57183705 (unexamined).
195. GAKKO HOJIN KINKI. (1985). European Patent 63194 (examined).
196. MORITA KAGAKU KOGYO. (1982). Japanese Patent 57206603 (unexamined).
197. KINKI DAIGAKU, G. H. (1984). Japanese Patent 59101408 (unexamined).
198. COMMISSION OF THE EUROPEAN COMMUNITIES. (1985). *Reports of the Scientific Committee for Food* (*Sixteenth Series*), Brussels.
199. YAMADA, A., OHGAKI, S., NODA, T. and SHIMIZU, M. (1985). *J. Food Hyg. Soc. Japan*, **26**(2), 169.
200. MEDON, P. J., PEZZUTO, J. M., HOVANEC-BROWN, J. M., NANAYAKKARA, N. P. D., SOEJARTO, D. D., KAMATH, S. K. and KINGHORN, A. D. (1982). *Fed. Proc. Fed. Amer. Soc. Exp. Biol.*, **41**(5), 1568.
201. TABARELLI, Z. and CHAGAS, A. M. (1984). *Braz. J. Med. Biol. Res.*, **17**, 531.
202. MORI, N., SAKANOUE, M., TAKEUCHI, M., SHIMPO, K. and TANABE, T. (1981). *J. Food Hyg. Soc. Japan*, **22**(5), 409.
203. MORENO, A. R. (1771). *La Medicina en 'El Paraguay Natural'*.
204. ANATA, P. N. (1898). *Botanica Medica Americana—Los Herbarios de Las Misiones de Paraguay*, Administracio de la Biblioteca, Buenos Aires.

205. RODRIGUEZ, P. M. (1915). *Plantas Medicinales del Paraguay*, Imp. La Mundial, Asuncion.
206. REPUBLICA DE PARAGUAY. (1924). *Plantas Medicinales usadas por el vulgo en el Paraguay*, Imp. Nacional, Paraguay.
207. BERTONI, M. S. (1927). *La Medicina Guarani*, Y Edicion 'Ex-Sylvis', Paraguay.
208. MICHALOWSKI, M. (1955). *Plantas Medicinales del Paraguay*, Ministerio de Agricultura y Ganaderia, Asuncion.
209. STEWARD, J. H. (ED.). (1946). *Handbook of the South American Indians*, Smithsonian Institute, Bureau of Ethnology, Washington, DC.
210. BETHELL, L. (1984). *The Cambridge History of Latin America*, Cambridge University Press, Cambridge.
211. STRAUSS, L. (1946). *Handbook of the South American Indians*, J. H. Steward (Ed.), Vol. 6, Smithsonian Institute, Bureau of Ethnology, Washington, 486.
212. WINGARD, R. E., BROWN, J. P., ENDERLIN, F. E., DALE, J. A., HALE, R. L. and SEITZ, C. T. (1980). *Experientia*, **36**, 5, 519–20.
213. NAKAYAMA, K., KASAHARA, D. and YAMAMOTO, F. (1986). *J. Food Hyg. Soc. Japan*, **27**(1), 1.
214. KELMER-BRACHT, A. M., ALVAREZ, N. S. and BRACHT, A. (1983). *Braz. J. Med. Biol. Res.*, **16**, 521.
215. KELMER-BRACHT, A. M., ALVAREZ, M., YAMAMOTO, N. S., KEMMELMEIER, F. S. and BRACHT, A. (1983). *Braz. J. Med. Biol. Res.*, **16**, 522.
216. KEMMELMEIER, F. S., YAMAMOTO, N. S., ALVAREZ, M. and BRACHT, A. (1983). *Braz. J. Med. Biol. Res.*, **16**, 521.
217. ISHII, E. L. and BRACHT, A. (1984). *Arq. Biol. Technol.*, **27**, 170.
218. PEZZUTO, J. M., NANAYAKKARA, N. P. D., COMPADRE, C. M., SWANSON, S. M., KINGHORN, A. D., GUENTHNER, T. M., LAM, L. K. T., SPARNINS, V. L. and WATTENBERG, L. W. (1985). *Amer. Ass. Cancer Res.*, **26**, 94.
219. PEZZUTO, J. M., NANAYAKKARA, N. P. D., COMPADRE, C. M., SWANSON, S. M., KINGHORN, A. D., GUENTHNER, T. M., SPARNINS, V. L. and LAM, L. K. T. (1986). *Mutation Res.*, **169**, 93.
220. SILK, D. A. (1986). Personal communication.
221. HENNO, K. (ED.). (1984). *Literature Survey on the Safety of Food Additives, No. 6*, Bureau of Citizens and Cultural Affairs of the Tokyo Metropolitan Government, Tokyo.
222. BERRY, C. W. and HENRY, A. (1981). *J. Dental Res.*, **60**, 430.
223. TABU, M., TAKASE, M., TODA, K., TANIMOTYO, K., YASUTAKE, A. and IWAMOTO, Y. (1977). *Hiroshima Daigaku Shigaku Zasshi*, **9**, 12.
224. GRENBY, T. H. and SALDANHA, M. G. (1986). *Caries Res.*, **20**, 7.
225. ANON. (1980). *Concessao de Regustro e Medicamento*, Nos 3875/80 and 3876/80, Diaro Oficial, Brasil.
226. JETRO (1985). *Your Market in Japan—Sweeteners*, No. 35, March.

Chapter 2

PROGRESS IN THE CHEMISTRY AND PROPERTIES OF REBAUDIOSIDES

B. CRAMMER and R. IKAN

Department of Organic Chemistry, Hebrew University of Jerusalem, Israel

SUMMARY

The leaves of the plant Stevia rebaudiana *contain a complex mixture of labdane diterpenes, triterpenes, stigmasterol, tannins, volatile oils and eight sweet diterpene glycosides. They are stevioside, steviolbioside, rebaudiosides A, B, C, D, E and dulcoside A. The* Stevia rebaudiana *plant is being cultivated in the Far East, Paraguay, Brazil, Israel and the United States. The leaves contain essentially stevioside and rebaudioside A as the principal diterpene glycosides. Rebaudioside A is more stable, much sweeter and with better taste characteristics than stevioside. Rebaudioside E is as sweet as stevioside, and rebaudioside D is as sweet as rebaudioside A, while the remaining diterpene glycosides are not as sweet as stevioside. Stevioside can be converted to rebaudioside A. Stevioside and rebaudioside A afford steviolbioside and rebaudioside B respectively on hydrolysis. The diterpene glycosides are non-toxic in man and a number of animal species. No proof has been reported that rebaudiosides are metabolised* in vivo *to the mutagen aglycone, steviol. The rebaudiosides also possess gibberellin-like activity in some plants.*

1. INTRODUCTION

The leaves of the plant *Stevia rebaudiana* Bertoni (Fig. 1) contain the sweetening principle known as stevioside (Chapter 1). The people of Paraguay had been using the sweet leaves of stevia long before the

FIG. 1. *Stevia rebaudiana* Bertoni (reproduced with permission from The Royal Society of Chemistry).

Spaniards colonized in the 16th century. In 1887 M. S. Bertoni, a natural scientist of Paraguay, discovered that stevia leaves were being used by the local inhabitants to sweeten Maté tea. Recently 184 stevia leaf samples taken from herbarium specimens, representing 110 species and 121 taxa, were screened for their taste sensation, but surprisingly only *S. rebaudiana* exhibited a potent and prolonged sensation of sweetness.[1]

The crude glycoside extract from the leaves of this plant tastes sweeter and less bitter than the purified crystalline stevioside. It was found about a decade ago to contain further sweetening principles known as rebaudiosides and dulcosides. The most important diterpene glycoside is the intense sweetener known as rebaudioside A.[2] The leaves of *S. rebaudiana* contain not only the eight sweet *ent*-kaurene glycosides stevioside,[3] steviolbioside,[2] rebaudiosides A, B,[2] C,[4] D,[4] E[4] and dulcoside A[2] (Table 1), but also seven flavanoid glycosides such as centaureidin and quercitrin;[5] labdane diterpenes such as jhanol and austroinulin;[6] triterpenes such as lupeol;[6] sterols such as stigmasterol;[6] and tannins which have not yet been identified.[7] Darise and his co-workers found that stevioside and rebaudioside A were also present in the flowers of *S. rebaudiana*.[8]

TABLE 1

DITERPENE GLYCOSIDES ISOLATED FROM STEVIA LEAVES

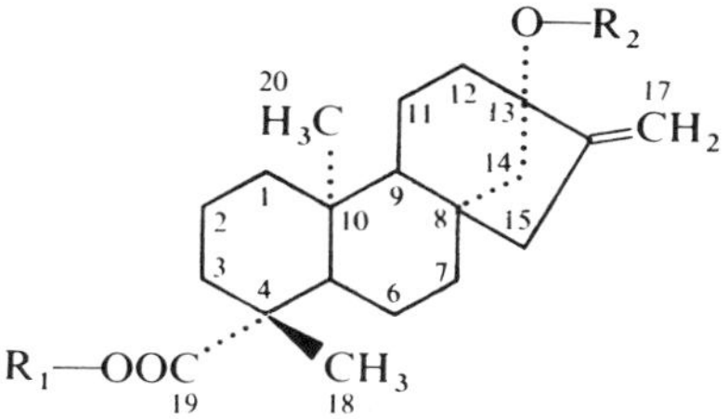

Diterpene glycoside	*R_1*[a]	*R_2*[a]	*Sweetening potency (sucrose = 1)*
Steviolbioside	H	glc^{2}—^{1}glc	100–125
Rubusoside	glc	glc	100–120
Stevioside	glc	glc^{2}—^{1}glc	150–300
Rebaudioside A	glc	glc$_{3}^{2}$—^{1}glc ^{1}glc	250–450
Rebaudioside B	H	glc$_{3}^{2}$—^{1}glc ^{1}glc	300–350
Rebaudioside C (dulcoside B)	glc	glc$_{3}^{2}$—^{1}rham ^{1}glc	50–120
Rebaudioside D	glc^{2}—^{1}glc	glc$_{3}^{2}$—^{1}glc ^{1}glc	250–450
Rebaudioside E	glc^{2}—^{1}glc	glc^{2}—^{1}glc	150–300
Dulcoside A	glc	glc^{2}—^{1}rham	50–120

[a] glc, β-D-glucopyranosyl; rham, α-L-rhamnopyranosyl.

2. THE REBAUDIOSIDES

The rebaudiosides, like stevioside, are diterpene glycosides, and have only been found in the leaves and flowers of *S. rebaudiana* Bertoni, which is a herb of the Compositae (daisy family). The aglycone, known as steviol or 13-hydroxy-*ent*-kaur-16-en-19-oic acid (Fig. 2) is the moiety found both in stevioside and the rebaudiosides. Steviol has been synthesised by several groups[9–11] but the low overall yields show the synthetic routes to rebaudiosides to be uneconomic.

Ruddat and his group demonstrated that steviol had gibberellin-like activity.[12] It was therefore not surprising that the rebaudiosides and stevioside also possess similar activity (see Section 10). Stevioside is the

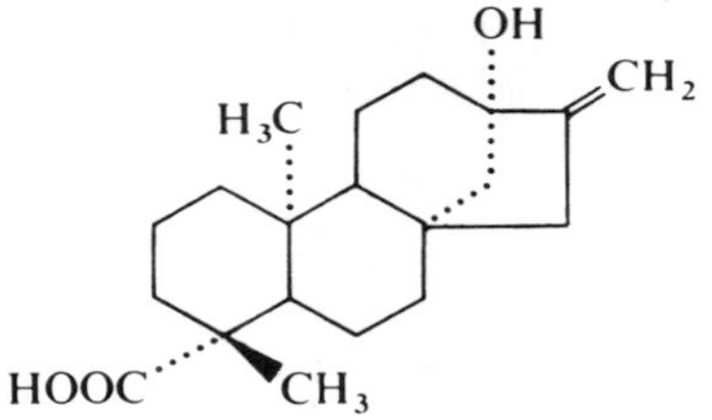

FIG. 2. Structure of steviol.

principal diterpene glycoside: up to 22% by weight has been detected in the dried leaves but only 1% in the dried flowers. The concentration of rebaudioside A in dried leaves varied from 25 to 54% relative to the concentration of stevioside.[13] Rebaudioside E has been found to be as sweet as stevioside, whilst the rebaudiosides A and D are 30% sweeter than stevioside. Rebaudioside C and dulcoside A are about half as sweet as rebaudioside E (Table 1). It has been recently observed that rebaudioside B is an artifact formed from rebaudioside A during the process of separation (O. Tanaka, Hiroshima University, Japan, personal communication). The overall yield of the minor rebaudiosides is less than 1% by weight in the leaves. No sweet diterpene glycosides have been found in the roots of the *S. rebaudiana* plant.

3. REBAUDIOSIDE A

The crude extract of the leaves of *S. rebaudiana* tastes much sweeter and has a more pleasant taste than pure stevioside because of the presence of the intensely sweet diterpene glycoside, rebaudioside A (Fig. 3, Table 1).

The sweetness of rebaudioside A has been evaluated to be as high as 450 times sweeter than sucrose, but the potency decreases, like that of stevioside, with increase in concentration. Further, the taste characteristic is closer to sucrose than to stevioside. A bitter aftertaste still persists, but it is again less intense than that of stevioside. The bitter aftertaste of rebaudioside A and the other diterpene glycosides is not a serious problem because these natural sweeteners are used in such small amounts in foods and beverages. The fact that rebaudioside A has superior taste qualities to stevioside has prompted several countries, in particular Japan, Israel and the USA, to develop this sweetener on a commercial scale. Unlike stevioside, rebaudioside A is water-soluble, and the sweet taste does not persist in the mouth for a long time.[14]

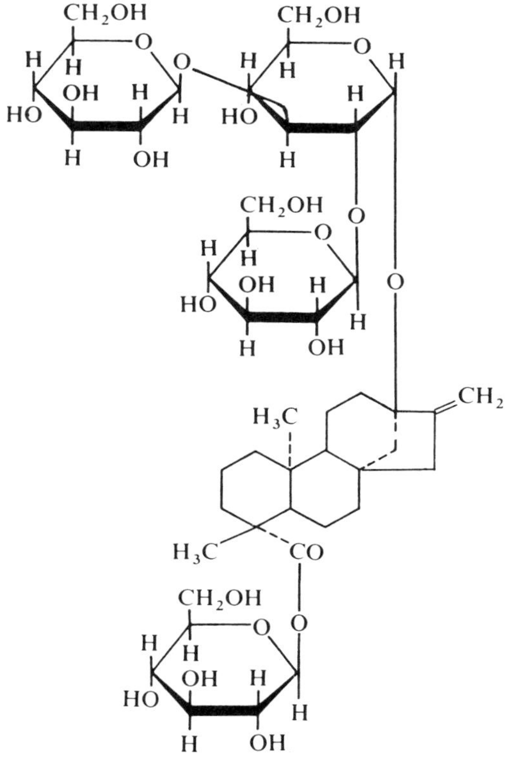

FIG. 3. Structure of rebaudioside A.

Rebaudioside A, together with the other diterpene glycosides, may be extracted from *S. rebaudiana* leaves, which may contain from 3 to 12% of this glycoside.[13] A more attractive method of obtaining pure rebaudioside A involves enzymic hydrolysis of stevioside to another natural sweet intermediate, rubusoside, which has been detected in the leaves of *Rubus suavissimus* (Rosaceae), a plant found in the Kwangchow region of China.[15] Rubusoside, a key intermediate, is converted chemically in a series of steps to rebaudioside A (see Scheme 1).

3.1. Extraction of Rebaudioside A from *S. rebaudiana*

During the past decade a considerable effort has been made to extract rebaudioside A on a commercial scale from the leaves of *S. rebaudiana*. At present there is no commercially attractive process to isolate this intensely sweet diterpene glycoside. In essence the known processes isolate the sweetening principles with a highly polar solvent such as water. The fact that stevioside is only slightly soluble in water whereas rebaudioside A is very soluble (Table 2) enabled Morita[14] to obtain rebaudioside A by concentrating the extract to remove the water, then extracting the concentrate with methanol, which preferentially extracts the rebaudioside A together with some stevioside. Removal of the methanol, followed by chromatographic separation on a silica gel column using a propanol–water–ethyl acetate solvent system, yields the desired compound. Pure rebaudioside A is finally obtained by recrystallisation from aqueous amyl alcohol. Morita obtained only 0·25% of the crystalline sweet glycoside from the dried stevia leaves. Realising that stevioside may be extracted from air-dried stevia leaves in yields up to 6·5% without the use of chromatographic separation techniques,[16] Japanese researchers have investigated the chemical conversion of stevioside to rebaudioside A.

O—glc^2—^{1}glc
H_3C
CH_2
glc—OOC
CH_3
Stevioside

Takadiastase
pH 4, 80 h, 37°C

O—glc
H_3C
CH_2
glc—OOC
CH_3
Rubusoside
Yield 95%

5% NaOH | CH_3OH

O—glc
H_3C
CH_2
HOOC
CH_3
Steviolmonoside
Yield 95%

C_6H_5CHO
98% HCOOH,
RT, 15 min

C_6H_5
O
O
O
O
HO
OH
H_3C
CH_2
HOOC
CH_3
Yield 87%

(i) 3,4,6-Tri-*O*-acetyl-α-D-glucopyranose/C_6H_5Cl/2 h
(ii) 30% CH_3COOH, 15 min, 80°C
(iii) 0·25M BaO/CH_3OH, 2 h, 0°C

Rebaudioside A
Yield 75%

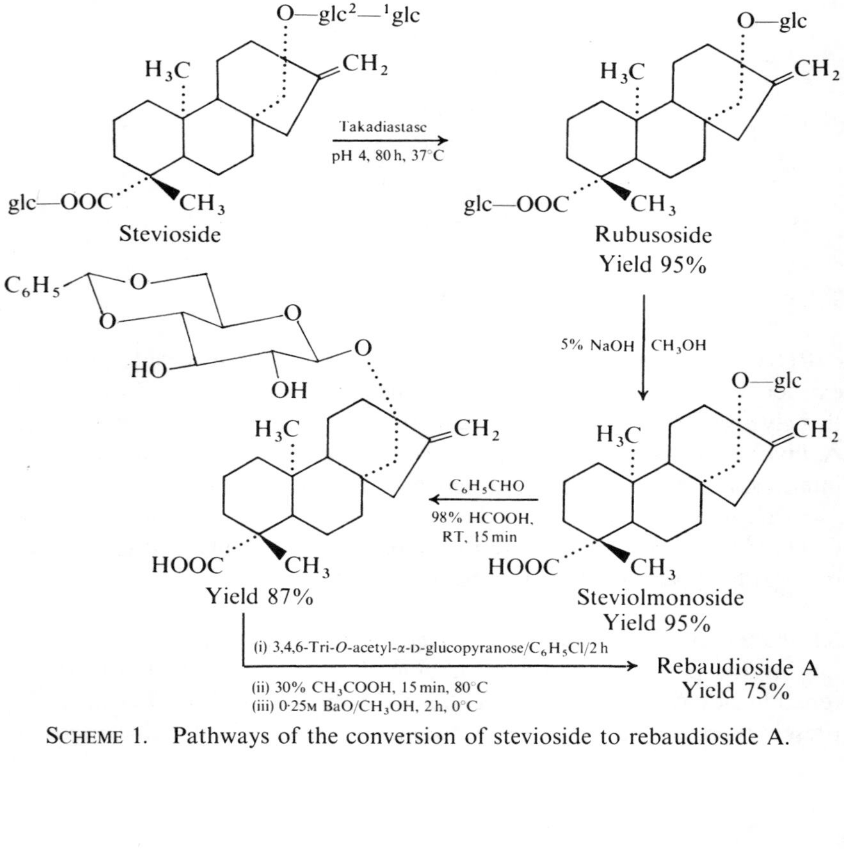

SCHEME 1. Pathways of the conversion of stevioside to rebaudioside A.

TABLE 2

COMPARATIVE PROPERTIES OF STEVIOSIDE AND REBAUDIOSIDES

Property	*Stevioside*	*Rebaudioside A (ref. 17a)*	*Rebaudioside B (ref. 2)*	*Rebaudioside C (ref. 4)*	*Rebaudioside D (ref. 20)*	*Rebaudioside E (ref. 20)*
Molecular weight	805	966	805	950	1128	966
Approximate content in dry leaves (%)	5–22	1·5–10	0·4	0·4	0·03	0·03
Melting point (°C)	196–198	242–244[a] 248–250[b]	193–195	215–217	283–286	205–207
Specific rotation $[\alpha]_D^{25}$ (deg.)	−39·3 $c = 5·7$ (H_2O)	−20·8 $c = 1$ (MeOH)	−45·4 $c = 0·96$ (MeOH)	−29·9 $c = 1$ (MeOH)	−22·7 $c = 1$ (MeOH)	−34·2 $c = 1$ (MeOH)
Solubility characteristics	0·13% in water. Slightly soluble in EtOH. Soluble in dioxane.	Soluble in water, MeOH, EtOH. Insoluble in acetone $CHCl_3$ or ether.	Slightly soluble in MeOH.	Slightly soluble in MeOH.	Slightly soluble in MeOH.	Slightly soluble in MeOH.

[a] Ref. 2.
[b] Ref. 14.

3.2. Chemical Conversion of Stevioside to Rebaudioside A

Tanaka and his co-workers in 1977 successfully converted stevioside by means of enzymic hydrolysis to another sweet natural diterpene glycoside known as rubusoside, which was converted after three steps to rebaudioside A in 75% yield[17] (Scheme 1).

The focal point of this process is the selective enzymic hydrolysis of the terminal glycosyl ether linkage of the β-sophorosyl moiety of the stevioside molecule to rubusoside (Table 1) by the digestive enzyme amylase (Takadiastase) derived from *Aspergillus oryzae.*

3.3. Interconversion of Rebaudiosides by Hydrolysis

Alkaline saponification of stevioside and rebaudioside E yields the same steviol glycoside, steviolbioside. It is sufficient to reflux the stevioside or rebaudioside E with 10% aqueous potassium hydroxide or sodium hydroxide solution for an hour.[18] DuBois improved the yields of steviolbioside by carrying out the reaction in aqueous–methanolic solution containing the alkaline hydroxide.[19]

Sakamoto saponified rebaudioside D to rebaudioside B which has also been obtained by enzymic hydrolysis of rebaudioside A and crude hesperidinase[20] (Scheme 2).

Stevioside[17,18] ⟶ Rubusoside[17] ⟶ Rebaudioside A[2] ⟶ Rebaudioside B
↓ (from Stevioside) ↑ (from Rebaudioside D)
Steviolbioside[18] ⟵ Rebaudioside E[20]   Rebaudioside D[20]

SCHEME 2. Hydrolytic conversion of steviol glycosides.

4. CHROMATOGRAPHIC SEPARATION OF THE REBAUDIOSIDES IN STEVIA

A number of different techniques have been reported for the separation and analysis of rebaudiosides and stevioside from the leaves of *S. rebaudiana.* These are gas–liquid chromatography (GLC),[21,22] thin layer chromatography (TLC), [23,24] droplet countercurrent chromatography (DCCC),[25] high-pressure liquid chromatography (HPLC),[13] enzymatic[26] and colorimetric determinations.[27,28] Some have disadvantages. For example, the GLC method requires acid or enzymic hydrolysis of the diterpene glycosides to steviol, followed by methylation to the methyl ester of either isosteviol[29] or steviol.[30]

TABLE 3

QUANTITATION OF STEVIA PLANT MATERIAL BY HPLC

Method	*Stationary phase*	*Moving phase*	*Detected constituents*	*Reference*
GLC	DEGS-5%		Stevioside, rebaudiosides A, C, dulcoside A	22
TLC	Silica gel F254	$CHCl_3$–CH_3OH (17:3)	Stevioside, rebaudiosides A, B (as phenacyl esters), steviolbioside	23
TLC	Kieselgel 60	$CHCl_3$–CH_3OH–H_2O (6:30:5)	Stevioside, rebaudiosides A, C, dulcoside A	24
TLC	Silica gel F254	$CHCl_3$–CH_3OH–H_2O (15:10:2)	Stevioside, rebaudioside A,	44
DCCD		$CHCl_3$–CH_3OH–$(CH_3)_2CHOH$–H_2O (19:9:4:8)	Stevioside, rebaudioside A dulcoside A	25
HPLC	Shodex OHpak M-414	CH_3CN–H_2O (4:1)	Stevioside, rebaudioside A	31
HPLC	Lichrosorb-NH_2	CH_3CN–H_2O (85:15)	Stevioside, rebaudioside A	33
HPLC	Unisil Q-NH_2	CH_3CN–H_2O	Stevioside, rebaudioside A	34
HPLC	Protein-1-125	$CH_3(CH_2)_2OH$	Stevioside, rebaudiosides A, C	35
HPLC	Protein-1-125	$(CH_3)_2CHOH$	Stevioside, rebaudiosides B, D, E, dulcoside A	36
HPLC	Zorbax NH_2	CH_3CN–H_2O (84:70)	Stevioside, rebaudiosides A, C, dulcoside A	37
HPLC	P-NH_2	CH_3CN–H_2O (80:10)	Stevioside, rebaudiosides A, C	38
HPLC	Bondpak	CH_3CN–H_2O (80:20)	Stevioside, rebaudioside A	44

It was found from the enzymic method with crude hesperidinase that glucose release by the hydrolysis of rebaudiosides A and C, as well as stevioside, was complete, and that the rebaudioside estimations correlated favourably with other methods of analysis.

Several investigators have applied HPLC analytical methods for the separation and quantitation of the rebaudiosides.[31–38] Its advantages are: (1) it is extremely rapid; (2) the recovered rebaudiosides are unchanged and can be used as standards for analysis or taste. HPLC has also been applied to the quantitation of stevia plant material[31–38] and to foods[39–43] and beverages[44] (Table 3).

Successful separations of the rebaudiosides, stevioside, steviolbioside and dulcoside have been achieved by two groups using HPLC as the desired method.[35,37]

5. TASTE CHARACTERISTICS OF THE REBAUDIOSIDES

Tanaka reported that the pure rebaudiosides A, D and E are not only sweeter than stevioside but they are also less bitter[45] (Table 4). Morita and coworkers did not agree entirely with Tanaka. They reported that stevioside has a slow latent sweetness coupled with a strong bitterness and an unpleasant aftertaste. In contradiction of Tanaka's evaluation of the taste characteristics of rebaudioside A, Morita described it as resembling sucrose in latent sweetness with no bitter or unpleasant aftertaste.[14] Unlike

TABLE 4
RELATIVE SWEETNESS[a] OF DITERPENE GLYCOSIDES TO SUCROSE[45]

Diterpene glycoside	*A (%w/v)*		
	0·1	*0·05*	*0·025*
Stevioside	89	149	143
Rebaudioside A	85	149	242
Rebaudioside D	89	163	221
Rebaudioside E	85	125	174
Rubusoside	84	63	114

[a] Relative sweetness = B/A, where A = concentration (%w/v) of aqueous solution of sample; B = concentration (%w/v) of aqueous solution of sucrose with same sweetness as the sample solution.

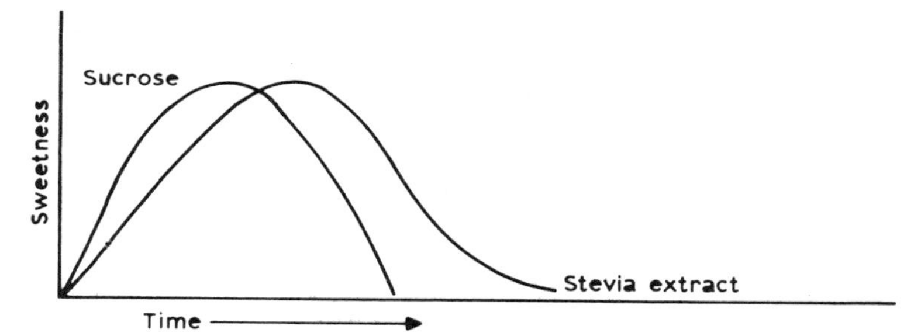

FIG. 4. Sweet taste characteristics of stevia extract.

stevioside, which is sparingly soluble in water, rebaudioside A is water-soluble. Schiffman did not agree with either Tanaka or Morita in evaluating the taste characteristics of both stevioside and rebaudioside A.[46] She reported that both stevioside and rebaudioside A had not only a strong bitter aftertaste, but also a metallic aftertaste resembling that of sodium saccharin and acesulfam-K. The general opinion is that rebaudioside A is closer in sweetness profile to sucrose than stevioside, with no disagreeable aftertaste. The Maruzen Pharmaceutical Company of Japan also said that the taste profile of a stevia extract containing both stevioside and rebaudioside A was very similar to that of sucrose[47] (Fig. 4). The only difference with the stevia extract is that there is a delay in the development of sweetness. Stevia extract containing a mixture of the rebaudiosides and 50% stevioside has been given the name Marumilon 50 by the Maruzen Pharmaceutical Company.

TABLE 5
SWEETENING POTENCY OF MARUMILON 50 IN AQUEOUS AND 5% SALT SOLUTION[47]

Concentration of sugar (%)	*Sweetening potency in aqueous solution*		*Sweetening potency in 5% salt solution*	
	Concentration of Marumilon 50 (%)	*Sweetening potency*	*Concentration of Marumilon 50 (%)*	*Sweetening potency*
2	0·014	× 143	0·006	× 333
4	0·036	× 111	0·013	× 308
6	0·067	× 90	0·022	× 273
8	0·133	× 60	0·040	× 200
10	0·200	× 50	0·080	× 125

Tama Biochemical Company, also of Japan, produces a similar stevia extract under the name 'Stevix'. It was observed that the sweetening potency of Marumilon 50 decreases with increase in concentration. Furthermore its taste intensity was potentiated quite remarkably in the presence of a 5% saline solution (Table 5).

Although no aftertaste characteristics have been reported of the rebaudiosides C, D and E, it is assumed that they would have similar properties to rebaudioside A.

6. TASTE CHARACTERISTICS OF MARUMILON 50 AND GLYCYRRHIZIN

Glycyrrhizin is a natural sweetener about 50 times sweeter than sucrose. It is found in certain rhizomes and licorice roots (*Glycyrrhiza glabra* L., and *G. hersuta*). The ammonium salt of glycyrrhizic acid is also 50 times sweeter than sucrose, and when it is mixed with sucrose a potentiating effect doubles its sweetening potency. Glycyrrhizin and its ammonium salt have an unpleasant licorice aftertaste.[48] Glycyrrhizin does not give an immediate sweet taste sensation, but the sweetness gradually increases and then slowly fades. Glycyrrhizin also potentiates the sweetening potency of Marumilon 50 (Table 6). It was noticed that a mixture of Marumilon 50 and glycyrrhizin, known as Marumilon A, improves the sweetening potency in 5% saline solution (Table 7). The advantage of adding glycyrrhizin to Marumilon 50 is that the unpleasant bitter aftertaste associated with

TABLE 6
SWEETENING POTENCY OF MARUMILON A IN AQUEOUS AND 5% SALT SOLUTION[47]

Concentration of sugar (%)	*Sweetening potency in aqueous solution*		*Sweetening potency in 5% salt solution*	
	Concentration of Marumilon A (%)	*Sweetening potency*	*Concentration of Marumilon A (%)*	*Sweetening potency*
2	0·10	×20	0·020	×100
4	0·25	×16	0·040	×100
6	0·40	×15	0·065	×90
8	0·57	×14	0·100	×80
10	0·77	×13	0·200	×50

TABLE 7
CHANGE OF THE SWEETENING POTENCY OF MARUMILON A WITH VARIATION OF CONCENTRATION OF SALT SOLUTION[47]

Concentration of salt (%)	*Concentration of sugar* (%)	*Concentration of Marumilon A* (%)	*Sweetening potency*
0·5	6	0·30	×20
1·0	6	0·20	×30
3·0	6	0·15	×40
5·0	6	0·065	×90

glycyrrhizin and Marumilon 50 is no longer perceived. Marumilon A had an improved taste much closer to sucrose than Marumilon 50. Manufacturers in Japan have been using Marumilon A in a variety of foods, including salty ones, in which the salt apparently increases the sweetness, while at the same time the salty taste is considerably reduced. Pure stevioside gave similar results with glycyrrhizin but the overall taste characteristics were not as good as Marumilon 50, which also contains rebaudioside A.[47]

7. STABILITY OF REBAUDIOSIDE A

Rebaudioside A, like stevioside, is stable at room temperature for indefinite periods. After 22 h at 100°C only 11·5% had hydrolysed to rebaudioside B and glucose, compared with 20·6% of stevioside under the same conditions.[44] Furthermore, Chang and Cook showed that rebaudioside A was more stable than stevioside in acidic conditions.[44] No degradation products were detected when either phosphoric acid or citric acid beverages containing rebaudioside A were stored for three months at room temperature. Even raising the temperature to 60°C did not appreciably increase the hydrolysis of rebaudioside A after five days.[44]

8. TOXICITY EVALUATION OF THE REBAUDIOSIDES

Only Japan, Brazil and Paraguay have approved the use of stevioside and stevia extracts in foods and beverages. Most of the toxicity studies were carried out in Japan by a consortium of 11 firms collectively known as the 'Stevia Konwakai' (Stevia Association)[58] (Table 8), assessing either stevia

TABLE 8
JAPANESE COMPANIES CONSTITUTING THE 'STEVIA KONWAKAI'[58]

Dainippon Ink Kagaku Kogyo Co. Ltd, Tokyo
Fuji Kagaku Co. Ltd, Tokyo
Ikeda Toka Kogyo Co. Ltd, Hiroshima
Maruzen Kasei Co. Ltd, Tokyo
Morita Kagaku Kogyo Co. Ltd, Osaka
Nikken Kagaku Co. Ltd, Tokyo
Sanyo Kokusaku Pulp Co. Ltd, Tokyo
Sekisui Kagaku Kogyo Co. Ltd, Osaka
Tama Seikagaku Co. Ltd, Tokyo
Tokiwa Shokubutsu Kagaku Laboratories Co. Ltd, Chiba
Yoto Foods Co. Ltd, Tokyo

extracts containing the sweet diterpene glycosides or pure stevioside. Very little toxicity evaluation has been done on any of the pure rebaudiosides.

The LD_{50} of extracts from *S. rebaudiana* leaves containing 50% stevioside and about 40% of the rebaudiosides, administered intraperitoneally to rats, was reported to be 3·4 g/kg.[49] Acute toxicity tests have been carried out on stevia extracts containing the sweetening principles, particularly stevioside and rebaudioside A. When these extracts were administered orally to mice it was found that the LD_{50} was in the range 17–42 g/kg.[50,51] Medon found that rebaudiosides A, B and C, as well as steviolbioside or dulcoside A, were not toxic when orally administered to mice.[52] Sub-acute toxicity studies on female rats using a water extract of *S. rebaudiana* leaves containing rebaudioside A and stevioside as the principal diterpene glycosides showed no significant effects after three months.[51]

There was no antifertility action when extracts of *S. rebaudiana* leaves containing essentially stevioside and rebaudiosides were fed to rats for a period of three weeks.[51] These findings appear to contradict the results of Planas and Kuć, who investigated the contraceptive effects of aqueous extracts of *S. rebaudiana* leaves for periods up to two months on rats,[53] supporting the claim that the Paraguayan Matto Grosso Indians have been using the leaves and stem of *S. rebaudiana* in Maté tea as a contraceptive for centuries. In our opinion the explanation of these conflicting results lies in the source material. The Japanese researchers used processed stevia extracts containing only the diterpene glycosides, whereas Planas and Kuć and the Paraguayan Indians used a decoction in water from dry powdered leaves and stems of the plant. Stevia leaves contain many other substances.[5–7] Some of these are flavanoids, including apigenin which is known to possess antifertility activity.[5,54]

Tanaka reported that no mutagenic activity was observed for rebaudioside A or stevia extracts using the Ames test.[45]

Recently Yamada and his colleagues published a chronic toxicity study of dietary stevia extracts containing rebaudioside A and stevioside in rats for 22 months. The highest level of stevia extracts which was non-toxic in rats was 550 mg/kg.[55] It was concluded that no chronic toxicity and carcinogenic effects ensued from feeding stevia extracts to rats over two years. There was no mention of whether the mutagen steviol was detected. Wingard (1980) showed that rebaudioside A was degraded to the aglycone steviol by rat intestinal microflora *in vitro*. It was found that rebaudioside A was converted to the extent of 65% to steviol, whereas stevioside under similar conditions was completely converted to steviol.[56]

In 1986 Nakayama and his team published their findings on the metabolic fate of [^{3}H]stevioside fed to rats and observed on analysing the intestinal contents that steviol was one of the metabolites.[57] Presumably rebaudioside A would undergo a similar fate but over a longer period if Wingard's work is taken into account. Kinghorn and Soejarto's work on the mutagenic effects of the metabolism of stevioside and the rebaudiosides offers cause for concern about the safety for human consumption of the diterpene glycosides either individually or as an extract.[58] There have been no reports to date of the metabolic fate of diterpene glycosides in man in spite of the fact that large quantities of stevia extracts are being consumed in the Far East and parts of South America.

It has been suggested that rebaudiosides and stevioside lower blood-sugar levels in man and animals, but according to Kinghorn further reliable data are required to substantiate this.[58]

9. SYNTHETIC ANALOGUES OF DITERPENE GLYCOSIDES

The sweetening principles stevioside and rebaudioside A are not entirely satisfactory as completely safe, non-nutritive sweeteners with taste characteristics similar to sucrose. Along with the bitter aftertaste, controversy is continuing over the safety of these two glycosides as well as the extract of the leaves of *S. rebaudiana*. DuBois and co-workers therefore prepared a number of synthetic analogues of stevioside and rebaudioside A[59] to obtain sweet analogues which were not only devoid of any unpleasant bitter aftertaste, but also resisted enzymic hydrolysis to steviol, recently discovered to be mutagenic in rats.[58]

Stevioside hydrolyses with ease to the less sweet steviolbioside which contains a carboxylic group at C-4. The object was to substitute the glucosyl ester moiety with other groups of similar polarity without loss of sweetness. In 1981 DuBois prepared the sulphopropyl ester of steviolbioside (Table 1, $R_1 = (CH_2)_3SO_3^-Na^+$, $R_2 = glc^2—^1glc$).[59] This ester was easily prepared from steviolbioside and propane sultone in the presence of an alkali metal carbonate such as potassium carbonate in dimethylformamide. The ester was 125 to 200 times sweeter than sucrose with taste characteristics closer to sucrose than stevioside. It was very stable, resistant to boiling alkali and virtually unaffected by hot dilute sulphuric acid. Furthermore it was highly resistant to bacterial degradation, with less than 0·1% being hydrolysed to steviol. Under the same conditions stevioside was hydrolysed with ease. The bitter aftertaste was considerably less than that of stevioside.

Four years later DuBois synthesised the sulphopropyl ester of rebaudioside B.[60] Just as rebaudioside A had improved taste characteristics compared with stevioside, so did the sulphopropyl ester of rebaudioside B with respect to the sulphopropyl ester of steviolbioside. This analogue was found to be as sweet as rebaudioside A with no bitter aftertaste and taste characteristics even closer to sucrose than the sulphopropyl ester of steviolbioside. The stability of the sulphopropyl ester of rebaudioside B towards alkali, acidic and enzymic hydrolysis has not yet been reported but it is expected to be as resistant as the sulphopropyl ester of steviolbioside. The polar sulphopropyl group of steviolbioside and rebaudioside B seems to minimise the usual bitter aftertaste of the natural members of the diterpene glycosides.

It is of interest to mention that the sulpho group $—SO^-Na^+$ is part of the molecule of the pleasant synthetic sweetener, cyclamate $R—NHSO_3^-Na^+$ (R = cyclohexyl). Sweetness is retained when R represents certain other alkyl and cycloalkyl groups.[48]

10. PLANT GROWTH-REGULATING PROPERTIES OF REBAUDIOSIDES

It is known that gibberellic acid (Fig. 5) promotes stem extension and fruit growth. Like the auxins (plant hormones), gibberellic acid also stimulates flowering in a number of plants and overcomes the seed dormancy of certain species. Gibberellin glycosides are inactive as plant growth regulators,[61] but it was discovered by Valio and Rocha that the aglycone

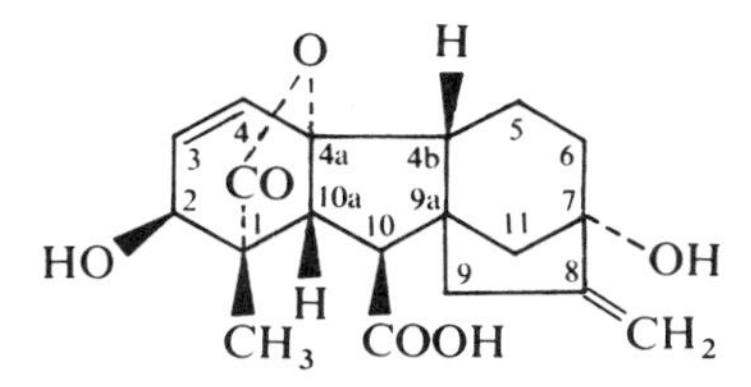

FIG. 5. Structure of gibberellic acid.

diterpene steviol has gibberellin-like activity in lettuce, cucumber and bean plants.[62]

Steviol glycosides such as stevioside, steviolbioside and the rebaudiosides have excellent plant growth-promoting activity. This was surprising, as the addition of a glycopyranosyl moiety to gibberellic acid inactivates it. Iwamura and Komai found that rice plant stems, after two weeks cultivation, increased in growth by as much as 50% in the presence of 500 ppm of a diterpene glycoside[63] (Table 9).

The rebaudiosides are more soluble in water than the gibberellins. Unlike the gibberellins, the sweet diterpene glycosides do not cause yellowing in plants and do not inhibit root growth. The diterpene glycosides also accelerate flowering in plants.[63] A 500 ppm aqueous solution of stevioside sprayed on 5-year-old Ginchoge trees (*Daphne japonica* Thunb.) increased blooming by as much as 42% and there was also an increase in root growth when Yaemugura trees [*Galium aparine* L.] were cultivated in the presence of 100 ppm stevioside solution.[63]

TABLE 9
GROWTH-PROMOTING EFFECTS OF DITERPENE GLYCOSIDES ON RICE SEEDLINGS

Diterpene glycoside	*Elongation of rice seedlings (%)[63] Concentration (ppm)*				
	Control	*Medicated*			
	0	*100*	*200*	*500*	*1000*
Stevia extract[a]	100	133·2	149·1	145·4	143·4
Stevioside	100	165·2	186·4	186·7	179·2
Rebaudioside A	100	119·5	126·2	138·3	136·7
Rebaudioside B	100	118·2	143·7	144·8	131·9
Rebaudioside C	100	114·6	121·1	135·7	144·8
Dulcoside A	100	107·5	126·7	130·3	128·5
Steviolbioside	100	125·2	134·6	143·1	137·1

[a] Containing (% by wt): steviolbioside (2), stevioside (75), rebaudioside A (3), Dulcoside A (2), rebaudioside C (3) and others (13).

11. CONCLUSIONS

It has been confirmed without any doubt that rebaudioside A is superior in sweetness, aftertaste and stability to stevioside. Opinions differ on the aftertaste characteristics of rebaudioside A, but the majority verdict is that its taste is close to that of sucrose with a weak bitter aftertaste. However, as stevia extracts and stevioside are widely used in South-East Asia and parts of South America, this implies that consumers there are satisfied with these natural sweeteners. Pure rebaudioside A is not yet commercially available but is a constituent of stevia extracts such as Stevix and Marumilon 50. The controversy over the safety of stevioside and rebaudiosides continues. No proof has been reported that sweet diterpene glycosides are metabolised to the mutagen, steviol, in man. It is of interest in this respect that the Delaney Amendment to the US Food and Drug Laws does not cover mutagenicity.

There have been several reports of plant tissue culture experiments with the purpose of increasing the stevioside content of *S. rebaudiana* plants.[58] A recent report from Morita Kagaku Kogyo of Japan apparently had successfully applied tissue culture techniques to the *S. rebaudiana* plant to obtain 2·56 times as much rebaudioside A as that of stevioside.[64] We are of the opinion that the sweet diterpene glycosides, in particular rebaudioside A, have a promising future as a substitute for sweeteners and sugar. The sweet principles of the stevia plant have been in use longer than any other sweetener except honey.

REFERENCES

1. SOEJARTO, D. D., KINGHORN, A. D. and FARNSWORTH, N. R. (1982). *J. Nat. Prod.*, **45**, 590.
2. KOHDA, H., KASAI, R., YAMASAKI, K., MURAKAMI, K. and TANAKA, O. (1976). *Phytochemistry*, **15**, 981.
3. (a) BRIDEL, M. and LAVIEILLE, R. (1931). *J. Pharm. Chim.*, **14**, 99.
   (b) WOOD, H. B., Jr and FLETCHER, H. G., Jr (1956). *J. Amer. Chem. Soc.*, **78**, 207.
4. SAKAMOTO, I., YAMASAKI, K. and TANAKA, O. (1977). *Chem. Pharm. Bull.*, **25**, 844.
5. RAJBHANDARI, A. and ROBERTS, M. F. (1983). *J. Nat. Prod.*, **46**(2), 194.
6. SHOLICHIN, M., YAMASAKI, K., MIYAMA, R., YAHARA, S. and TANAKA, O. (1980). *Phytochemistry*, **19**, 326.
7. SUZUKI, H., IKEDA, T., MATSUMOTO, T. and NOGUCHI, M. (1976). *Agric. Biol. Chem.*, **40**, 819.
8. DARISE, M., KOHDA, H., MIZUTANI, K., KASAI, R. and TANAKA, O. (1983). *Agric. Biol. Chem.*, **47**(1), 133.
9. MORI, K., NAKAHARI, Y. and MATSUI, M. (1970). *Tetrahedron Lett.*, **28**, 2411.

10. COOK, I. F. and KNOX, J. R. (1970). *Tetrahedron Lett.*, **47**, 4091.
11. ZEIGLER, F. E. and KLOEK, J. A. (1977). *Tetrahedron*, **33**, 373.
12. RUDDAT, M., LONG, A. and MOSETTIG, E. (1963). *Naturwiss.*, **50**, 23.
13. HASHIMOTO, Y. and MORIYASU, M. (1978). *Shoyakugaku Zasshi*, **32**, 209.
14. MORITA, T., FUJITA, I. and IWAMURA, J. (1978). US Patent 4 082 858.
15. DARISE, M., MIZUTANI, K., KASAI, R., TANAKA, O., KITAHATA, S., OKADA, S., OGAWA, S., MURAKAMI, F. and FENG-HUAI CHEN. (1984). *Agric. Biol. Chem.*, **48**(10), 2483.
16. PERSINOS, G. J. (1973). US Patent 3 723 410.
17. (a) KANEDA, N., KASAI, R., YAMASAKI, K. and TANAKA, O. (1977). *Chem. Pharm. Bull.*, **25**(9), 2466; (b) TANAKA, O. (1980). In *Chemistry of* Stevia rebaudiana *Bertoni—New Source of Natural Sweeteners*, Recent Advances in Natural Products Research, Seoul National University Press, 111.
18. WOOD, H. B., Jr, ALLERTON, R., DIEHL, H. W. and FLETCHER, H. G., Jr. (1955). *J. Org. Chem.*, **20**, 875.
19. DUBOIS, G. E. (1984). US Patent 4 454 290.
20. SAKAMOTO, I., YAMASAKI, K. and TANAKA, O. (1977). *Chem. Pharm. Bull.*, **25**(12), 3437.
21. MITSUHASHI, H., UENO, J. and SUMIDA, T. (1975). *Yakugaku Zasshi*, **95**, 1501.
22. KOBAYASHI, M., HORIKAWA, S., DEGRANDI, I. H., UENO, J. and MITSUHASHI, H. (1977). *Phytochemistry*, **16**, 1405.
23. AHMED, M. S., DOBBERSTEIN, R. H. and FARNSWORTH, N. R. (1980). *J. Chromatogr.*, **192**, 387.
24. JUNICHI, I., RYOHIKO, K., TOSHIO, I., OSAMU, K., TOYOSHIGE, M. and NENOKICHI, H. (1982). *Nippon Nogei Kagaku Kaishi*, **56**, 87.
25. KINGHORN, A. D., NANAYAKKARA, N. P. D., SOEJARTO, D. D., MEDON, P. J. and KAMATH, S. (1982). *J. Chromatogr.*, **237**, 478.
26. MIZUKAMI, H., SHIIBA, K. and OHASI, H. (1982). *Phytochemistry*, **21**, 1927.
27. SUGISAWA, H., KASAI, T. and SUZUKI, H. (1977). *Nippon Nogei Kagaku Kaishi*, **51**, 175.
28. MIYAZAKI, Y., WATANABE, H. and WATANABE, T. (1978). *Eisi Shikensho Hokoku*, **96**, 86.
29. MITSUHASHI, H., UENO, J. and SUMIDA, T. (1975). *Yakugaku Zasshi*, **95**, 127.
30. SAKAMOTO, I., KOHDA, H., MURAKAMI, K. and TANAKA, O. (1975). *Yakugaku Zasshi*, **95**, 1507.
31. HASHIMOTO, Y., MORIYASU, M., NAKAMURA, S., ISHIGURO, S. and KOMURO, M. (1978). *J. Chromatogr.*, **161**, 403.
32. CHEN, W. S. and YEH, C. S. (1978). *T'ai-wan T'ang Yeh Yen Chiu So Yen Chiu Hui Pao*, **79**, 43; *Chem. Abstr.* , **89**, 211995d.
33. YOHEI, H. and MASATAKA, M. (1978). *Shoyakugaku Zasshi*, **32**, 209.
34. HIROKADO, M., NAKAJIMA, I., NAKAJIMA, K., MIZOIRI, S. and ENDO, F. (1980). *Shokuhin Eiseigaku Zasshi*, **21**, 451.
35. AHMED, M. S. and DOBBERSTEIN, R. H. (1982). *J. Chromatogr.*, **236**, 523.
36. AHMED, M. S. and DOBBERSTEIN, R. H. (1982). *J. Chromatogr.*, **245**, 373.
37. MAKAPUGAY, H. C., NANAYAKKARA, P. D. and KINGHORN, A. D. (1984). *J. Chromatogr.*, **283**, 390.
38. HONGYONG, N. and YU, S. H. (1985). *Shengwu Huaxue Yu Shengwu, Wuli Yinzhan*, **64**, 63.

39. MINAMISONO, H. and AZUNO, K. (1978). *Nenpo-Kagoshima-ken Kogyo Shikenjo*, **24**, 66.
40. SHIRAKAWA, T. and ONISHI, T. (1979). *Kagawa-ken Hakko Shokuhin Shikenjo Hokoku*, **71**, 35.
41. NAKAJIMA, I., HIROKADO, M., USAMI, H., MIZOIRI, S. and ENDO, F. (1979). *Tokyo-toritsu Eisei Kenkyusho Kenkyu Nempo*, **30–31**, 153.
42. NAKAJIMA, I., HIROKADO, M., NAKAJIMA, K., MIZOIRI, S. and ENDO, F. (1980). *Tokyo-toritsu Eisei Kenkyusho Kenkyu Nempo*, **31–32**, 180.
43. TEZUKA, S., YAMANO, T., SHITOU, T. and TADAUCHI, N. (1980). *Shokuhin Kogyo*, **23**(18), 43.
44. CHANG, S. S. and COOK, J. M. (1983). *J. Agric. Food Chem.*, **31**, 409.
45. TANAKA, O. (1982). *Trends Anal. Chem.*, **1**(11), 246.
46. SCHIFFMAN, S. S., REILLY, D. A. and CLARK III, T. B. (1979). *Physiology and Behaviour*, **23**, 1.
47. *Utilization of Stevia Extracts to Food Industry* (1980). Internal publication of Meruzen Pharmaceutical Co., Ltd, Onomichi, Japan.
48. CRAMMER, B. and IKAN, R. (1977). *Chem. Soc. Rev.*, **6**, 431.
49. LEE, S. J., LEE, K. R., PARK, J. R., KIM, K. S. and TCHAI, B. S. (1979). *Hanguk Sikp'um Kwahakhoe Chi*, **11**, 224.
50. FUJITA, H. and EDAHIRO, T. (1979). *Shokuhin Kogyo*, **22**(20), 66.
51. AKASHI, H. and YOKOYAMA, Y. (1975). *Shokuhin Kogyo*, **18**(20), 34.
52. MEDON, P. J., PEZZUTO, J. M., HOVANEC-BROWN, J. M., NANAYAKKARA, N. P. D., SOEJARTO, D. D., KAMATH, S. K. and KINGHORN, A. D. (1982). *Fed. Proc. Amer. Soc. Exp. Biol.*, **41**, 1568.
53. PLANAS, G. M. and KUĆ, J. (1968). *Science*, **162**, 1007.
54. WENIGER, B., HAAG-BERRURIER, M. and ANTON, R. (1982). *J. Ethnopharmacol.*, **6**(1), 67.
55. YAMADA, A., OHGAKI, S., NODA, T. and SHIMIZU, M. (1985). *J. Food Hyg. Soc. Japan*, **26**(2), 169.
56. WINGARD, R. E., Jr, BROWN, J. P., ENDERLIN, F. E., DALE, J. A. HALE, R. L. and SEITZ, C. T. (1980). *Experientia*, **36**, 519.
57. NAKAYAMA, K., KASAHARA, D. and YAMAMOTO, F. (1986). *Shokuhin Eiseigaku Zasshi*, **27**(1), 1.
58. KINGHORN, A. D. and SOEJARTO, D. D. (1985). Current status of stevioside as a sweetening agent for human use, in *Economic and Medicinal Plant Research*, Vol. 1, H. Wagner, H. Hikino and N. R. Farnsworth (Eds), Academic Press, London, 2.
59. (a) DUBOIS, G. E., DIETRICH, P. S., LEE, J. F., MCGARRAUGH, G. V. and STEPHENSON, R. A. (1981). *J. Med. Chem.*, **24**, 1269; (b) DUBOIS, G. E. (1982). US Patent 4 332 830.
60. DUBOIS, G. E. and STEPHENSON, R. A. (1985). *J. Med. Chem.*, **28**, 93.
61. CROZIER, A., KUO, C. C., DURLEY, R. C. and PHARIS, R. P. (1970). *Can. J. Bot.*, **48**, 867.
62. VALIO, I. F. M. and ROCHA, R. F. (1976). *Z.Pflanzenphysiol.*, **78**(1), 90.
63. IWAMURA, J. and KOMAI, K. (1984). US Patent 4 449 997.
64. MORITA KAGAKU KOGYO (1986). Japanese Patent 61 202 667.

Chapter 3

TECHNICAL AND COMMERCIAL ASPECTS OF THE USE OF LACTITOL IN FOODS AS A REDUCED-CALORIE BULK SWEETENER

C. H. DEN UYL

CCA Biochem BV, Gorinchem, The Netherlands

SUMMARY

Lactitol (trade name Lacty®, manufactured by CCA Biochem, The Netherlands) is a new sweetener which can substitute for sucrose in many food applications. It is a sugar alcohol derived from lactose by hydrogenation of the glucose moiety. It can be crystallized from solution under special conditions, and it provides foods with the bulk and texture normally given by sucrose, but with only half of the calories. Lactitol does not affect blood-glucose or insulin levels, so it is valuable for diabetics. It is very slowly converted into acid by oral plaque bacteria and is thus suitable for products that are safe for the teeth. Although it is less sweet than sucrose (30–40% of the sweetness), this is an advantage in some processed foods. In other products the sweetness can be increased by adding intense sweeteners. Dietetic foods containing lactitol are virtually identical in appearance, taste and texture with conventional foods. One particular application is in the manufacture of slimming products.

1. PREPARATION AND SPECIFICATIONS

1.1. Preparation

The preparation of lactitol (Fig. 1) by hydrogenation of lactose with a nickel catalyst has been known for about 50 years.[1]

A 30–40% lactose solution in water at about 100°C is subjected to a

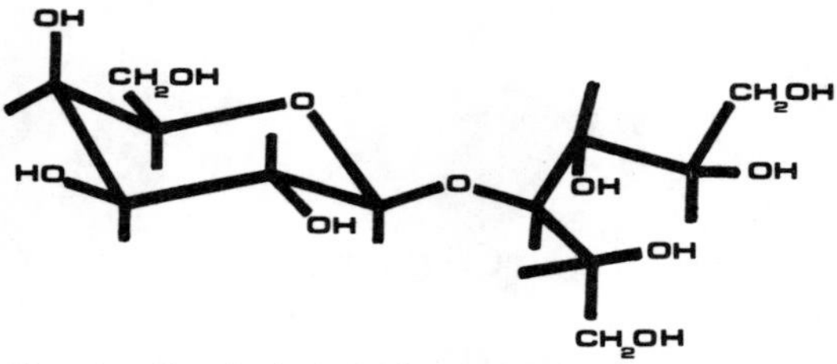

FIG. 1. Lactitol, 4-*O*-(*β*-galactosyl)-D-glucitol.

hydrogen pressure of 40 bar or more in the presence of a Raney nickel catalyst, according to the following reaction:

$$C_{12}H_{22}O_{11} + H_2 \xrightarrow{\text{Raney nickel}} C_{12}H_{24}O_{11}$$

After purification, lactitol can be crystallized in a mono- or dihydrate form, depending on the crystallization conditions.[2]

1.2. Specifications

The purity of the food grade of lactitol is specified in Table 1. There are two types available: monohydrate and dihydrate crystalline forms. The dihydrate is marketed by CCA Biochem BV under the trade name Lacty®.

TABLE 1
SPECIFICATIONS OF LACTITOL FOOD GRADE

Specification	*Lactitol monohydrate food grade*	*Lactitol dihydrate food grade Lacty®*
Molecular weight	362	380
Description	White crystalline powder	White crystalline powder
Taste	Sweet	Sweet
Odour	Odourless	Odourless
Lactitol content	Min. 97·5% d.s.[a]	Min. 97·5% d.s.
Water (Karl Fischer)	4·5–5·5%	9·0–10·5%
Other polyols	Max. 2·5%	Max. 2·5%
Reducing sugars	Max. 0·1%	Max. 0·1%
Specific rotation $[\alpha]_D^{25}$	+13·5–15·0°	+13·5–15·0°
pH of 10% solution	4·5–8·5	4·5–8·5
Chloride	Max. 30 ppm	Max. 30 ppm
Sulphate	Max. 30 ppm	Max. 30 ppm
Ash content	Max. 0·1%	Max. 0·1%
Heavy metals	Max. 10 ppm	Max. 10 ppm
Arsenic	Max. 1 ppm	Max. 1 ppm
Nickel	Max. 1 ppm	Max. 1 ppm

[a] Abbreviation: d.s., dry solids.

2. SENSORY EVALUATION

The relative sweetness of polyol sweeteners is normally determined by comparison with sucrose solutions at different concentrations. The relative sweetness of lactitol, compared with sucrose solutions of 2, 4, 6 and 8%, was determined using triangle tests and untrained participants. The results shown in Table 2 were obtained.

TABLE 2
RELATIVE SWEETNESS OF LACTITOL AT DIFFERENT CONCENTRATIONS

Concentration of sucrose (%)	*Relative sweetness* (*sucrose = 1·0*)
2	0·30
4	0·35
6	0·37
8	0·39

Lactitol has a clean sweet taste, closely resembling sucrose in sweetness-character.

Although it is not as sweet as sugar, in certain processed foods a reduced sweetness may even contribute to an attractive taste, in which the other flavours have more chance of being perceived. A lactitol-containing hard-boiled sweet, for instance, has a very good taste without the need of an intense sweetener.

When the sweetness of lactitol-containing food products has to be increased, however, artificial sweeteners can be added. Aspartame and acesulfam-K have proved to be especially suitable as they do not have such an unpleasant, bitter aftertaste as saccharin. It was found that a 10% lactitol solution containing 0·03% aspartame or acesulfam-K or 0·013% sodium saccharinate was of equal sweetness to a 10% sucrose solution.

3. PHYSICAL PROPERTIES

Lactitol is available in a white crystalline form with good storage stability. There are two physical forms available commercially: a monohydrate with a melting point of about 120°C, and a dihydrate with a melting point of about 75°C.

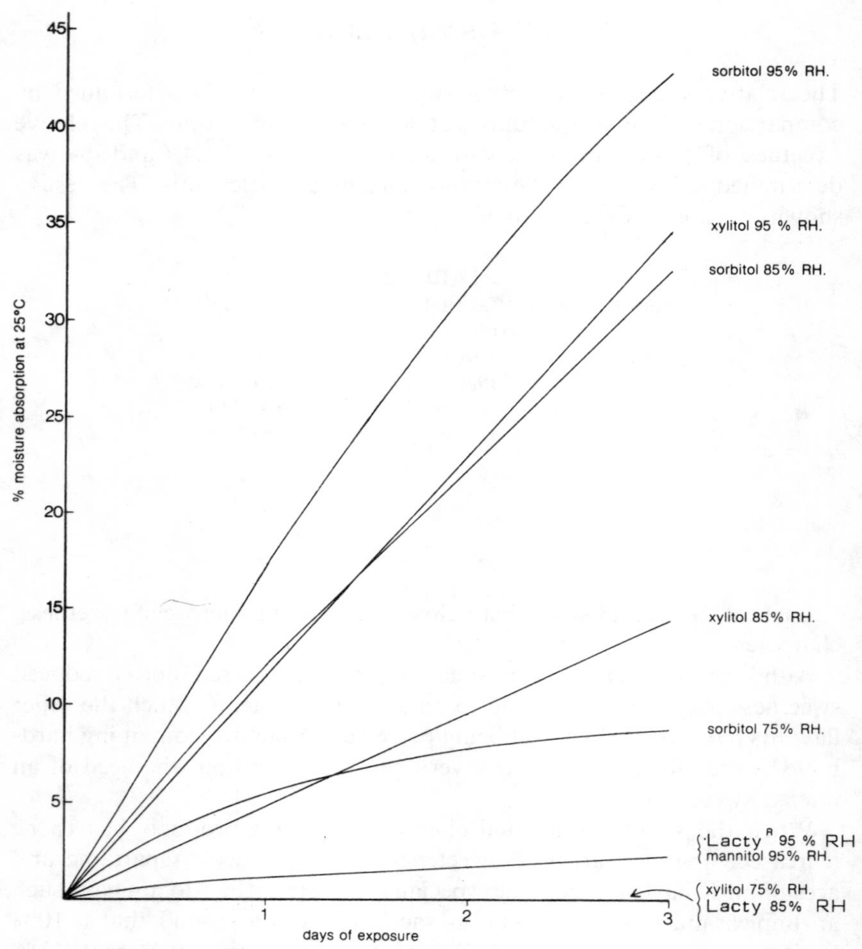

FIG. 2. Moisture absorption of crystalline polyols at different relative humidities.

When heated at temperatures of 179–240°C lactitol is partly converted into anhydro derivatives (lactitan), sorbitol and lower polyols.

In solid and melted conditions, at temperatures up to 270°C, it is not possible to ignite lactitol with a flame.

3.1. Hygroscopicity and Water Activity

Lactitol mono- and di-hydrate crystals are non-hygroscopic, far less hygroscopic than sorbitol and xylitol, and similar to mannitol (Fig. 2).

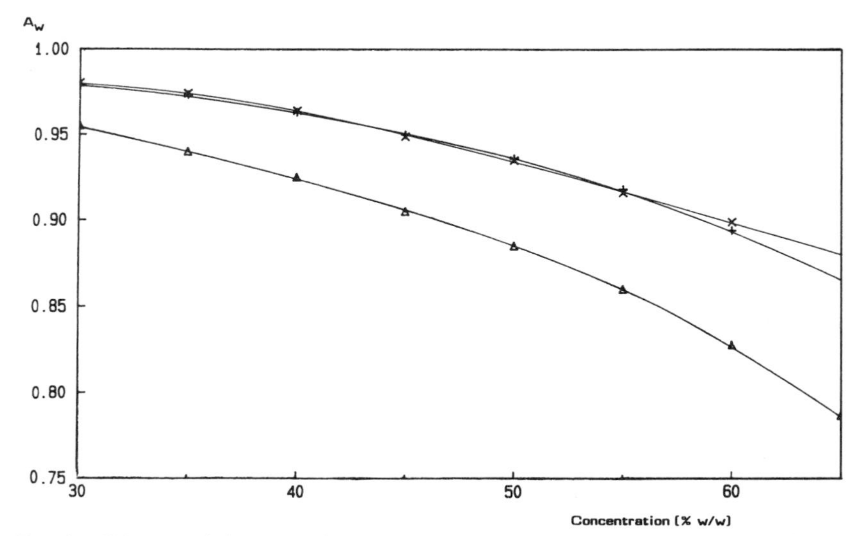

FIG. 3. Water activity (A_w) of sucrose, sorbitol and lactitol solutions at different concentrations: +, sucrose; ×, lactitol; △, sorbitol.

Hygroscopicity gives an indication of its ability to absorb water, as it is closely related to water activity (A_w) and equilibrium relative humidity (ERH):

$$A_w = \frac{P}{P_0} = \frac{\text{vapour pressure of solution}}{\text{vapour pressure of water}} = \frac{\text{ERH}}{100}$$

The water activity is important in the regulation of the moisture pick-up and release. It also influences enzyme activity, Maillard reaction, fat oxidation, microbial stability and texture. As water activity is related to molar concentration, it can be easily understood that the water activity of lactitol in solution is comparable with that of sugar solutions, whereas the A_w of sorbitol is lower. This is shown in Fig. 3. The lower water activity of sorbitol shows that sorbitol solutions are more hygroscopic than lactitol solutions.

3.2. Solubility

Lactitol has good solubility in water. At 25°C, 150 g of lactitol monohydrate or 140 g of lactitol dihydrate will dissolve in 100 ml water. These solutions have a dry solids (d.s.) content of 60% and 58% respectively.

The solubility increases greatly with increasing temperature. Figure 4

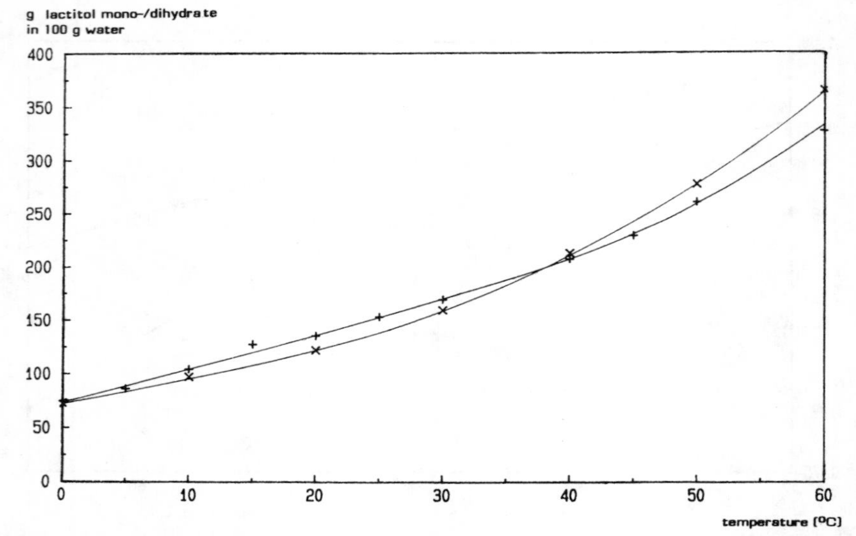

FIG. 4. Solubility of lactitol mono- and di-hydrate in water at different temperatures: +, monohydrate; ×, dihydrate.

shows the solubility of lactitol mono- and dihydrate at different temperatures, demonstrating that lactitol can be used in the preparation of solutions for all kinds of applications. Only when making products with a high dry solids content can crystallization occur.

For example, jams with a high dry solids content (67% d.s.) cannot be prepared by using only lactitol; a crystallization inhibitor has to be used. For hard-boiled sweets there are comparable problems: another polyol has to be used to 'doctor' it and inhibit the crystallization of lactitol.

3.3. Heat of Solution

When lactitol is dissolved in water, the crystals need energy to dissolve, which gives an endothermic cooling effect. The heat of solution of lactitol monohydrate is −52·1 J/g. The value for lactitol dihydrate is −58·1 J/g.

3.4. Viscosity

The viscosity of lactitol at different concentrations is shown in Fig. 5. The viscosity of aqueous lactitol solutions is quite similar to that of sucrose solutions of equal concentration (Table 3).

The viscosity is of importance in food products with a high dry solids content in which a high amount of lactitol is used. In hard-boiled sweets the

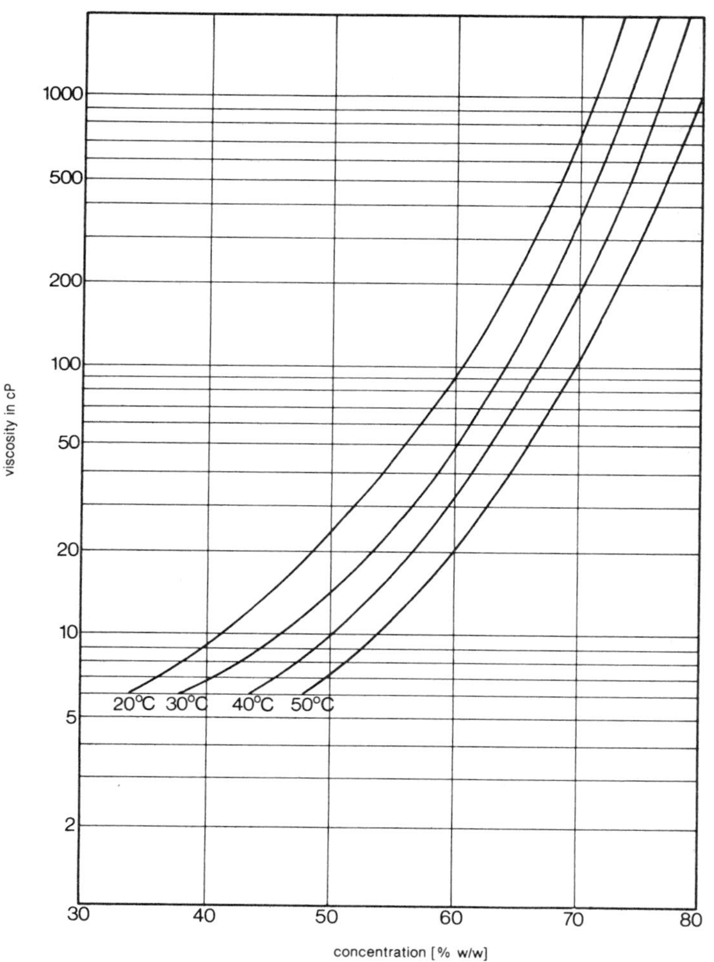

FIG. 5. Viscosities of lactitol solutions.

TABLE 3

VISCOSITIES OF LACTITOL AND SUCROSE SOLUTIONS AT EQUAL CONCENTRATION (60% d.s.)

Temperature (°C)	*viscosity (cP)*	
	Lactitol	*Sucrose*
30	47	45
50	21	19

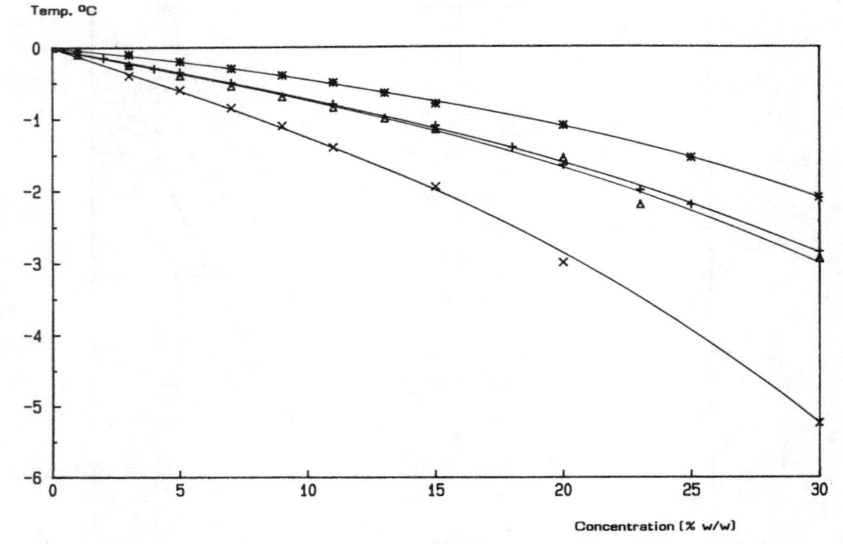

FIG. 6. Freezing points of sucrose (△), sorbitol (×), lactitol (+) and polydextrose (*) solutions at different concentrations.

viscosity has a strong influence on the processability of the boiled mass and on the shape-stability of the boilings during storage (cold flow). In this respect lactitol functions very well.

3.5. Effect on Freezing Point

Lactitol has an influence on the freezing point of solutions. Freezing point depression is related to the molecular weight of the solute. A comparison was made between solutions of lactitol, sucrose, sorbitol and polydextrose. The results are shown in Fig. 6, from which it appears that lactitol and sucrose are again quite similar. This property is of importance when making ice-cream: it influences the freezing point of ice-cream and also other properties such as sweetness perception, hardness and melting characteristics.

4. CHEMICAL PROPERTIES

Lactitol is a polyol with nine —OH groups, which can be esterified with fatty acids resulting in emulsifiers,[3] or it can react with propylene oxide, forming polyurethanes.

TABLE 4
DECOMPOSITION OF A 10% LACTITOL SOLUTION UNDER EXTREME CONDITIONS OF pH AND TEMPERATURE

pH	*Recovery after 24 h at 105°C (% w/w)*			
	Lactitol	*Galactose*	*Sorbitol*	*Not identified*
2·0	42·9	28·1	21·4	7·6
10·0	98·0	—	—	2·0 (acids)
12·0	98·3	—	—	1·7 (acids)

Owing to the absence of a carbonyl group, lactitol is chemically more stable than related disaccharides such as lactose. Its stability in the presence of alkali is higher than that of lactose, but is similar to that of lactose in the presence of acids.

The absence of a carbonyl group also means that lactitol does not take part in non-enzymic browning (Maillard) reactions which are important in bakery products to give these products a browned appearance.

Lactitol solutions have excellent storage-stability in the pH range 3·0–7·5 and at temperatures up to 60°C.

No decomposition is shown by 10% lactitol solutions in the pH range 3·0–7·5 at 60°C after 1 month. After 2 months only at pH 3·0 some decomposition (15%) is detected. With increasing temperature, and especially with increasing acidity, a hydrolytic decomposition of lactitol is observed, sorbitol and galactose being the main decomposition products.

At high pH lactitol is stable even at 105°C. This is shown in Table 4.

Under the severe conditions of baking cookies and manufacturing hard candies (pH 3, 145°C), the lactitol content remains unchanged.

When heated at temperatures of 179–240°C, lactitol is partly converted into anhydro derivatives (lactitan), sorbitol and lower polyols.

5. METABOLISM OF LACTITOL BY MICRO-ORGANISMS

5.1. Tooth Protective Properties

Sugars are the main dietary factor in dental caries: oral bacteria convert sugars into polysaccharides that contribute to dental plaque deposited on the teeth, and in this plaque several sugars are fermented into acids. The acid demineralizes the enamel and causes cavities.

At the University of Zürich, a special method has been developed for the

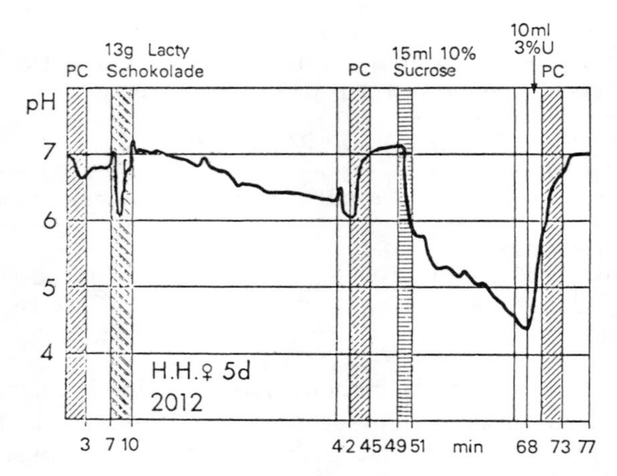

FIG. 7. Telemetrically recorded pH of 5-day interdental plaque in subject H.H. during and after consumption of 13 g Lacty® plain chocolate. A 10% sucrose solution was used as a positive control. PC, 3 min paraffin chewing; U, rinsing with 10 ml of 3% carbamide.

determination *in vivo* of plaque pH in humans. After consumption of chocolate or confectionery in which sucrose is replaced by lactitol, changes in plaque acidity are detected by electrodes and transmitted electronically to a graph recorder. The term 'Zahnschonend' (safe to teeth/friendly to teeth) is used officially in Switzerland when the pH of a dental plaque does not fall below 5·7 during a 30-min period. By this means Mühlemann demonstrated that in chocolates lactitol was 'safe to teeth'.[4]

In Fig. 7 it is shown that during and after eating 13 g of Lacty®-chocolate (plain lactitol-containing chocolate), the plaque pH was very little influenced, whereas with 15 ml of a 10% sucrose solution the plaque pH was significantly reduced to about 4·5, which means that sucrose is easily fermented into acids by the oral bacteria.

In another study it was demonstrated that *Streptococcus mutans*, and other bacteria such as *Lactobacillus* and *Bifidobacterium* isolated from the dental plaque, form acid from lactitol, but the acid formation is slow, comparable with that from sorbitol. Moreover, no dental plaque is formed from lactitol.[5]

The cariogenicity of lactitol was also tested in an experiment with programme-fed rats.[6] For this purpose, lactitol was incorporated in a powdered diet, consisting of 50% of a basal diet (SPP, Trouw & Co., Putten, The Netherlands), 25% wheat flour and 25% of test substance. Lactitol was

TABLE 5
AVERAGE OF CARIOUS FISSURE LESIONS AND WEIGHT GAINS IN FIVE DIETARY GROUPS OF 12 RATS

Diet	*Number of fissure lesions*[a]		*Weight gain*
	B	*C*	*(g)*
Sucrose	5·75	2·58	82·0
Sorbitol	1·33*	0·42*	46·0**
Lactitol	1·17*	0·33*	72·2
Wheat flour	0·67*	0·25*	80·8
Xylitol	0·42*	0*	48·6**
Pooled standard error	0·56	0·43	5·4

[a] Twelve fissures at risk. B-lesion: enamel lesion with progression into the dentin.[7] C-lesion: advanced lesion of the dentin whereby enamel destruction has occurred.[7]
* $P<0{\cdot}01$, ** $P<0{\cdot}001$ } significantly different from the sucrose group (Tukey's test).

compared with sorbitol, xylitol, sucrose and a control consisting of 50% wheat flour and 50% basal diet.

The rats were programme-fed. None of the animals suffered from diarrhoea, and no significant adverse effects to general health were observed. The results are shown in Table 5. Obviously, substitution of sucrose by lactitol significantly reduced the dental caries figures. The caries results for sorbitol and xylitol were in accordance with other studies on these polyols. Further reports are now beginning to appear on the favourable dental properties of lactitol.[8,9]

5.2. General Fermentability of Lactitol

Many bacteria which are able to ferment lactose (such as Enterobacteriaceae e.g. *E. coli*) cannot ferment lactitol.

6. PHYSIOLOGICAL PROPERTIES

6.1. Metabolism and Metabolizable Energy of Lactitol

The enzyme that brings about breakdown of lactose by hydrolysis can operate on lactitol only very slowly. From studies based on six human intestinal biopsies,[10] it was demonstrated that activities towards lactitol and Palatinit® were only 1–3% of those towards lactose and isomaltose.

The activity towards maltitol was much higher, approximately 10% of that for maltose. These findings indicate that lactitol and Palatinit® are poorly digested, while significant amounts of maltitol might be digested and utilized by man.

Consequently, lactitol can pass the small intestine without being hydrolysed or absorbed, but it will be converted in the large intestine by the flora there into organic acids, carbon dioxide (CO_2) and biomass. These factors will have an effect on the metabolizable energy.

The metabolic energy (ME) was determined in several studies. First, in rats,[11] lactitol had an ME of 68% in comparison with starch. Second, in mini-pigs,[12] the ME value was $56 \pm 9\%$ compared with starch. Third, a study with human volunteers was carried out.[13] They were kept for 4-day periods in a respiratory room. In one period they had a diet with 49 g of sugar a day, and in the second period this sugar was replaced by lactitol. It was shown that the ME of lactitol was about 60% lower than that of sucrose. Because of experimental variation, the conclusion was that the ME of lactitol (anhydrous) is at most 50% of that of sucrose, or 2 kcal/g. This study also proved that the volunteers had hardly any discomfort from eating 50 g of lactitol monohydrate divided into 4–6 portions throughout the day. There was no diarrhoea, but in the lactitol period the bulk of the faeces was greater and the dry matter content was lower (13·7%). Tolerance to lactitol is dependent on the manner of consumption and adaptation.

6.2. Laxative Properties

The laxative effects of lactitol, xylitol and sorbitol were compared with those of lactose in feeding tests in young male rats. In the initial stages of the study, lactitol (10%) induced considerably less diarrhoea than xylitol and sorbitol. The rats showed rapid adaptation to the diets. It was concluded that each of the sugar alcohols, and xylitol in particular, when fed at the 10% level, was more laxative than lactose. When fed at the 5% level, lactitol was considerably less laxative than xylitol and sorbitol, and comparable with 10% lactose.[14]

To determine laxative thresholds in man, 21 healthy volunteers entered a single-blind, randomized cross-over trial, taking, in divided doses, increasing amounts of lactitol, sorbitol or placebo. The laxative thresholds of lactitol ($74{\cdot}7 \pm 6{\cdot}3$ g/day) and sorbitol ($71{\cdot}9 \pm 4{\cdot}9$ g/day) were similar and the incidence of gastrointestinal effects was not significantly different comparing sorbitol and lactitol.[15]

In man disturbing physiological side-effects of lactitol intake (osmotic diarrhoea, flatulence) will diminish when it is regularly used. The intake of lactitol in sweets, in portions of 20 g, does not cause inconvenient side-effects in people accustomed to lactitol. When smaller doses, spread over the day, are taken, far larger doses can be consumed, without side-effects worth mentioning.

6.3. Dietary Fibre

From the foregoing on the metabolic pathways of lactitol, we may conclude that lactitol is not metabolized as a carbohydrate:

—It is not absorbed or hydrolysed in the small intestine.
—It is not absorbed in the large intestine.
—It functions as a carbon source for the intestinal microflora, giving more stool plus gas formation.

So lactitol, chemically not a carbohydrate but a hydrogenated carbohydrate, does not function metabolically as a carbohydrate, but has certain similarities to dietary fibre.

Figure 8 shows the pathways of metabolism of dietary fibres: there is no absorption in the small intestine, but part or total fermentation in the large intestine by bacteria, giving fatty acids and a large bacterial mass which increases faecal mass. Thus lactitol is, like lactulose, a non-absorbable carbohydrate with metabolic effects which are comparable with dietary fibre. From this point of view the physiological side-effects of lactitol are not unfavourable, but may help in protection against a range of diseases prevalent in Western communities and blamed on a diet low in dietary fibre.

6.4. Suitability for Diabetics

Lactitol, 24 g/day orally, was well tolerated by healthy or diabetic persons and it did not influence blood-glucose or blood-insulin levels. It did not induce diarrhoea in diabetic patients.[17]

Loading tests with equal amounts of sucrose, lactose, lactitol and lactitol + sucrose were carried out on eight healthy adults.[18] The average maximal increases in blood-glucose concentration after the different loadings were 63 ± 26, 43 ± 19, 6 ± 3 and 40 ± 14 mg% respectively. The increase in blood glucose level on lactitol was within the normal fluctuation.

Lactitol is therefore suitable for both insulin-dependent (type I) diabetics and non-insulin-dependent (type II) diabetics. This second group can limit diabetes by reducing sugar intake, as well as taking dietetic measures, so that lactitol also fits into their diet because of its reduced calorie value in comparison with sugar.

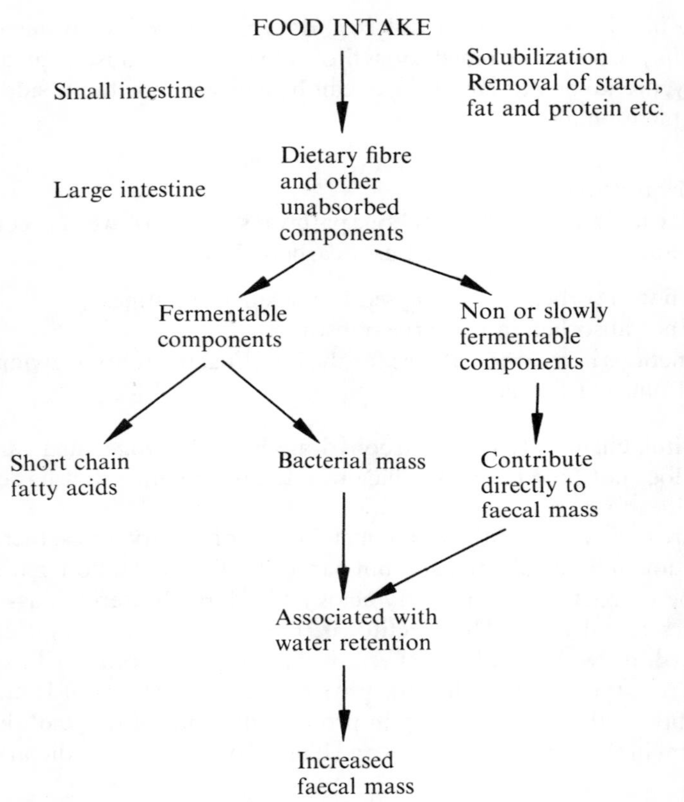

FIG. 8. Involvement of dietary fibre in production of faecal mass.[16]

6.5. Toxicology

All the required toxicity studies for a food additive petition have been carried out. The results did not reveal deleterious effects of feeding lactitol at levels up to 10% of the diet.

The Joint FAO/WHO Expert Committee on Food Additives evaluated all the biological data at its meeting in Geneva in April 1983, and allocated an ADI (Acceptable Daily Intake) 'not specified' for lactitol.[19]

In the EEC, the Scientific Committee on Food evaluated lactitol and other polyols and identified it as a safe product with the remark (for all polyols) that: 'Laxation may be observed at high intakes. Consumption of the order of 20 g/person/day of polyols is unlikely to cause undesirable laxative symptoms'.[20]

7. APPLICATIONS

Because of its biological properties, lactitol can replace sugar as a sweet-tasting bulking or texturizing agent in a variety of applications such as

—diabetic products;
—products which are safe for teeth;
—slimming products (low-calorie);
—table-top sweeteners (low-calorie).

In these applications lactitol may be used alone or in combination with intense sweeteners. By using lactitol instead of sucrose, foods can be prepared that are equal in palatability and eating qualities to their sucrose-containing counterparts. The lactitol-containing products have a sweet taste and definitely no aftertaste.

Lactitol can be used in several ways in different products, described below.

7.1. Table-top Sweeteners and Compressed Sweets

Since lactitol is not hygroscopic and since it has a reduced caloric value, it is a very suitable bulking agent for intense sweeteners in table-top use.

7.2. Bakery Products

Lactitol-containing bakery products have essentially the same properties as regards texture, specific volume and shelf-life as conventional sugar-containing products (e.g. cakes and pastries).

Crispness, very essential for products like biscuits, can be obtained by using lactitol as a sugar substitute. When sorbitol or xylitol is used these products are subject to very rapid deterioration of crispness and become soft within hours of production.

7.3. Chocolate

The preparation of lactitol-containing chocolate is similar to that of its sugar-containing counterpart, since lactitol fine powder is only slightly hygroscopic and has little impact on the viscosity of the chocolate mass during conching. Typical formulations are listed in Table 6.

Other important properties of the product are:

—it is easy to wrap because of the low hygroscopicity of lactitol;
—it has a good taste, but an additional intense sweetener such as aspartame is needed;
—it has a reduced calorie content because of the caloric value of lactitol and a lower fat content than sorbitol-containing chocolate.

TABLE 6
TYPICAL CHOCOLATE FORMULATIONS

	Milk chocolate	*Bitter chocolate*
Cacao	20	45
Cacao butter	20	10
Lecithin	0·3	0·5
Lactitol	51·23	44·35
Sodium caseinate	4·4	—
Butter oil	3·9	—
Aspartame	0·10	0·15

7.4. Chewing Gum

Sugarless chewing gum has received considerable interest since it is suitable as a diabetic foodstuff and it does not cause dental caries.

Chewing gums with lactitol comply with these characteristics. Moreover, they have a reduced calorie content, because the caloric utilization of lactitol in man is only about 2 kcal/g.

Using lactitol powder instead of sorbitol powder gives a chewing gum which is more flexible and not hygroscopic.

The amount of gum-base, which is often 27–30% in sorbitol-containing chewing gum, can be reduced to about 25% when lactitol powder in combination with Lycasin® is used. Because of its non-hygroscopicity, lactitol may also be used for the surface dusting of sugar-free chewing gum at the rolling-out stage.

7.5. Hard-Boiled Sweets

The viscosity of molten lactitol (2·4 Pa s at 135°C), resembles that of sucrose–glucose (60:40) mix, enabling the production of hard candy by traditional kneading and forming procedures. As the viscosity of a sorbitol melt (0·1 Pa s at 135°C) is far lower than lactitol, sorbitol hard-boilings can only be produced using the more difficult depositing process.

When lactitol is used in preparing a hard candy, a glassy structure is obtained which remains stable over a long period. During storage, crystallization of lactitol may occur, but this can be inhibited, e.g. by using 30–40% Lycasin in the product. This does not affect the overall quality. The product can be made without the need for an additional sweetener.

7.6. Sugar-Free Fruit Gums and Pastilles

Lactitol can be used in sugar-free pastilles giving a good taste. Another polyol (up to about 50%) is also needed to prevent the lactitol from crystallizing.

7.7. Ice-Cream

Ice-cream made with lactitol has good melting characteristics and a very good structure. It has a bland sweet taste with no aftertaste. The low sweetness can be increased by using an intense sweetener. This offers the possibility of producing reduced-calorie ice-creams. The method of preparation of lactitol-containing ice-cream is similar to that of the conventional product, using non-fat milk-solids, stabilizers and emulsifiers. The level of lactitol in a low-fat ice-cream is about 14%.

7.8. Other Applications

Other possible applications of lactitol include:

—sugar-free and low-calorie instant beverages;
—low-calorie jam;
—toothpaste.

REFERENCES

1. KARRER, P. and BÜCHI, J. (1937). *Helv. Chim. Acta*, **20**, 86.
2. CHEMIE COMBINATIE AMSTERDAM CCA. (1981). European Patent 0039981.
3. VAN VELTHUIJSEN, J. A. (1979). *J. Agric. Food Chem.*, **27**, 680.
4. MÜHLEMANN, H. R. (1977). Unpublished report to CCA.
5. HAVENAAR, R. (1976). Unpublished report, University of Utrecht, The Netherlands.
6. VAN DER HOEVEN, J. S. (1986). *Caries Research*, **20**, 441.
7. KÖNIG, K. G. (1966). Möglichkeiten der Kariesprophylaxe beim Menschen und ihre Untersuchung im kurzfristigen Rattenexperiment, Huber, Bern.
8. GRENBY, T. H. and PHILLIPS, A. (1987). *Caries Res.*, **21**, 170.
9. GRENBY, T. H. and DESAI, T. (1987). *J. Dent. Res.*, **66**, 856.
10. NILSSON, U. and JÄGERSTAD, M. (1987). *Br. J. Nutrition* (in press).
11. VAN BEEK, L. (1977). Unpublished report, CIVO-TNO, Zeist, The Netherlands.
12. BIRD, S. (1984). Unpublished report, Department of Nutrition, National Institute for Research in Dairying, Reading, UK.
13. VAN ES, A. J. H. (1986). *Br. J. Nutrition*, **56**, 545.
14. DE GROOT, A. P. and ADRINGA, M. (1976). Unpublished report, CIVO-TNO, Zeist, The Netherlands.
15. PATIL, D. H., GRIMBLE, G. K. and SILK, D. B. A. (1985). *Gut*, **26**, A1114.
16. SOUTHGATE, D. A. T. and PENSON, J. M. (1983). In: *Dietary Fibre*, G. G. Birch and K. J. Parker (Eds), Applied Science Publishers, London.
17. DOORENBOS, H. (1977). Unpublished report, University of Gröningen, The Netherlands.
18. ZAAL, J. and OTTENHOF, A. (1977). Unpublished report, CIVO-TNO, Zeist, The Netherlands.
19. JOINT FAO/WHO EXPERT COMMITTEE ON FOOD ADDITIVES. (1983). *International Programme on Chemical Safety*. WHO Food Additives Series no. 18, 82–94.
20. ANON. (1984). EC-Document III/1316/84/CS/EDUL/27 rev.: *Report of the Scientific Committee for Food on Sweeteners.*

Chapter 4

MALBIT® AND ITS APPLICATIONS IN THE FOOD INDUSTRY

Ivan Fabry
Zentralfachschule der Deutschen Süsswarenwirtschaft, Solingen-Gräfrath, Federal Republic of Germany

SUMMARY

Data on the composition and characteristics of the various forms of Malbit® are given, plus a review of what is known about its biological properties. This is followed by detailed information on the applications of Malbit® in the production of a wide range of confectionery, including chocolate, candies, caramels, toffees, chewing-gum, pastilles, jellies and dragées. The formulation and manufacture of Malbit®-containing jams and ice-creams are also described.

1. INTRODUCTION

It is generally accepted that in industrialized countries food consumption far exceeds actual caloric requirements, with an important contribution to this over-consumption from the traditional bulk sweeteners. In fact there is a suggestion that this over-supply of calories from carbohydrate may be partially responsible for an increase in diabetes, dental caries and obesity.

A consequence of this is the increased market demand for foods in which the classical carbohydrates are substituted by alternative sweeteners. To meet this demand, the sweeteners should have the characteristics of safety for health, reduced calorie content, non-cariogenicity, organoleptic properties equal to those of sucrose and similar technological properties to those of sucrose.

In addition to these features, and to permit the production of sweets with close, equivalent or superior organoleptical properties to those of sucrose and with acceptable production costs, the alternative sweeteners should have the following additional properties:

—good stability to high temperature, a wide range of pH and micro-organisms;
—high water solubility;
—low hygroscopicity;
—rheological properties close or equivalent to those of sucrose and to the usual mixtures of sucrose and glucose syrup;
—needing a minimum change in the usual technology;
—equivalent shelf-life properties of the final products to those of the standard products;
—relative sweetness close or equivalent to that of sucrose and without need of intense sweeteners like saccharin, cyclamate or aspartame;
—good flavour-releasing properties without the extra cost of additional flavouring.

2. TYPES OF ALTERNATIVE BULK SWEETENERS

Most of these are polyhydric alcohols obtained by hydrogenation of reducing carbohydrates under raised temperature and pressure in the presence of nickel as catalyst. Their technological properties are related to those of the carbohydrates. The chief differences of the polyhydric alcohols used in sugarless food products, compared with traditional carbohydrates, are:

—relative sweetness mostly lower,
—more hygroscopic,
—lower viscosity,
—no participation in the Maillard reaction,
—a certain laxative effect, a limit of intake being recommended.

Commercially and technologically they can be divided into three main groups:

First-generation alternative sweeteners:
—sorbitol,
—mannitol,
—xylitol.

Second-generation alternative sweeteners:
—maltitol,
—lactitol,
—palatinol,
—Palatinit®,
—hydrogenated glucose syrups (Lycasin).

Non-sweet bulking agents:
—polydextrose, which is obtained by melt polycondensation of dextrose in the presence of small amounts of sorbitol and citric acids.

3. MALBIT®

Recent work at the Central College of the German Confectionery Trade (ZDS Solingen) as well as industrial experience demonstrate that Malbit® largely fulfils the above requirements.

3.1. Definition

Malbit® is available in two basic forms:

1. Malbit® liquid is produced by catalytic hydrogenation of a high-maltose syrup and subsequent purification and concentration.
2. Malbit® crystalline, produced from a solution with high maltitol content by concentration under mild conditions and crystallization and cooling. The crystalline powder is obtained in two particle-size ranges: 20 mesh and 50 mesh.

As shown in Table 1, Malbit® liquid is a clear, stable, highly concentrated aqueous polyhydric alcohol solution which besides maltitol, the main component, contains set amounts of sorbitol, maltotriitol and higher hydrogenated oligo- and poly-saccharides.

Malbit® liquid is produced by enzymic hydrolysis of edible potato or corn starch to yield a 'high-maltose syrup' which is hydrogenated in a

CH_2OH CH_2OH O OH H H H H H CH_2OH OH H O OH H HO H OH H OH

FIG. 1. The maltitol molecule (4-*O*-α-D-glucopyranosyl-D-glucitol, $C_{12}H_{24}O_{12}$, mol. wt 344·37).

TABLE 1
ANALYTICAL COMPOSITION OF MALBIT® LIQUID

Solids	Min. 74·0%
Maltitol	73·0–77·0%
D-Sorbitol	2·5–3·5%
Maltotriitol	9·5–13·5%
Hydrogenated oligo- and poly-saccharides	6·5–13·0%
Reducing sugars	Max. 0·3%
Sulphated ash	Max. 0·1%
Chlorides	Max. 20 ppm
Nickel	Max. 2 ppm
Arsenic	Max. 3 ppm
Heavy metals (total)	Max. 10 ppm
Lead	Max. 1 ppm
pH of 40% aqueous solution	5·0–7·0
Average content of higher hydrogenated oligosaccharides (%)	
Tetrasaccharide alcohols	1·5
Pentasaccharide alcohols	0·8
Hexasaccharide alcohols	0·6
Heptasaccharide alcohols	0·6
Higher saccharide alcohols	5·7

second step to the corresponding 'high-maltitol syrup' (79–80% maltitol, which is the main component). As shown in Fig. 1, maltitol is a 1,4-linked molecule of glucose and sorbitol with a molecular weight of 344·37.

Malbit® crystalline is produced from a high-maltose syrup with a greater amount of maltose and less dextrose, oligo- and poly-saccharides. Its average composition is shown in Table 2.

3.2. Technological Properties

The properties of Malbit® liquid and Malbit® crystalline are largely determined by their polyol character. In Table 3 a simple comparative survey of their technological properties is given.

3.2.1. Sweetness

Both products are odourless, and their relative sweetness compared with sucrose is as follows:

Sucrose	1·0
Malbit® crystalline	0·8–0·9
Malbit® liquid	0·6

TABLE 2
ANALYTICAL COMPOSITION OF MALBIT® CRYSTALLINE

Moisture content	Max. 1·0%
Maltitol	86·0–90·0%
D-Sorbitol	1·0–3·0%
Maltotriitol	5·0–8·0%
Hydrogenated oligo- and polysaccharides	2·0–6·0%
Reducing sugar	Max. 0·3%
Sulphated ash	Max. 0·1%
Chlorides	Max. 20 ppm
Nickel	Max. 2 ppm
Arsenic	Max. 3 ppm
Lead	Max. 1 ppm
pH of a 40% aqueous solution	5·0–7·0

Trials have shown that the sweetness of confectionery products based on Malbit® is very close to that of traditional products based on pure sucrose (chocolate) or on sucrose/glucose syrup (high-boiled candies). No aftertaste has been detected, and the cooling effect on the tongue observed with sorbitol and xylitol, due to their negative heat of solution, is much less distinct. It is remarkable that no intense sweeteners were needed to obtain sweetness and flavour release properties very close to those of the sucrose products.

3.2.2. Solubility

In food technology the solubility of the various carbohydrates and polyols plays a very important role. For example, if the relative solubilities of the sugar in a hard candy or of the polyols in a sugarless candy are not taken

TABLE 3
FUNCTIONAL PROPERTIES OF MALBIT® IN RELATION TO THEIR COMMERCIAL FORM

Property	*Malbit® liquid*[a]	*Malbit® crystalline*[a]
Sweetness	←	→
Solubility	→	←
Viscosity of solution	→	←
Hygroscopicity[b]	←	→
Boiling point of solution	←	→
Depression of freezing point	←	→
Osmotic pressure	←	→

[a] The direction of the arrows indicate the relative trends in the qualities.
[b] See Fig. 2.

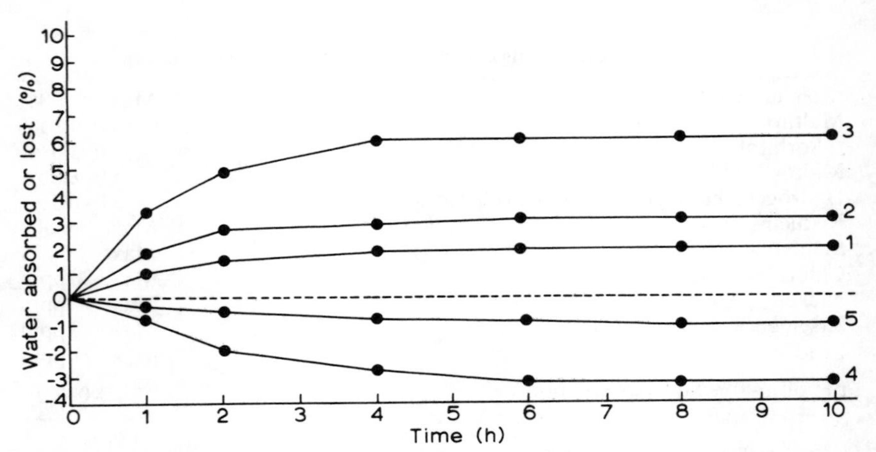

FIG. 2. Hygroscopicity of Malbit® liquid and crystalline. 1–3, Malbit® powder; 4,5, Malbit® liquid. Relative humidity: 1 and 4, 30% 2 and 5, 50%; 3, 90%. Source: Laboratories of EC Sintesi (May 1986).

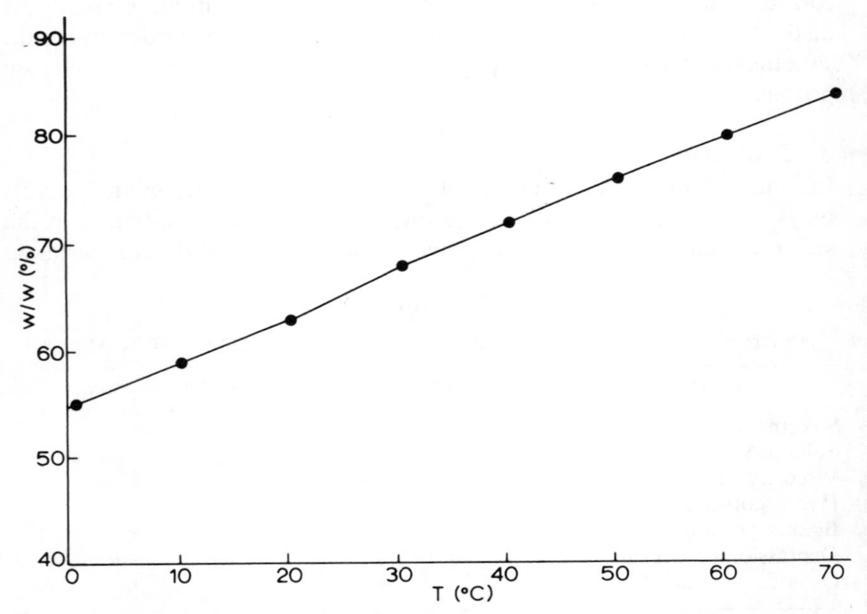

FIG. 3. Solubility of Malbit® crystalline in water. Source: Laboratories of EC Sintesi (May 1986).

TABLE 4
SOLUBILITY OF MALBIT® CRYSTALLINE AND OF SUCROSE IN WATER[a]

Saturation temperature (°C)	*Malbit® crystalline* (% w/w)	*Sucrose* (% w/w)
20	63	67
30	68	68
40	72	70
50	76	72
60	80	74

[a] Source: Laboratories of EC Sintesi (May 1986).

into account, an unwanted crystallization may occur and the finished product will lose its glassy texture.

Table 4 and Fig. 3 show that Malbit® crystalline is less soluble than sucrose at 20°C but above 30°C it has a greater solubility than sucrose.

In aqueous ethanolic solutions the solubility of Malbit® decreases as the ethanol content rises (Fig. 4).

Malbit® liquid has a higher solubility than Malbit® crystalline. At 20°C the solution is saturated at 74–76% w/w. This increase of solubility is due to the presence of a higher amount of maltotriitol as well as to more hydrogenated oligo- and poly-saccharides.

This greater proportion of higher-molecular-weight polyols increases the viscosity and has an effect on the so called 'doctor-sugars' or 'doctor-polyols', which inhibit undesirable crystallization and raise the overall level of polyols in solution, in a similar mode of action to glucose syrup.

3.2.3. Viscosity

The viscosity of Malbit® solutions is lower than that of pure sucrose or sucrose/glucose syrup solutions, which influences the rheological properties both of pure Malbit syrup and highly concentrated masses. This means that the forming temperature of high-boiled candies has to be changed (see Fig. 5 for temperature dependence).

3.2.4. Boiling Point

The boiling point of a Malbit solution is higher than that of equivalent solutions of pure sucrose or sucrose/glucose syrup. Furthermore, the boiling points of solutions prepared from Malbit® crystalline are higher than those of solutions prepared from Malbit® liquid. This difference is related to the lower amount of high-molecular polyols contained in Malbit powder.

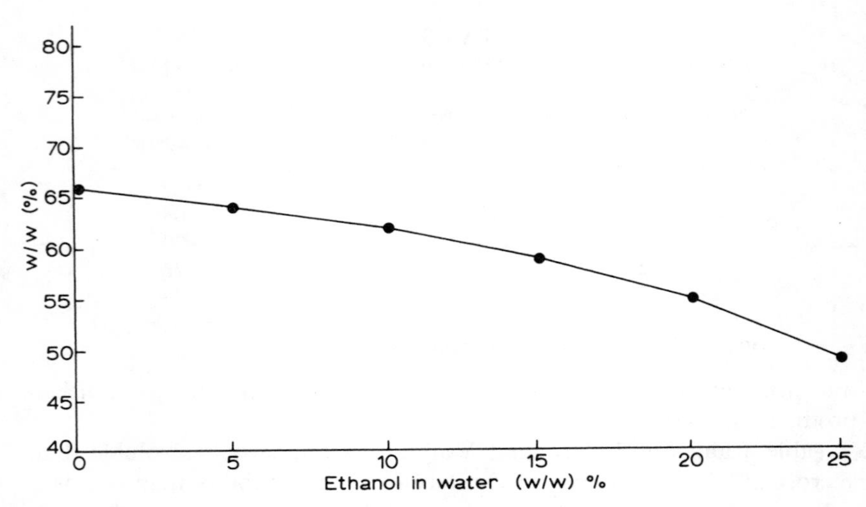

FIG. 4. Solubility of Malbit® crystalline in aqueous ethanol solutions at 25°C.

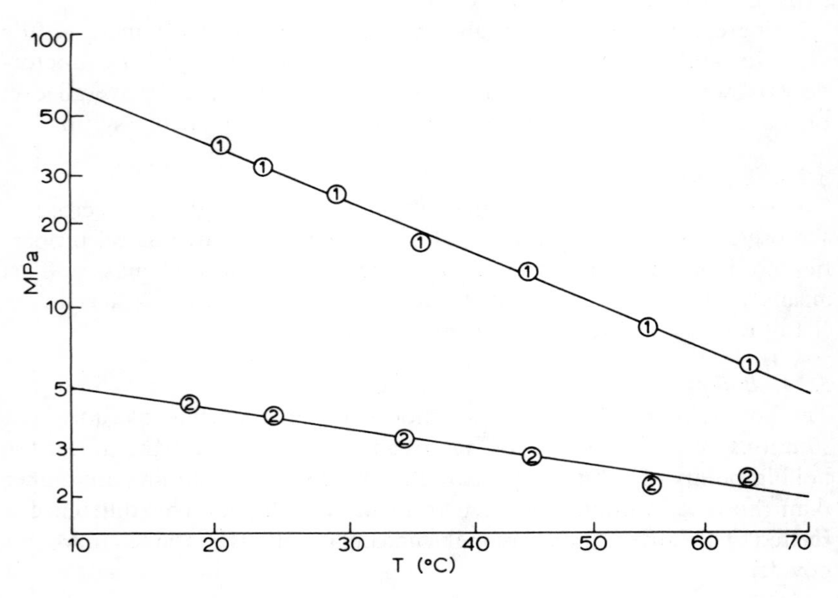

FIG. 5. Viscosity of Malbit® solution: 1, Malbit 60; 2, Malbit 30.

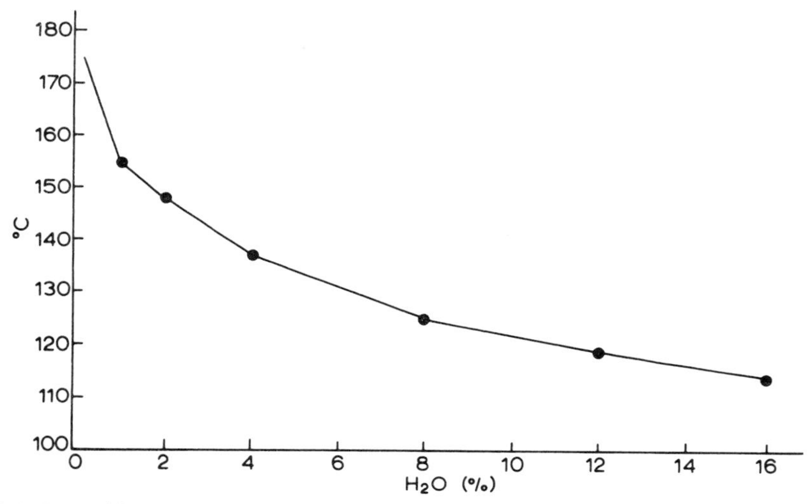

FIG. 6. Boiling point of Malbit® solution, using Malbit® crystalline. Source: Laboratories of EC Sintesi (May 1986).

In practice it is very important to take this into consideration. For example, to obtain a sugarless Malbit hard candy by using a vacuum cooker, the boiling temperature must be increased at least to 160°C to obtain a finished product with a residual water content of 2%. For classical high-boiled candies the cooking temperature used is within the range 140–145°C.

Figure 6 shows the boiling point of Malbit solutions produced with Malbit powder.

3.2.5. Melting Point

Malbit® crystalline melts at 135–140°C. Trials have shown a good heat-stability and no degradation at this temperature. This relatively low melting point causes problems when Malbit® crystalline is pulverized, and special mills are necessary to prevent the risk of melting by frictional heat.

3.3. Health Safety*

Extensive work on acute[1,2] and chronic toxicity,[3,4] on carcinogenesis[3,4] as well as on mutagenesis and teratogenesis,[5–7] have proved the safety of Malbit®. A daily intake of 50 g does not cause any intestinal trouble.[8] The data presented to the European health authorities have permitted registration of Malbit® in Denmark, Norway, Finland, the UK, Switzerland,

* Sections 3.3 and 3.4 are contributions from Dr G. Celia, Melida SpA, Milan, Italy.

Sweden, Belgium, France, Austria and Italy. Other registrations are expected in the near future in The Netherlands and Spain.

3.4. Biological Properties of Malbit® (Claims)

3.4.1. Non-cariogenic

Malbit® alone and in manufactured products is not hydrolysed or transformed into acids in the mouth. Therefore the product(s) have excellent non-acidogenic and non-plaque forming properties[9,10] and have been recognized in Switzerland as *zahnschonend* (non-caries producing). Furthermore the 5% maltotriitol content inhibits maltase activity and thus represents a further positive factor in this respect.[11]

3.4.2. Reduced Calorie Utilization—Safety for Diabetics

Japanese studies[12,13] and recent work done in Italy on rats[14] showed low weight gains in rats fed on Malbit®. New experimentation in Switzerland[15] has shown low absorption of Malbit® and its components; and reduced metabolic utilization of Malbit® in humans, with a caloric value of 50% (2 kcal/g) of that of sucrose. Furthermore, reduced glycaemic and insulinaemic levels have been recorded. Work in humans showed similar results.[16] The reason is the low hydrolysis as well as the incomplete absorption of Malbit® hydrolysis products.

These data have been interpreted as permitting claims for hypocaloric properties and safety for diabetics, and further experimentation is under way.

3.4.3. Final Metabolic Fate of Malbit®

The part of Malbit® not utilized in the small intestine enters the caecum where natural absorption is small or zero. There Malbit® is partly metabolized by micro-organisms and transformed into microbial biomass.

4. APPLICATIONS

Based on the above properties Malbit® offers a range of applications in the food industry. Products are already on the market in several European and Asian countries as a result of trials at the Zentralfachschule der Deutschen Süsswarenwirtschaft, Solingen, Federal Republic of Germany.[17]

4.1. Chocolate

Dark and milk chocolate can be produced with Malbit® crystalline by applying normal manufacturing processes as in the formulations listed in Table 5.

To obtain a Malbit® chocolate with good properties the following precautions must be taken:

(1) As with sorbitol, lactitol, Palatinit and other sugar substitutes, a temperature of 46°C must not be exceeded during the kneading, refining and conching processes, as well as in storage tanks and during the preheating phase when tempering the Malbit® chocolate. Higher temperatures increase very rapidly the viscosity giving the chocolate an unmanageable, sandy texture.

Control of the local temperature so that it does not increase in the conching is especially important.

(2) The sandy effect is aggravated by the presence of moisture, either in the Malbit® crystalline, or in other ingredients such as milk powder, or by absorption during kneading, refining and conching.

(3) The tempering temperature should be 31°C for the dark chocolate and 28°C for the milk.

When the above precautions are observed, the well-tempered moulded chocolate is non-hygroscopic and presents an excellent gloss, good hardness as well as good melting properties in the mouth. Furthermore, the sweetness is similar to that of sucrose chocolate; artificial sweeteners are not necessary; and no unpleasant aftertaste, nor cooling effect, is detectable. Compound coatings with cocoa butter equivalents and substitutes as well as fat fillings, pastes based on hazelnuts and other nuts, etc., can also be produced with Malbit® when these precautions are observed.

4.2. Hard-boiled Candies

Hard-boiled candies can be produced with Malbit® crystalline and liquid. They have a good glossy texture, good sweetness without addition of artificial sweeteners and a pleasant fruit taste. Due to the excellent heat-stability of Malbit® there is no loss of colour during boiling. Malbit® boiled sweets can be manufactured with stamping machines and thus filled boiled sweets can be produced.

The formulations shown in Table 6 have been tested with good results.

With 2% residual moisture the formulations show the polyol composition listed in Table 7.

TABLE 5

MALBIT® CHOCOLATE: RECIPES AND PRODUCTION STEPS

Raw materials	*Dark chocolate*		*Milk chocolate*	
	Wt (kg)	(%)	*Wt (kg)*	(%)
Cocoa liquor	21·000	42·00	4·500	9·00
Cocoa butter	5·750	13·50	13·920	27·84
Whole milk powder	—	—	7·250	14·50
Skimmed milk powder	—	—	2·000	4·00
Malbit® crystalline	22·100	44·20	22·080	44·16
Lecithin	0·140	0·28	0·240	0·48
Vanillin	0·010	0·02	0·010	0·02
Total	50·000	100·00	50·000	100·00
Total fat		36·20		36·20

Processing step	*Processing method*	
	Dark chocolate	*Milk chocolate*
1. Mixing/kneading	Malbit® cryst., cocoa liquor and 1·5 kg of the cocoa butter amount are mixed batchwise in a Z-kneader to a homogeneous chocolate ground paste.	Malbit® cryst., cocoa liquor, skimmed and whole milk powder are mixed batchwise in a Z-kneader to a homogeneous chocolate ground paste.

	Mixing time	10 min	Mixing time	10 min
	Mixing temperature	44°C	Mixing temperature	44°C
	Fat content	25·7%	Fat content	30·36%
2. Refining	The chocolate paste is refined on a traditional five-roll refiner to a chocolate ground powder. Particle size is measured by micrometer: 28μm/27μm/27μm/28μm/28μm/22μm		The milk chocolate paste is refined on a traditional five-roll refiner to a chocolate ground powder. Particle size is measured by micrometer: 22μm/24μm/26μm/28μm/21μm	
3. Conching	Total conching time	22 h	Total conching time	22 h
	Dry conching time	20 h	Dry conching time	20 h
	Liquid conching time	2 h	Liquid conching time	2 h
	Addition of the rest cocoa butter	After 20 h	Addition of cocoa butter	After 20 h
	Addition of lecithin and vanillin	After 21 h	Addition of lecithin and vanillin	After 21 h
	Conching temperature	46°C	Conching temperature	46°C
	Conche type	Petzholdt rotary conche type PVS 100	Conche type	Petzholdt rotary conche type PVS 100

TABLE 6

MALBIT® HARD BOILED CANDIES: RECIPES AND PRODUCTION STEPS

Raw materials	*Malbit® liquid*	*Malbit® crystalline*
Malbit® liquid (76% solids)	30·000 kg	—
Malbit® crystalline	—	25·000 kg
Water	—	8·000 kg
Citric acid monohydrate	0·255 kg	0·280 kg
Fruit flavour	0·035 kg	0·040 kg
Colour solution (10%)	0·025 kg	0·030 kg

Processing steps in the batchwise manufacturing process

1. Cooking	Cook the Malbit® liquid at 160°C. Bring the cooked mass under vacuum. Conditions: Vacuum level Maximum Vacuum time 4 min Output temperature of the cooked mass ±140°C	Dissolve the Malbit® crystalline in the water. Cook at 160°C. Bring the cooked mass under vacuum. Conditions: Vacuum level Maximum Vacuum time 4 min Output temperature of the cooked mass ±140°C
2. Kneading and tempering Cooker type: discontinuous vacuum cooker	Pour on to a lightly oiled cooling table with retaining bars. Add acid, flavour and colour. Temper until required forming plastic temperature is obtained.	
3. Forming/stamping	Forming temperature 65°C. *Remark* In comparison with the production of a classical boiled sweet, the batch roller, size rollers and stamping machine of the forming line must be preheated to a lower temperature, below 65°C.	
4. Cooling	Cooling conditions: Temperature 16–18°C Relative humidity 40–45%	

TABLE 7
FINAL COMPOSITION OF MALBIT® HARD-BOILED CANDIES

Ingredient	*Hard-boiled candies with:*	
	Malbit® liquid	*Malbit® crystalline*
Residual moisture	2·00	2·00
Maltitol	73·50	85·26
D-Sorbitol	2·94	2·45
Maltotriitol	11·27	6·37
Higher-molecular polyols	10·29	3·92
Total	100·00	100·00

Because of the higher amount of maltotriitol and higher-molecular polyols the hard candies produced with Malbit® liquid are of low hygroscopicity and reasonably stable against undesirable crystallization.

In any case, as for all hard candies made with sugar substitutes, both types of Malbit® hard candies must be wrapped with a material with good water-vapour barrier properties for long shelf-life.

4.3. Soft Caramels, Toffees, Chewy Fruits

Malbit® soft caramels, toffees and chewy fruits can be produced by using the usual manufacturing processes. The flow diagram in Fig. 7 shows the production steps of chewy fruits with both Malbit® qualities.

As in the case of hard-boiled candies, the production of Malbit chewy fruits does not require any reinforcement of sweetness by intense sweeteners, no higher aromatization level is required and no off-taste is detectable.

The following changes from the normal sucrose procedure are necessary:

(1) to obtain a final product with 7–9% residual moisture the cooking temperature of the boiled Malbit mass must be increased to 135–140°C. A sucrose chewy fruit mass is normally cooked at 120–124°C.

(2) the forming temperature should be lower. At temperatures of 30–35°C the mass does not stick to the size rollers and the cutting knife. The amount of hardened fat used must be higher than for sucrose chewy fruits.

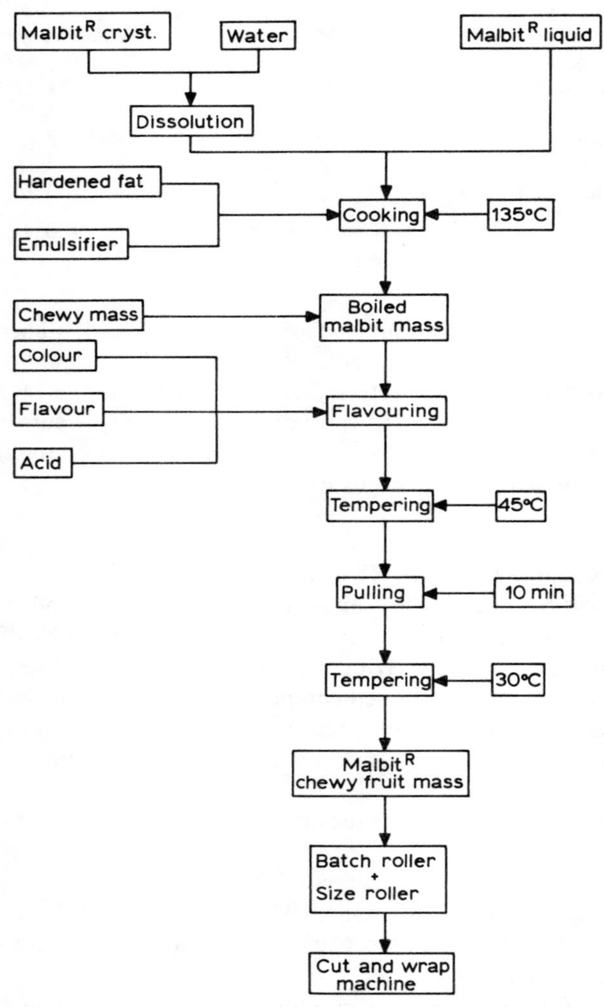

FIG. 7. Production steps of chewy fruits with both Malbit® qualities.

The formulations in Table 8 have been tested successfully. The composition of the 'chewy mass' is as follows:

(1)	Gelatin, 130 Bloom	0·500 kg
	Water	1·000 kg

—Dissolve the gelatin in hot water.

(2)	Gum arabic powder	0·500 kg
	Water	0·500 kg

—Dissolve the gum arabic in the water.
—Mix the gelatin and gum arabic solution together.

(3)	Malbit® liquid	2·500 kg
	Hardened palm kern fat, m pt 34–36°C	2·500 kg
	Lecithin	0·100 kg

—Add Malbit® liquid, melted fat and lecithin to the gelatin–gum arabic solution. To obtain a homogeneous chewy mass use a high-speed mixer.
—Store the obtained chewy mass in a cold room for 12 h before use.

4.4. Gum Arabic Gums/Pastilles

Fruit pastilles and liquorice gums based on gum arabic can be produced without processing modifications, except that the amount of gum arabic must be increased to 45–55% to obtain a texture comparable with that of the sucrose pastilles. The following formulation is a basic one, but the final texture can be modified by varying the amount of gum arabic used and the final moisture content (10–14%).

Basic Formulation

(1)	Gum arabic solution	
	Gum arabic powder (Kordofan, instant powder)	15·000 kg
	Water	10·000 kg
(2)	Malbit® liquid	15·000 kg
(3)	Flavour, colour and acid	to taste

Working Method

(1) Dissolve the gum arabic powder in hot water (70°C).
(2) Cook the Malbit® liquid to 92% solids and cool the cooked mass under vacuum to 90°C.
(3) Add gum arabic solution to the cooled, boiled Malbit® mass and mix well.

TABLE 8

MALBIT® CHEWY FRUITS: RECIPES AND PRODUCTION STEPS

Raw materials	*Malbit® liquid*	*Malbit® crystalline*
Malbit® liquid	33·000 kg	—
Malbit® crystalline	—	25·000 kg
Water	—	—
Hardened fat (Palm kern fat, m pt 34–36°C)	3·500 kg	3·500 kg
Lecithin	0·100 kg	0·100 kg
Chewy mass	3·500 kg	3·500 kg
Citric acid	0·400 kg	0·400 kg
Cherry flavour	0·060 kg	0·060 kg
Colour solution (10%)	0·030 kg	0·030 kg
Processing steps in the batchwise production		
1. Cooking	Heat the Malbit® liquid at 80°C. Add the premelted fat and the lecithin. Mix this premix for 5 min at high speed. Cook the mass at 135°C. Turn off the steam, add the chewy mass and mix at high speed for a short period.	Dissolve the Malbit® crystalline in the water. Add the premelted fat and the lecithin. Mix this premix for 5 min at high speed. Cook the mass at 135°C. Turn off the steam, add the chewy mass and mix at high speed for a short period.
2. Forming	Pour the mass on to an oiled cooling table and add acid, flavour and colour. Temper at 45°C and pull the mass during a period of 10 min. Temper the pulled mass at 30°C. Process the mass on a cut and wrapping machine.	

(4) Add the colour, flavour and acid.

(5) Deposit in dry (5–7% moisture) and warm (40–50°C) starch moulds.

(6) Transfer the starch trays to a drying stove (temperature 50–55°C; relative humidity, 20–25%) and hold until the desired residual moisture and hardness are reached.

(7) Thoroughly clean the gum pastilles and glaze with oil.

4.5. Gelatin Gums and Jellies

Malbit® gelatin gums and jellies can be obtained by using a pig-skin gelatin obtained by acid treatment. This gelatin, type A, has a relatively high iso-electric point. The amount required to produce a satisfactory gel varies between 7 and 10% depending upon the final texture and the Bloom strength used. The best results are obtained with a high-Bloom gelatin (200–240 Bloom). Besides the use of a pig-skin gelatin, the following rules must be adhered to for satisfactory results.

(1) Do not add the gelatin solution to the boiled Malbit® syrup at a temperature above 80°C.
(2) Add the acid solution just before the depositing process.
(3) Deposit the Malbit®-gelatin syrup in dry starch (5–7% moisture).

The following basic formulation has been tested with satisfactory results:

(1) Gelatin solution	
Pig-skin gelatin, 240 Bloom	1·500 kg
Cold water	2·700 kg
(2) Malbit® syrup	
Malbit® liquid	15·000 kg

Working Method

(1) Prepare the gelatin solution in advance, by soaking the gelatin in cold water for 1 h and subsequently dissolving it at a temperature below 65°C.
(2) Cook the Malbit® liquid at 92% solids and then cool the syrup to 80°C.
(3) Add the gelatin solution to the cooled Malbit® syrup and stir in gently to prevent the incorporation of air bubbles. Finally add flavour, colour and acid solution to taste.
(4) Deposit in dry and tempered (25–30°C) moulding starch.
(5) Transfer the starch trays to a climatized room and hold until the required residual moisture content and texture are obtained.
(6) Thoroughly clean the gelatin gums and glaze with oil.

4.6. Chewing Gums and Bubble Gums

Sugarless chewing gums and bubble gums can be easily produced using a combination of Malbit® liquid and crystalline. As in other sugarless chewing gums the amount of gum base required to produce a good quality varies between 25 and 30% depending upon the required chewiness and the type of gum base.

The following formulation has been tested with good results:

Chewing Gum Sticks

Gum base	26·6%
Malbit® liquid	17·8%
Malbit® crystalline	54·3
Glycerol	0·5%
Mint oil	0·8%

Working Method

(1) Heat the chewing gum in a Z-kneader to 60°C.
(2) Add the gum base and knead for 1 min.
(3) Add the Malbit® liquid and knead for 4 min. During this the gum base must be completely melted ensuring that the temperature does not increase above 60°C.
(4) Add one-third of the amount of Malbit® crystalline and mix for 8 min.
(5) Add the second one-third of the amounts of Malbit® crystalline and mix again for 8 min.
(6) Add the last part Malbit® crystalline as well as the flavour and the glycerol. Continue the kneading process for a further 8 min.
(7) On completion of kneading, discharge the Malbit chewing gum and temper the mass at 35–40°C.
(8) Transfer the tempered Malbit® chewing gum mass to the hopper of the chewing gum extruder and extrude through staggered reducing rollers.
(9) After passing through transverse and longitudinal cutting rolls wrap the chewing gum sticks.

To prevent problems during the manufacturing process the following rules must be observed.

(1) The total kneading time should be less than 30 min.
(2) The temperature of the chewing gum mass must be kept below 60°C during the kneading process.

Higher temperatures and longer kneading times reduce the quality of the finished products; the mass shows a higher tendency to stickiness, and the chewing gum loses elasticity and chewiness.

The above formulation can also be applied for the production of a hard-panned chewing centre. In this case the amount of gum base must be increased to 30% to secure sufficient mechanical stability.

4.7. Panned Confectionery/Dragées

Malbit® chocolate dragées can be produced without difficulty. The manufacturing process is the same as usual. A temperature of 46°C must not be exceeded during the melting of the chocolate mass. Higher temperature rapidly increases the viscosity of the Malbit® chocolate mass and a sandy texture occurs.

Malbit® hard dragées of good quality can also be produced by applying a combination of hard and soft panning technology. The flow diagram in (Fig. 8) shows the different production steps.

Basically, 40 g of Malbit engrossing syrup should be added to 5 kg of centres which are revolving in the pan, and Malbit® crystalline powder sprinkled on until they dry. This must be repeated until the desired percentage of coating has been achieved.

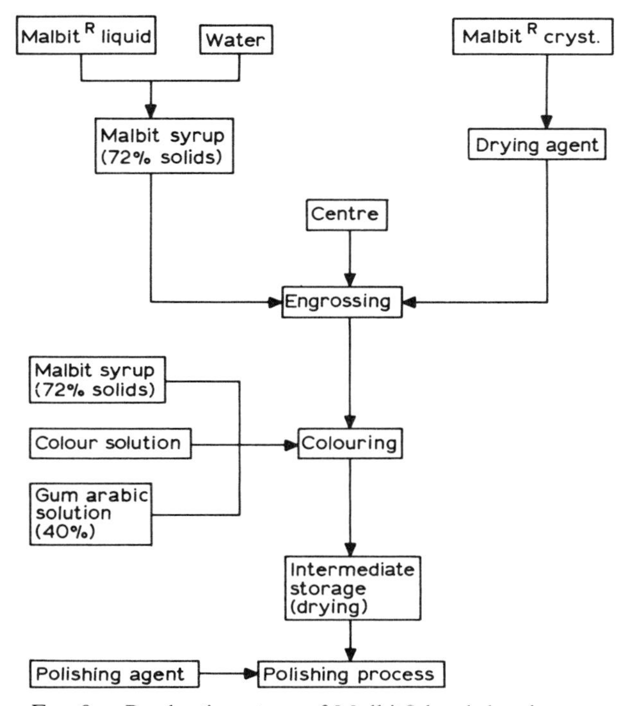

FIG. 8. Production steps of Malbit® hard dragées.

To obtain a smooth coloured surface, use:

Malbit syrup (72%)	0·900 kg
Gum arabic solution (40%)	0·100 kg
Colour solution (10%)	as wanted

After the colouring process the panned and coloured centres must be removed from the pan, sieved and dried for one night at a temperature of 25–30°C and a relative humidity of 40–45%.

To obtain a satisfactory gloss the panned dragées must be free from any dust and kept under very dry conditions. As for sucrose dragées, beeswax, carnauba wax, or another usual polishing agent can be applied.

4.8. Other Types of Confectionery Products

Malbit® can also be used for other types of confectionery:

—compressed sugarless tablets;
—sugarless hard croquante with hazelnuts, almonds, peanuts etc.;
—sugarless aerated enrobed candy bars;
—sugarless cereal bars such as 'granola-bars' and 'muesli-bars'.

The following basic formulation can be applied for a 'muesli-bar':

(1)	Cereals and dry fruit mixture	
	Oat flakes	3·000 kg
	Rice crispies	2·500 kg
	Corn flakes	1·000 kg
	Chopped roasted hazelnuts	2·000 kg
	Roasted sesame seeds	0·500 kg
	Dried apricot slices	1·500 kg
(2)	Binding syrup	
	Malbit® crystalline	2·000 kg
	Malbit® liquid	7·500 kg
	Water	1·000 kg

Cook the syrup at 86% solids

Gum arabic solution (50% solids)	1·000 kg
Honey	2·500 kg
Hardened vegetable fat (m pt 34–36°C)	2·000 kg
Lecithin	0·150 kg

Add this to the cooked syrup and mix well. To bind the above dry mixture add 4·5–5·5 kg of binding syrup (90°C) and mix for 1 min.

TABLE 9
BASIC FORMULATIONS FOR JAM (%)

Ingredients	*Formulation 1*	*Formulation 2*
Malbit® crystalline	53·50	—
Malbit® liquid	—	58·40
Apricot pulp	41·00	—
Strawberry pulp	—	34·30
Water	5·00	—
Pectin HM slow set	0·25	—
Pectin HM rapid set	—	0·20
Tartaric acid	0·25	—
Citric acid	—	0·20
°Brix	61·00	60·60
pH	3·2	3·2

4.9. Jams

Malbit® jams can be prepared without difficulty. The shelf-life of the final product is excellent due to the good resistance of Malbit to micro-organisms and moulds. No intense sweetener must be used. Sucrose can be substituted at a ratio of 1:1.

The basic formulations in Table 9 have been tested successfully.

4.10. Ice-Cream

On account of its sweetness, which is very similar to that of sucrose, and its good solubility, Malbit® can be used to produce sugarless ice-cream without the addition of intense sweeteners. The formulations in Table 10 have been developed by Enichem.

TABLE 10
SUGARLESS ICE-CREAM FORMULATIONS (%)

Ingredients	*Formulation 1*	*Formulation 2*
Malbit® crystalline	17·60	—
Malbit® liquid	—	22·00
Cream (40%)	17·10	17·00
Milk	64·30	60·00
Carboxymethylcellulose	0·40	0·40
Emulsifier 'GMS'	0·30	0·30
Vanilla	0·30	0·30

TABLE 11

LEGAL EUROPEAN STATUS ON THE APPROVAL OF MALBIT® FOR USE IN FOOD

Country	Table-top	Candies	Chewing gum	Chocolate	Bakery products	Ice-cream	Soft drinks	Confitures	Other products	Claims[a]	Notes
Italy[b]	+	+	+	+	+	+	+	+	+	A, B, C	Every product must be approved by the Health Ministry. Claims approved with final product Law 1983 No. 1211
UK	+	+	+	+	+	+	+	+	+		
Switzerland[b]	+	+	+	+	+	+	+	+	+	A, B, C	Admitted as additive
Belgium		+	+		+			+		B, C	Admitted as additive. Claims approval with products. Law 13 Nov. 1985
The Netherlands											Approval as additive expected
France[c]	+	+	+	+	+	+	+	+	+		Claims approval with final products. Law 4 July 1987.
Sweden		+	+	+						B, C	Claims approval with final products
Norway		+	+							B, C	Claims approval with final products
Finland		+	+							B, C	Statute of sweeteners 14 Feb. 1986. Possible use also in other food stuffs upon permission of the Commerce and Industry Board
Denmark	+	+	+	+	+	+	+	+	+	B, C	Final product must be submitted to National Food Institute
Austria	+	+	+	+	+	+	+	+	+	B, C	Malbit is classified as common food. Law 1975 BGBL No. 86
Germany										A, B, C	Petition under examination
Spain										B, C	Petition under examination

[a] A = safe for diabetics, B = safe for teeth, C = reduced calories.
[b] Only for dietetic products.
[c] Only for dietetic products except soft drinks.
Source: Melida SpA, Milan, Italy, 31 July 1987.

5. CONCLUSIONS

Due to its physiological, organoleptic and technological properties, Malbit® in liquid or crystalline form can be used for the development of a new generation of food products which could be termed 'health products', or 'specialty products', including dietary, diabetic, tooth-protective and slimming products.

One of its important organoleptic properties is its level of sweetness, similar to that of sucrose. Reinforcement of the intensity of sweetness by intense sweeteners is not required in most cases. If necessary an adjustment of the sweetness can be obtained by the use of a combination of Malbit® with, for example, xylitol or fructose.

Sugarless Malbit® confectionery and chocolate products, as well as ice-cream and other milk desserts, have been on the Japanese market for about 10 years. Similar products are now on the market of several European countries, such as Switzerland, the UK, Finland, Italy, Sweden and Norway.

Experimentation in two European institutes (ZDZ Solingen, West Germany;[17] TNO Wageningen, The Netherlands[18]) has shown the possibilities in manufacturing food products (sweets) comparable with conventional sucrose products and employing the same industrial techniques.

As with other sugar substitutes, special attention should be paid to the packaging material. Due to higher hygroscopicity in comparison with sucrose products, packaging material with excellent water-vapour barrier properties should be used to assure a long shelf-life.

Finally Table 11 gives a detailed picture of the registration status of Malbit® in several European countries.

REFERENCES

1. Yamasaki, M. *et al.* (1973). *J. Yonago Med. Ass.*, **24**, 1.
2. Nishibori, K. (1968). *Study on Acute Toxicity Test of Maltitol*, Notre Dame Selshin Women's University, Okayama.
3. Shimo, K. *et al.* (1977). *J. Tox. Sci.*, **2**, 4.
4. Mito, A. (1973). *Long-Term Successive Malbit Feeding Test with Wistar-Derived Rats*, Intermediate Report, University of Hiroshima.
5. Anon. (1979). *Study of the Mutagenic Activity of Marvie with* Schizosaccaromyces pombe, RBM Institute of Biomedical Researches A. Marxer, Ivrea, Italy.

6. ANON. (1979). *Study of the Mutagenic Activity of Marvie with* Salmonella typhimurium, RBM Institute of Biomedical Researches A. Marxer, Ivrea, Italy.
7. ANON. (1980). *Stimulating Test of Non-programmed DNA Synthesis by Marvie in Human Cultivated Cells*, RBM Institute of Biomedical Researches A. Marxer, Ivrea, Italy.
8. ABRAHAM, R. R. and DAVIS, M. (1981). *J. Hum. Nutr.*, **35**, 165.
9. IMFIELD, T. (1983). In: *Cariology Today, Int. Congr. Zurich*, Karger, Basel, 147.
10. SAXER, U. P. (1984). *Swiss Dent.*, **4**, 33.
11. WÜRSCH, P. and KOELLREUTTER, B. (1982). *Caries Res.*, **16**, 90.
12. INOUE, Y. *et al.* (1971). *J. Japanese Soc. Food Nutr.*, **24**, 399.
13. OKU, T., HIM, S. H. and HOSOYA, N. (1981). *J. Japanese Soc. Food Nutr.*, **32**, 2.
14. MARANESI, M., GENTILI, P. and CARENINI, G. (1984), *Acta Vit. Enzym.*, **6**(1), 3–9, 11–15.
15. FELBER, J. P. (1984). *Etude Comparative du Maltitol et du Sucrose par Calorimetrie Indirecte et Continu*, Conclusive Report, Centre Hospitalier Universitaire Vaudois, Switzerland.
16. SECCHI, A., PONTIROLI, A. E. and POZZA, G. (July 1982). *Influence of Maltitol (Malbit) on Glycemia and Insulinemia after Oral Administration*, Clinica Medica V, Univ. Milano.
17. CAPPELMANN, K., FABRY, I. and WEBER, H. (July 1986). *Versuchsprotokoll—Herstellung von Malbithaltigen Zuckerschokoladenwaren*, Zentralfachschule der Deutschen Süsswarenwirtschaft, Solingen.
18. FIELLIETTAZ GOETHART, R. I. (Oct. 1982). *Evaluation of Malbit in Eight Food Products: A Comparison with Six Commercial Sweeteners*, TNO Division of Nutrition and Food Research, Netherlands Organization for Applied Scientific Research, Zeist.

Chapter 5

THE METABOLISM AND UTILIZATION OF POLYOLS AND OTHER BULK SWEETENERS COMPARED WITH SUGAR

SUSANNE C. ZIESENITZ and GÜNTHER SIEBERT

Division of Experimental Dentistry, University of Würzburg Medical School, Würzburg, Federal Republic of Germany

SUMMARY

The development of polyols and other bulk sweeteners has its main purpose in replacing sugar in human foods. The extent of substitution depends on the goal to be achieved: in diets for diabetic patients, a balanced exchange of sucrose and other easily metabolizable dietary sugars by sugar substitutes is required; in non-cariogenic foods the replacement of sucrose and other easily fermentable sugars in sweets and snacks is essential; in calorie-reduced foods, sugar substitution must fit into a dietary scheme providing less energy over a period of time for obese individuals.

These different purposes require a variety of properties: calorie-reduction depends on certain gastrointestinal processes; dietetic measures in the treatment of diabetes involve some gastrointestinal functions; while non-cariogenicity manifests itself in the mouth, implications in the gastrointestinal tract also require attention. This review is therefore focused on the digestion, absorption, water balance, and microbial degradation in the lower gut.

Gradual conceptual development has led to a systematic approach in the categorization, ranking and assessment of polyols and other bulk sweeteners, which is capable of still further expansion.

1. INTRODUCTION

1.1. Definitions

Polyols (sugar alcohols) are defined on a chemical basis: they are derived from carbohydrates whose carbonyl group has been hydrogenated into a primary or secondary hydroxyl group. The original carbohydrate may have been a mono-, di- or higher oligo-saccharide.

Bulk sweeteners are defined by their applications: they are used to replace sucrose, both for food manufacture and in the household. They are sweet-tasting and give body (bulk) to foods and beverages. Chemically, bulk sweeteners are in most cases carbohydrates in the strict sense or polyols; their sweetness lies within the order of magnitude of that of sucrose and they belong to the group of nutritive sweeteners, but with a reduced energy content in some cases. The term sugar substitute is generally used when no distinction is necessary between polyols and other bulk sweeteners.

Intense (non-nutritive) sweeteners, e.g. acesulfam-K, cyclamate or saccharin, cannot give bulk to foods due to their high sweetness and are therefore not discussed in this review. Also not discussed is the intense sweetener of carbohydrate nature, 4,1′,6′-trichloro-4,1′,6′-trideoxygalacto-sucrose (sucralose®).[1]

Bulking agents are important constituents of many foods but lie beyond the limits of this review if they lack sweetness. Many of such bulking agents are oligo- or poly-saccharides or their derivatives, and may be mentioned later in passing.

1.2. Applications

There are several reasons for the replacement of sucrose (saccharose; 'sugar') by polyols or other bulk sweeteners:

—In diabetic patients, glucose concentrations in the body ought to be kept at normal levels which are usually monitored as blood-glucose concentrations. Also, the insulin system ought to be relieved of transient heavy loads of glucose. Polyols and other bulk sweeteners may yield less glucose to the body and may cause flattened and lengthened glucose kinetics in the organism and thus successfully replace sugar. It should be kept in mind that dietetic foods for diabetics serve therapeutic purposes and thus are distinguished from foods for healthy people.

In industrialized countries, diabetics comprise nearly 3% of the population. In designing suitable dietetic schemes, total dietary carbohydrates need to be accounted for, but only readily metabolizable carbohydrates like sucrose and free glucose are primary candidates for replacement by polyols or other bulk sweeteners.

—In obesity, calorie-reduced polyols and other bulk sweeteners serve to supply bulk with less calories to the consumer. Not all polyols and bulk sweeteners are calorie-reduced, and naturally the total energy content of the daily food intake needs to be controlled in order to achieve any weight reduction. In Western countries 15–25% of the population are over-weight or obese and may thus benefit from calorie-reduced diets whose sweetness may be maintained with the use of polyols and other bulk sweeteners.

—Dental caries afflicts up to 99% of the population of many Western countries, demonstrating a large need for sugar substitution by non-cariogenic polyols and other bulk sweeteners. Some of their aspects will be dealt with in Chapters 6 and 7 of this volume.

The cariogenic action of sucrose, glucose and fructose on teeth may be modified by a number of factors, one of them of great relevance for non-cariogenic sweeteners: the use-pattern of snacks and other 'in-between-meals' items. This is of major significance in whether and how fast caries develops after the intake of easily fermentable sugars. Accordingly, the primary target in caries prevention by polyols and other bulk sweeteners is their inclusion in all sweets and snacks consumed between meals. Generally, 10–20% of the total sucrose consumption stems from such sweets and snacks, and represents the amount to be substituted in safe-for-teeth foods.

In principle, non-cariogenicity of polyols and other bulk sweeteners is based on the inability of oral micro-organisms to ferment the modified carbohydrates anaerobically into acids. Under these circumstances, lack of utilizability by the oral microflora may have consequences for the metabolic fate of the sugar substitute in the host organism as well. This will be discussed in detail in Sections 2–5 of this review.

Besides these health-related considerations, there are also some aspects of food technology which may form a major reason for using polyols and other bulk sweeteners in food manufacturing, including prevention of fungal and microbial growth, moisturizing effects of some polyols, and the control of browning reactions. These technological aspects of sugar substitutes are not dealt with in this review.

2. GENERAL ASPECTS OF THE PHYSIOLOGY OF POLYOLS AND OTHER BULK SWEETENERS

According to the definitions given above, a monomeric, dimeric or oligomeric carbohydrate is modified in its structure in such a way that it becomes a sugar substitute.

In the case of monosaccharides, the whole burden of structural modification rests on the hydrogenation of the carbonyl function—as in the case of sorbitol (D-glucitol) formed from D-glucose or D-fructose—or on stereoisomerism—as in L-sorbose or perhaps L-glucose.

Hydrogenation of disaccharides leads to glycosylalditols, e.g. maltose to maltitol; lactose to lactitol, which may be used as sugar substitutes. In Lycasin® 80/55, hydrogenation also is performed conserving the existing glycoside bonds.

In di- or oligo-saccharides, a shift of the glycoside bond(s) may partially or totally replace the alterations of the monosaccharide structure. Examples are an $\alpha(1 \rightarrow 5)$- or $\alpha(1 \rightarrow 6)$-glycoside bond instead of $\beta(1 \rightarrow 2)$- in D-glucosyl-fructose; a number of polyols and other bulk sweeteners contain glucose bound in this way, which is slowly or not at all available by digestive and metabolic steps, so that the intended effect of lowered blood-glucose or of nondegradability by oral bacteria is easily achieved.

Those polyols and other bulk sweeteners that are modified monomeric carbohydrates are usually called first-generation sugar substitutes. But when a sugar substitute contains glycoside bonds its metabolic fate is mainly determined by the cleavability of such bonds. Compounds with glycoside bond(s) are termed second-generation sugar substitutes. In a chemical sense, products of glycosyl transfer like neosugar or coupling sugars belong to this category.

The most frequently-found building-blocks in polyols and other bulk sweeteners are D-glucose and its reduction product D-glucitol (sorbitol). D-Mannitol, D-xylitol and D-galactose are much less frequent. All five compounds, whether first- or second-generation, have been well studied as to their metabolic fate in the human body. All of them are constituents of the human body, products of human biosynthetic pathways and/or food components. Except for a few instances discussed below, the intermediary metabolism, including energy utilization, has been extensively studied and does not need a detailed description here, since it is to be found in many textbooks of biochemistry or nutrition.

This review will therefore concentrate on metabolic effects which arise from modified glycoside bonds and from the symbiosis of microbial cells and mammalian tissues in the lower gut.

2.1. The Mouth

The oral mucosa as well as adjacent structures of the human mouth are supplied with nutrients essentially through the systemic route, not by local uptake and subsequent metabolism.

However, a special local role of the oral microflora deserves attention: dental caries is initiated by the fermentation products of dietary sugars, mainly lactic and lower fatty acids. Their formation is only possible under anaerobiosis since aerobic conditions would lead preferentially to carbon dioxide and water. Anaerobic conditions in dental plaque undoubtedly exist, with ethanol, formate and methane as their unambiguous indicators, but it is still a matter of speculation how anaerobiosis is achieved and maintained in the dental plaque.

A second prominent feature is the relatively high specificity of oral micro-organisms in choosing carbohydrates for fermentation. If oral bacteria were as omnivorous as intestinal bacteria, there would be no prospect of developing non-fermentable and thus non-cariogenic sugar substitutes.

The high sugar specificity shown by oral micro-organisms depends on—amongst other factors—the availability of highly specific proteins in sugar transport systems and of carbohydrases secreted by the micro-organisms. A systematic approach relating sugar specificity to the genetic characteristics of oral micro-organisms is only just beginning. The frequently-discussed problem of adaptation of the oral microflora to previously non-utilizable polyols and other bulk sweeteners may find an answer in the genetic constitution of the individual strains of the bacteria. The investigations by Birkhed *et al.*[2] of the fermentation of sorbitol point in this direction.

Sugar fermented in the mouth comprises only a minor fraction in absolute terms of total sugar intake, and does not significantly contribute, through eventual absorption of the acids formed, to the metabolism of sugar in the host organism.

This review is not specifically concerned with the details of non-cariogenicity (see Chapter 6 in this volume) and the microbial fate of polyols and other bulk sweeteners. However, the metabolism of sugar substitutes also has important whole-body aspects.

2.2. Small Intestine

If a polyol or other bulk sweetener is modified far enough from the structure of sucrose to become non-fermentable by oral bacteria, its subsequent utilization in the small intestine may also be impaired, and there are many examples of this. When sugar substitutes are present in the small intestine, delayed digestion and incomplete absorption may generate shortened intestinal transit times.

2.2.1. Enzymic Cleavage

The enzymic cleavage of glycoside bonds in polyols and other bulk sweeteners[3] by digestive carbohydrases is brought about mainly by the sucrase–isomaltase[4,5] and the glucoamylase–maltase[6] complexes of the intestinal brush border.

With maltose or sucrose as reference substrates, rates (v_{max}) as well as affinities (k_m) with polyols and other bulk sweeteners[3] are diminished (Table 1). It appears that the hydrogenation of a carbonyl group is of lesser influence on the cleavage rates than a shift in the position of glycoside bonds. However, one exception may be noted, in that maltitol almost consistently has been characterized as a high-k_m (low-affinity) substrate.[3,7] Maltotriitol, which is frequently admixed with maltitol, exerts an inhibitory

TABLE 1

CLEAVAGE OF DISACCHARIDES AND HYDROGENATED DISACCHARIDES BY A POOLED SAMPLE ($n = 8$) OF HUMAN JEJUNAL MUCOSA[3]

Substrate		k_m (mM)	v_{max} (nmol/min per mg protein)	v_{max} (%)
glc α(1 → 1) fru	Glucosylfructose	21	136	15
glc α(1 → β2) fru	Saccharose	22	280	31
glc α(1 → 3) fru	Turanose	20	120	13
glc α(1 → 4) glc	Maltose	9	900	100
glc α(1 → 5) fru	Leucrose	11	315	35
glc α(1 → 6) glc	Isomaltose	7	234	26
glc α(1 → 6) fru	Isomaltulose (Palatinose®)	9	73	8
glc α(1 → 1) mtl	Glucosylmannitol	11	32	4
glc α(1 → 1) mtl, glc α(1 → 1) gut	Hydrogenated trehalulose (glucosyl-α(1 → 1) fructose)	3	35	4
glc α(1 → 3) mtl, glc α(1 → 3) gut	Hydrogenated turanose	27	22	2
glc α(1 → 4) gut	Maltitol	13	112	12
glc α(1 → 4) gut	Malbit®	50–70	250–310	28–34
glc α(1 → 6) gut	Glucosylglucitol	8	67	7
glc α(1 → 1) mtl, glc α(1 → 6) gut	Palatinit®	5	35	4

action against the cleavage of maltitol by intestinal carbohydrases.[8] It has not been stated so far with certainty whether the high apparent k_m of maltitol is an intrinsic property of this disaccharide alcohol, or follows from the simultaneous presence of maltotriitol. In any case, saturation of small intestinal disaccharidases with maltitol may require concentrations (at least 2–3 times k_m) of 120–270 mmol/litre, which would correspond to 4·2–9·3% (w/v) solutions of maltitol. Such high concentrations are hardly ever attained in the small bowel, with the consequence that the digestion of maltitol should take place at much below the saturating concentrations, and thus hardly ever at v_{max} conditions.

This observation stresses the necessity for high purity of the sugar substitutes under investigation. Older literature on maltitol without data on the presence of maltotriitol cannot be fully evaluated. Another example is the contamination of some samples of isomaltulose with traces of sucrose which may conceal its non-cariogenic properties.

In the experience of the authors, cleavage rates *in vitro* of polyols or other bulk sweeteners which are above 25% of that of maltose, suffice for complete hydrolysis of most sugar substitutes in the small intestine. If rates are lower than 10% of that of maltose, cleavage may be incomplete. Below 2% it certainly will. The uncleaved fraction of the sugar substitute under consideration is then transported into the large bowel (see below).

The small-intestinal digestion of polyols and other bulk sweeteners is therefore of major significance: slow hydrolysis will result in damped blood-glucose kinetics.

—This is a desirable effect for diabetic patients (see Section 1.2).
—Since undigested sugar substitutes reach the large bowel, a mandatory condition for caloric reduction is met (see Sections 2.4 and 2.5).
—Incomplete digestion in the small intestine is not a requirement for non-cariogenic sugar substitutes, but is frequently a consequence of their non-fermentability by oral bacteria, and thus a determinant of their wholesomeness.

2.2.2. Absorption

Amongst polyols, other bulk sweeteners and their building blocks, only D-glucose and D-galactose are actively transported from the small intestinal lumen into the blood stream.[9,10] At the usual amounts of a sugar substitute consumed—between 1 and 20 g per meal or snack—and with the provision that only a part of this consists of D-glucose or D-galactose, their absorption from the small intestine will be rapid and complete. Diurnal profiles of plasma glucose may be altered to smaller peak height and longer duration if the enzymic release of glucose is slower than about one-quarter of that of maltose. In this way, polyols and other bulk sweeteners with bound glucose may act as donors of retard glucose, representing a kind of slow release system.

Detailed mechanisms of the active transport of monosaccharides in the small intestine are found in textbooks of physiology and some recent reviews.[9,10] The inhibitory action of polyols, especially of D-glucitol,[11] on the absorption of glucose have not been studied in a conclusive manner; some scattered observations are suggestive of such an effect but of course only direct measurements of glucose absorption permit the study of the inhibitory effects of polyols on the absorption of glucose.

No polyol, whether free or bound, is absorbed through active transport but rather by 'passive transport' in the small intestine, by mechanisms that may include facilitated diffusion. In principle, the laws of diffusion apply to D-fructose, D-xylitol, D-sorbitol and D-mannitol. Theoretically, their absorption will only be complete at infinity; in practice, rates of absorption of for example D-fructose in the small intestine are fast enough to ensure almost complete absorption before the end of the ileum is reached. With polyols, however, rates of absorption are slow enough to let some non-absorbed material into the large bowel.

As examples, 7·5 mg of D-glucitol, administered by stomach tube to rats of 280 g body weight, give rise to a highly significant participation of the caecum in the utilization of sorbitol.[12] While this dosage in rats corresponds to 27 mg D-glucitol/kg body weight, 5 g of Palatinit®, given in solution to volunteers of 60 kg body weight (45 mg hexitols/kg), can be shown to need the participation of the colon for utilization (by measurement of exhalation of H_2).[13]

In general, incomplete absorption of monomeric polyols and of the building blocks of higher polyols and other bulk sweeteners must be assumed to be the rule.

Two factors thus contribute to the transfer of material into the large intestine: incomplete enzymic cleavage and incomplete absorption in the small intestine. In a report on pigs with an ileo-caecal fistula, a detailed analysis of the intestinal contents after oral intake of Palatinit® revealed the presence of uncleaved D-glucosyl-$\alpha(1 \rightarrow 6)$-D-glucitol and D-glucosyl-α-$(1 \rightarrow 1)$-D-mannitol, together with free D-glucitol and D-mannitol, while D-glucose was absent.[14] At 10% Palatinit® in their diets, pigs transferred about 55% of their intake into the large intestine. From the analytical composition it may be concluded that about 65% of the doses were cleaved in the small intestine; thus 35% entered the colon unsplit; the remaining part consisted of the hexitols D-glucitol and D-mannitol. With 65% of the dose split, and D-glucose absent, it must be concluded the only 20% of the hexitols which were released enzymically in the upper bowel were absorbed before the end of the ileum was reached.

The rapid absorption of D-glucose compensates for the increase of osmotic activity due to enzymic hydrolysis, so that osmotic pressures in the small intestine remain essentially unchanged.[14]

2.2.3. Water Balance

Normal functions of digestive processes in the small intestine require that solutes there achieve isotonicity. Amongst other properties, gastric emptying is dependent on isotonic conditions in the duodenum. To achieve isotonicity 7–9 litres of water per day are secreted from the body fluids into different parts of the human gastrointestinal tract, where they are mixed with about 1·5–2·5 litres of water from beverages and food. Any nutrient which is not effectively absorbed in the small intestine, e.g. polyols and other bulk sweeteners, will therefore carry osmotically-bound water into the lower parts of the gut. Consequences such as laxative effects will be discussed in Section 2.5. Under normal circumstances, 99% of the water in the gastrointestinal tract is reabsorbed in the large bowel (8·5–11·5 litres/day).

With 0·33 mol/litre as the isotonic concentration of carbohydrates, 6% (w/v) solutions of a monosaccharide or monomeric polyol are isotonic; for disaccharides, 11·4% (w/v) solutions are isotonic. If fermentation in the lower gut sets in, a monosaccharide will yield 6–8 times as many osmotically-active molecules:

$$1 \text{ Glucose} \rightarrow 2 \text{ acetate}^- + 2HCO_3^- \text{ (or osmotic equivalents)} + 4H^+$$

A tetrasaccharide yields accordingly, e.g. >30 osmotically-active particles if dissociation of $CO_2(HCO_3^- + H^+)$ is included

$$1 \text{ Nystose} \rightarrow 4{\cdot}8 \text{ acetate} + 2 \text{ propionate} + 1{\cdot}2\,n\text{-butyrate} + 9CO_2.$$

These considerations point to the enormous capacity of water flow through the digestive tract for the maintenance of isotonicity during digestion and fermentation of unabsorbed sugars.

2.3. Large Intestine

The outstanding feature of the lower gut is the symbiosis between host colonic tissues and its anaerobic microflora.[15–25] Only a few methods are available to discriminate between host (eukaryotic) and microbial (prokaryotic) contributions to total carbohydrate utilization, and to measure the metabolic fate of polyols and other bulk sweeteners in this part of the digestive tract.

2.3.1. Caecal and Colonic Mucosa

Irrespective of different anatomical details in various animals, only scattered data are available on the metabolic capacities of these tissues. Germ-free and gnotobiotic techniques have been applied to the study in live animals of the fate of Palatinit®, an equimolar mixture of D-glucosyl-$\alpha(1 \rightarrow 6)$-D-glucitol and D-glucosyl-$\alpha(1 \rightarrow 1)$-D-mannitol. When fed to germ-free Sprague Dawley rats at 5 and 10% levels, no excretion of unsplit Palatinit® in the faeces occurred. Thus intact glucosyl-glucitol and glucosyl-mannitol coming from the upper gut must have been cleaved in the caecum and colon of the germ-free animals.[26-31]

No universally reliable data are available on carbohydrase activities in colonic and caecal mucosa of germ-free animals. In an unpublished study, duodenal, jejunal, ileal, caecal and colonic mucosae were therefore studied with maltose, sucrose, lactose and Palatinit® as substrates.[27] They were cleaved by the large intestinal mucosa homogenates of both conventional and germ-free rats. The enzyme activities in caecum and colon were between a few per cent and a fraction of one per cent of those in jejunal homogenates.[32-34] However, since the intestinal contents are in contact with the mucosa for a longer period in the large than in the small bowel, a rough calculation demonstrated that the observed disaccharidase activities in the lower gut should suffice to effect the cleavage of about one-half of the amount of Palatinit® fed to germ-free animals—the other half certainly being split during the passage of the small intestine, mainly in its jejunal part.

Thus the absence of glucosyl-hexitols in the faeces of germ-free animals can easily be explained by the enzymic activities of carbohydrases in the caecal and colonic mucosa. Interestingly enough, nutritive factors like the nature of the dietary carbohydrates, age factors like the change in activity from the third to the eighth week of life, and microbial factors (comparing conventional and germ-free animals), had basically similar effects on both small-intestinal and large-intestinal carbohydrases. These data[32-34] are suggestive of common regulatory mechanisms of disaccharidase activities in the small- and large-intestinal mucosa.

Besides these digestive aspects, the study of the excretion of products of Palatinit® in germ-free rats[28-31] permitted also a first approach to the absorptive capacity for hexitols and glucose in the lower gut. Since glucose

TABLE 2

EXCRETION OF PALATINIT® AND HEXITOLS UNDER GERM-FREE, GNOTOBIOTIC AND CONVENTIONAL CONDITIONS[27]

	Amount excreted (% Palatinit® fed) in 24 h				
	As Palatinit®		*As hexitols*		*Totals*
	Urine	*Faeces*	*Urine*	*Faeces*	
Germ-free rats					
Basal diet	0	0	0	0	0
5% Palatinit	0·29 ± 0·13	0·13 ± 0·07	6·6 ± 4·4	11·8 ± 3·0	19
10% Palatinit	0·1 ± 0·08	0·24 ± 0·12	18·3 ± 5·4	7·6 ± 1·6	45
Gnotobiotic rats					
10% Palatinit	0·50 ± 0·29	0·08 ± 0·04	1·2 ± 0·6	0·4 ± 0·2	2·2
5% Palatinit	0·94 ± 0·58	0·14 ± 0·09	1·5 ± 0·5	0·4 ± 0·3	3·0
Basal diet	0	0·04 ± 0·07[a]	0·04 ± 0·07[a]	0·12 ± 0·3[a]	0·2
Conventional rats					
Basal diet	0	0	0	0	
5% Palatinit	0·99 ± 0·58	0·01 ± 0·02	0·67 ± 0·20	0·01 ± 0·03	1·7
10% Palatinit	0·51 ± 0·23	0·002 ± 0·005	0·43 ± 0·12	0·002 ± 0·004	1·0

[a] Percentage excreted of fed 5% Palatinit® in diet.

was never found in the faeces of germ-free animals, its complete absorption after release in the colon must be deduced. Hexitols, however, were excreted by these animals in higher amounts in the urine than via faeces. Accordingly, a limited capacity for the absorption of hexitols prevails in the mucosa of the lower gut, the extent of which cannot be assessed exactly. Urinary excretion of glucitol and mannitol, when added to faecal excretion, did not match the total taken in. With the reasonable assumption that a fraction of the hexitols, absorbed in the caecum or colon, was metabolized in the body of these rats and was not available for excretion by the kidneys, a table may be constructed showing the distribution of hexitols between absorption and intestinal excretion in germ-free rats (Table 2).

When these germ-free animals were inoculated with *Klebsiella* and *Enterobacter* spp. known to utilize Palatinit®, a dramatic change of the excretion pattern of Palatinit® constituents occurred within 1–2 days. Glucitol and mannitol levels in urine and faeces fell to those low values observed in conventional rats and in germ-free rats without Palatinit® in their food.[28–31]

With enzymic cleavage and absorption well-established for the lower gut of rats, the question arises of how much of the utilization of polyols and other bulk sweeteners can be ascribed in the large intestine to the host tissues, and how much then depends on microbial activity.

2.3.2. Large-Bowel Microflora

The microflora of man and animals in the large intestine have been reviewed recently.[35–37] For the purpose of this review, only a few points will be discussed in more detail.

The pattern of acids formed anaerobically has been described as similar in man and the pig: the molar ratio of acetic:propionic:*n*-butyric acids is about 60:25:15.[15,16,38] Seven to nine different intestinal bacteria ferment either glucitol or mannitol, but only two or three degrade both.[39] In the absence of polyols and other bulk sweeteners in the human diet, intestinal bacteria live on carbohydrates from dietary fibre and mucous material flowing down the intestinal tract as their main source of energy. The daily amount of carbohydrate supply lies between 16 and 25 g.[16] With sugar substitutes added to the diet, much more carbohydrate is made available to the intestinal microflora and results in a considerable increase of bacterial mass, excreted via faeces. The best data available for sugar substitutes are those of Kirchgessner *et al.*[40] with Palatinit®, showing that the enhanced faecal nitrogen loss is compensated in rats by a decrease of urinary nitrogenous compounds.

There is considerable regularity in the utilization of Palatinit® in man. One of the microbial pathways is the formation of H_2 and eventually CH_4; small fractions of these gases diffuse into the bloodstream and are exhaled via the lungs. Analysis of the expired air has shown that doses of Palatinit® between 10 and 50 g/day result in a linear correlation with $H_2 + CH_4$ exhaled over 10 h (Fig. 1). Thus, $6{\cdot}4 \pm 1{\cdot}6$ ml ($H_2 + CH_4$) are expired in 10 h per g of Palatinit®.[41] Since it is known that other polyols[12,42,43] as well as *β*-lactulose[43,44] or lactose in lactase-deficient children yield expiratory H_2,[45] a generalization is possible that all polyols and all other bulk sweeteners in whose degradation the intestinal microflora participates will contribute to increased intestinal gas formation. Flatulence is indeed a frequent side-effect with the intake of some sugar substitutes.

Obligate anaerobes of the gut microflora may be found in the rat[46] and also in man for studies of anaerobic metabolic processes *in vitro*.[47] In the case of nystose, its fermentation rate with caecal bacteria of lab-chow-fed rats was determined in a pH-stat at neutral pH; while sucrose was fermented at a rate of about 0·5 μeq acid/min per g caecal contents, nystose

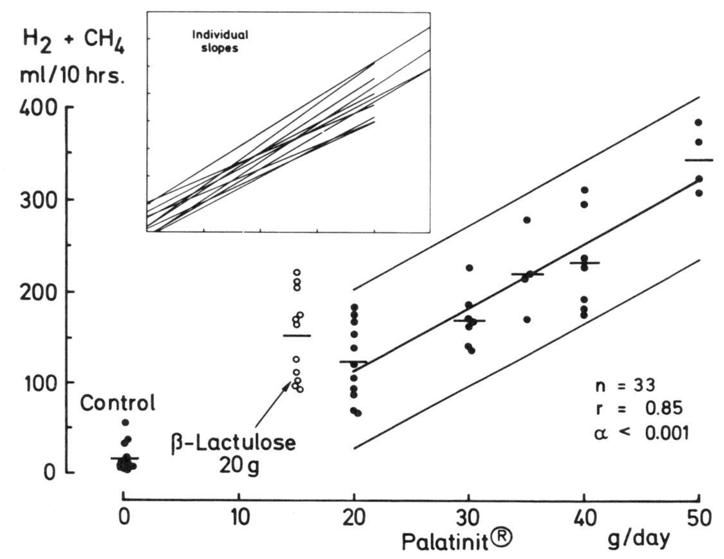

FIG. 1. Dose–effect relation of the integral of the 10 h total volume of hydrogen + methane in 11 healthy volunteers on a low-fibre control diet with 20 g β-lactulose and with various doses of Palatinit® (an equimolar mixture of D-glucosyl-α(1 → 1)-D-mannitol and D-glucosyl-α(1 → 6)-D-glucitol [Isomalt®]). Regression line with 95% confidence interval; $r = 0{\cdot}85$; $P < 0{\cdot}001$. Mean values are represented by horizontal bars. Inset: individual regressions.

gave about 60% of this rate.[48] In another example, the disappearance of D-glucose by fermentation was determined analytically to be near 1 μmol/min per g caecal contents.[49]

2.3.3. Symbiotic Cooperation Between Host Tissues and Microbes

This seems to be self-evident (see above), but any dietary component which, like polyols and other bulk sweeteners, may perturb established equilibria, deserves attention.

In much the same way as brain tissues depend on glucose as their fuel, or striated muscle on long-chain fatty acids, *n*-butyrate has been demonstrated by Roediger[50] as the preferred source of energy for colonocytes. The exact amount of ATP derived by colonocytes from *n*-butyrate oxidation, however, needs to be determined. In terms of symbiosis, this observation would suggest that colonocytes obtain their major fuel supply from the colonic contents where *n*-butyrate comprises 15% of total acids formed.[15,38]

The extraction of *n*-butyrate by colonocytes during the absorption of short-chain fatty acids, however, will eventually result in an experimental difficulty. The quantities found in the portal venous blood may not reflect the total amount of absorbed acids, due to partial colonic utilization, and this leads to an underestimation of absorptive efficiencies.

The host/microbial symbiosis in the large bowel may also be studied in a quite different way, by the determination of the bioavailability of glucose from dietary carbohydrates. This assay[51] is based on the fact that the growth of weaned rats, kept on a ketogenic diet, depends linearly on the glucose concentration in the diet between 0 and 3%.[52] It has been shown for a number of polyols and dietary carbohydrates that glucose bioavailability may be smaller than that calculated theoretically when digestion is impaired, e.g. with modified starches,[52] or gluconeogenesis is incomplete, e.g. with xylitol[51] or glucitol.[11,51] In an investigation of Palatinit®, the glucose moiety demonstrated only about 70% bioavailability.[11] From the data on fistulated pigs[14] and germ-free rats[27-31] it may be concluded that about 65% of the bound glucose was absorbed after enzymic cleavage in the small intestine while 35% of the bound glucose reached the lower gut and was degraded in this symbiotic environment. The 70% bioavailability of glucose from Palatinit® should then be interpreted as showing the overwhelming role of the caecal and colonic microflora in the lower-gut utilization of Palatinit®.[11]

2.4. Reduced Energy Content of Polyols and Other Bulk Sweeteners

Much in the same way as for proteins, carbohydrates may have a diminished caloric value for the organism which creates, although for other reasons, a distinction between physical and physiological energy values. The sole mechanism by which a diminished caloric value may be elicited in carbohydrates stems from the participation of the intestinal microflora in the utilization of the carbohydrate. Intestinal microbes get access to such carbohydrates if cleavage and/or absorption in the upper gut are incomplete. It is therefore obvious that a polyol or bulk sweetener carries its full energy content of about 4 kcal/g if, as in the case of leucrose, for example,[53a,54] the small-intestinal utilization is complete. Very simple, qualitative indicators are the absence of caecal enlargement in the rat[53a] and the absence of flatulence and related symptoms in man.[55]

Several mechanisms exist in the lower gut by which the caloric value of a polyol or other bulk sweetener is diminished.[56]

(i) The anaerobic conversion of carbohydrates into lower fatty acids implies that only a fraction—in most cases below 50%—of the original energy value is conserved in the acids formed, the remainder being given off as heat of reaction. This thermogenesis dissipates its heat within the body of the host, without any chance of contribution of useful energy to the host's energy balance.

(ii) Intestinal microbes use products of carbohydrate fermentation for the synthesis of microbial biomass, with about 0·1–0·2 mol ATP consumed for the synthesis of 2–6 g biomass of bacteria, the normal daily excretion in human stool. Any additional bacterial biomass formed in the colon by the energy derived from sugar substitutes would require ATP from about 45 g glucose equivalents per 20 g bacterial dry mass, approximately.[15,57–59] A corresponding loss of energy available for the host results from this process.

(iii) Gases like H_2 and CH_4 carry energy but their heat of combustion represents an energy loss for the host.

(iv) Host tissues compete by absorption for the utilization of volatile fatty acids in microbial synthesis. The appearance of acetate, propionate and *n*-butyrate in the portal venous blood[60] indicates this with some certainty. Whether both host and microbial utilization of lower fatty acids result in complete consumption ought to be checked by analyses of the faeces for these acids; in some instances, a loss of utilizable energy may be found.

(v) The creation of anaerobic conditions in the intestinal lumen may be brought about—according to the most plausible, yet unproven hypothesis—by a layer of oxygen-consuming micro-organisms close to the mucosal lining of the colon. If this is the case, carbohydrates are the most probable fuel, resulting in $CO_2 + H_2O$ formation, and thus representing another loss of energy whose extent is unknown, due to the lack of any suitable methodology.

While most of these processes are measurable by appropriate techniques, the determination of the overall loss of energy requires whole-body analyses with carbon and nitrogen balances, and is thus confined to animal studies. Reliable data on the energetic value require a feeding protocol in which foods with varied contents of the polyol or other bulk sweetener are given at different total energy levels: one or two above maintenance, another very close to maintenance, and at least one below maintenance levels.

Among sugar substitutes studied in this context, Palatinit®,[61–68] by far the most intensively investigated substance, and maltitol[18,31,43] show a caloric reduction of about 50%. This value agrees well with indirect calorimetric data in man, and also fits well with energy loss calculations derived from the biochemical reactions involved in the utilization of Palatinit®.

From the above considerations it follows that the physiological caloric value of some polyols and other bulk sweeteners may be severely diminished, but intense experimental work is needed to obtain definite data. From the mere fact that a substance like nystose passes undegraded from the upper into the lower gut,[69] convincing data on its caloric utilization cannot be deduced.[48] Rates of small-intestinal absorption, regardless of their validity,[65] cannot replace a direct measurement of the energy value.[66]

2.5. Laxative Effects and Safety

Depending on local food laws, polyols and other bulk sweeteners may be regarded as food additives, and may thus require a full toxicological evaluation, with three generations of rats fed on them, two kinds of mammals investigated, etc. In cases where these legal requirements have been fully met, e.g. xylitol and Palatinit®, the proof of food safety, i.e. the absence of any adverse effects, has been convincingly demonstrated. This was expected in these cases—for instance, Palatinit® has as its components glucose, glucitol and mannitol and as its glycoside bond an [$\alpha(1 \rightarrow 6)$] link, which are found in the body anyway.

Because of the close structural similarity between normal carbohydrates and some of the bulk sweeteners, it is not surprising that requests for a limited application of classical toxicological methods are made,[70] although a more detailed investigation of the digestive and metabolic fate is called for (see above).

Diarrhoea may be a consequence of the intake of some polyols and other bulk sweeteners, especially if the substance is given in too high doses to non-adapted individuals while fasting. Whenever observed, it is transient (for the osmotic mechanisms, etc., involved see Section 2.2 above). A comprehensive evaluation of the laxative potential of each material requires individual investigation and cannot be predicted with certainty by analogy.

3. GENERAL ASSESSMENT OF POLYOLS AND OTHER BULK SWEETENERS

Based on experience with a number of compounds, and looking for a rational approach to the assessment of the many new substances which will eventually emerge, a system of screening has been developed which rests on a theory of assays of general applicability. Tiers of testing allow the interruption of a screening programme as soon as any adverse effects are observed. Furthermore, the system places animal studies towards the end of tests and may thus serve to spare animals when possible.[71]

3.1. Prescreening

These assays are purely *in vitro*, and are therefore fast and inexpensive once they are set up.

3.1.1. Enzymic Procedures

A first characterization of new compounds, if glycoside bonds are present, consists of the use of α-glucosidase (maltase) from yeast.[72] Maltose usually serves as reference substrate. Characteristics to be assayed include rate (given as v_{max} from Lineweaver–Burk or similar plots) and k_m as well as eventual inhibitory activities (k_i and type of inhibition). An example is given in Table 3.[72]

A second step consists in the use of β-fructosidase (invertase) from yeast.[73] If the test material is not a suitable substrate a check on possible inhibitory

TABLE 3

CLEAVAGE OF DISACCHARIDES AND HYDROGENATED DISACCHARIDES BY α-GLUCOSIDASE FROM YEAST[72]

	Substrate	k_m (*m*M)	v_{max}	
			(*μmol/min per mg protein*)	(%)
glc α(1 → 1) fru	Glucosyl-α(1 → 1)-fructose	14	1·3	6
glc α(1 → β2) fru	Sucrose	37	23	105
glc α(1 → 3) fru	Turanose	23	21	95
glc α(1 → 4) glc	Maltose	30	22	100
glc α(1 → 5) fru	Leucrose	17	0·22	1
glc α(1 → 6) glc	Isomaltose	35	0·6	3
glc α(1 → 6) fru	Isomaltulose	14	0·8	4
glc α(1 → 1) mtl	Glucosylmannitol	27	0·29	1
glc α(1 → 1) gut, glc α(1 → 1) mtl	Hydrogenated glucosyl-α(1 → 1)-fructose	25	0·5	2
glc α(1 → 4) gut	Maltitol	80	8·7	40
	Malbit®	102	10·5	48
glc α(1 → 6) gut	Glucosylglucitol	9	0·56	3
glc α(1 → 1) mtl, glc α(1 → 6) gut	Palatinit®	11	0·40	2

TABLE 4

DISACCHARIDES AND HYDROGENATED DISACCHARIDES AS SUBSTRATES AND INHIBITORS OF β-FRUCTOSIDASE FROM YEAST[73]

		k_m (*m*M)	v_{max} (*μmol/min per mg enzyme*)	k_i (*m*M)	*Type of inhibition*
Substrate					
Sucrose		10–12	140		
No substrates					
glc $\alpha(1 \rightarrow 1)$ fru	Glucosylfructose				
glc $\alpha(1 \rightarrow 3)$ fru	Turanose				
glc $\alpha(1 \rightarrow 4)$ glc	Maltose				
glc $\alpha(1 \rightarrow 5)$ fru	Leucrose				
glc $\alpha(1 \rightarrow 5)$ fru	Isomaltulose				
glc $\alpha(1 \rightarrow 1)$ mtl glc $\alpha(1 \rightarrow 6)$ gut	Palatinit®				
Inhibitors					
glc $\alpha(1 \rightarrow 5)$ fru	Leucrose			10–12	Non-competitive
glc $\alpha(1 \rightarrow 6)$ fru	Isomaltulose (Palatinose®)			18	Uncompetitive
glc $\alpha(1 \rightarrow 1)$ mtl glc $\alpha(1 \rightarrow 6)$ gut	Palatinit®			60	Non-competitive
No inhibition					
glu $\alpha(1 \rightarrow 1)$ fru	D-Glucosyl-$\alpha(1 \rightarrow 1)$-D-fructose				

effects may be carried out to demonstrate in some instances an interference with the hydrolysis of sucrose (e.g. Table 4). In the experience of the authors, assays with β-fructosidase possess predictive power for fermentability by oral bacteria since streptococcal invertases, for example, seem to be quite similar to the yeast enzyme.[74] Furthermore, inhibition of β-fructosidase by some sugar substitutes is reflected in their inhibition of anaerobic sugar fermentation by *S. mutans* NCTC 10449.[75]

Another different source of carbohydrases[3] is found in the homogenate of human jejunal mucosa prepared from a pool of 8–10 individual scrapings in cases where death had occurred from non-intestinal causes. Again, kinetic data may be derived, and possible mutual inhibitions (maltose versus leucrose or Palatinose® and these two versus maltose) may be measured. From such data, certain predictions as to the extent of small-intestinal digestion (see Section 2.2.1 of this review) and the eventual

TABLE 5
EFFECT OF SOME POLYOLS AND OTHER BULK SWEETENERS ON THE ACTIVITY OF GLUCOSYLTRANSFERASE FROM *Streptococcus mutans* AHT No. 620 (SEROTYPE a)[76]

No substrates	
Leucrose	
Isomaltulose (Palatinose®)	
Polyglucose PL-3	
Palatinit®	
Nystose	
Inhibitors (at 20 mmol sucrose/litre ≅ $5k_m$)	
Polyglucose PL-3	Enhances efficiency
Palatinit®	No effect on efficiency
No effect	
Leucrose	
Nystose	
Activator	
Isomaltulose (Palatinose®)	Enhances efficiency

transfer of substances into the large bowel are possible. Some examples obtained are given in Table 1.[3]

Another enzyme system to be used is the mixture of glucosyltransferases secreted by some strains of *S. mutans* into the culture medium.[76] These enzymes are responsible for the synthesis of plaque polysaccharides from sucrose and thus contribute to the development of carious lesions. In a continuous assay system the release of fructose and glucose from sucrose is measured differentially; the amount of glucose not matching up with fructose will have been polymerized into a glucan.[77] Besides the substrate properties of some sugar substitutes, their inhibitory activity (Table 5) also deserves attention.[76] Interpretation of the data should include the type of glycoside bond synthesized [$\alpha(1 \rightarrow 3)$- or $\alpha(1 \rightarrow 6)$-] as well as the interplay with oligosaccharides which function as a primer for the polymerization of glucose. This type of assay for enzymic activity also permits an estimation of the efficiency with which the glucose moiety of sucrose is polymerized into a glucan instead of being released as free glucose. The efficiency of the glucosyl transferase-catalysed reaction depends to some extent on the sucrose concentration. Inhibitors may (Polyglucose PL-3) or may not (Palatinit®) enhance the efficiency, while an activator like Palatinose® also enhances the efficiency of the reaction.[76]

3.1.2. Assays with Micro-organisms

These are used for a screening of the eventual cariogenic potential of a test substance, by measuring growth rates in culture, and by determining anaerobic acid formation *in vitro*. The latter procedure is used with many variations in methodology. The following points are important.

(i) Culturing conditions for the strain under investigation may determine its fermentative capacity. The preparation of bacterial cell suspensions for incubations *in vitro* should be made with an appropriate salt medium which serves to stabilize the membrane potential.

(ii) Incubations for 15–20 min[78] represent a dentally relevant time: the oral clearance of sugars, unless they are contained in very sticky foods, is almost complete in 15–20 min. Depending on the purpose of the experiment, acidification (pH drop) or pH-stat (pH constant) measurements should be made under anaerobic conditions, and expressed—when pH values are recorded, by conversion into proton activities—on a molar basis, e.g. as nmol H^+ formed per min under given conditions.

Acid formation can then be calculated as initial velocity from the steepest linear part of the proton production curve ($v_i \cong$ nmol H^+/min) and total proton production can be derived from planimetry of the area under curve ($A \cong$ nmol H^+/18 min, for example). This approach to quantify acid formation is more informative than plain end-point pH data only.

(iii) The strength of the acid attack on tooth enamel depends on the total amounts of acids formed, and can not be read from pH values, which in the more acidic range reflect only the dissociated portions of the different acids. Electrode readings at pH 4·5, for example, lead to an underestimation by about 20–30% of the total acids formed. This depends on knowledge of the types and amounts of the individual acids formed.

(iv) The fermentation rates permit by their kinetic nature also the detection, quantification and qualitative characterization of inhibitors of acidogenesis.[79] A number of polyols and other bulk sweeteners have been characterized by this procedure with *S. mutans* NCTC 10449, *Lactobacillus casei* LSB 132, and *Actinomyces viscosus* Ny 1 #30.[78] In addition, a sample of 3–4 day-old dental plaque may be collected and assayed in exactly the same way as described above for pure bacterial strains. Results with a

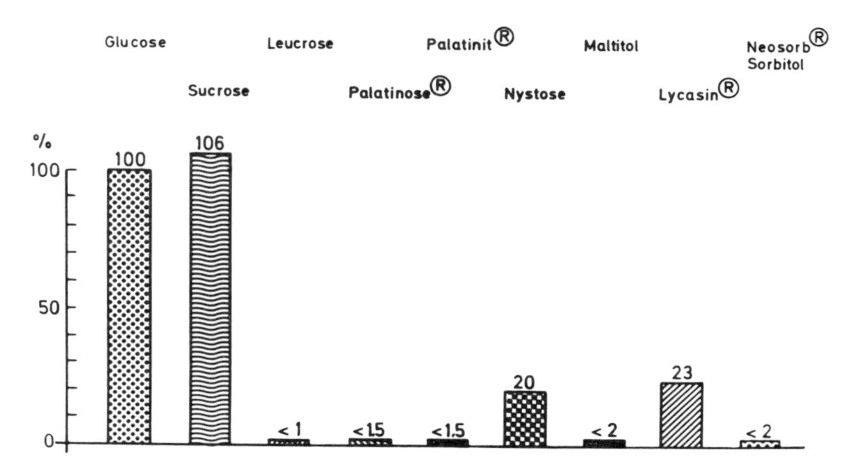

FIG. 2. Acid formation by mixed human plaque.[54] Acidification assay: pH drop; 15 mg plaque/2 ml; anaerobic; 37°C; all sweeteners at 20 mmol/litre; values relative to glucose (~100%).

sample of pooled dental plaque are usually quite similar to those with a single strain of bacteria, as can be seen from Fig. 2 for polyols and other bulk sweeteners.[80]

With this methodology, various polyols and other bulk sweeteners were found to be inhibitors of anaerobic sucrose fermentation (Table 6). However, except for D-xylitol, inhibiting concentrations are high and therefore probably not of immediate practical importance.[75]

3.2. pH Telemetry in Human Volunteers

Systems for the assay of acid formation *in vitro* lack a number of features which govern the fate of sugars in the human mouth. Salivary flow is of foremost importance, supplying buffering capacity and diluting the postprandial sugar concentrations in the fluid of the oral cavity. Swallowing allows the mouth to be an open system, in contrast to any of the above systems.

Recording of pH from an indwelling electrode in a denture, combined with telemetric transmission of the signals, therefore constitutes a major step in the assessment of polyols and other bulk sweeteners for their cariogenic potential. A recent monograph[81] describes methods and results in detail.

TABLE 6
INHIBITION OF ACIDOGENESIS BY SUGAR SUBSTITUTES[a]

Sugar substitute	ED_{25} (*m*M)	ED_{50} (*m*M)	k_i (*m*M)	*Type of inhibition*
Leucrose (at 0·5 mM-sucrose)	390	2 000	125	Competitive
Palatinose® (at 0·5 mM-sucrose)	450	4 500	125	Competitive
Palatinit® (at 0·5 mM-sucrose)	675	5 000	Not determinable	Not determinable
D-Glucosyl-α(1 → 1)-D-mannitol	1 200	15 000	2 000	Uncompetitive
D-Glucosyl-α(1 → 6)-D-glucitol (at 0·5 mM-sucrose)	540	4 000	375	Mixed, with competitive contribution
Maltitol (at 0·5 mM-sucrose)	200	1 500	320	Competitive
Xylitol	2·6	7·5	5	Uncompetitive
Lycasin® 80/55 (at 0·5 mM-sucrose)	200	1 250	410	Mixed

[a] Glucose-grown *S. mutans* NCTC 10449; 37°C, anaerobic; pH-stat at pH 7·0.[80]

Less sophisticated procedures reflect less well the anaerobic formation of acids in dental plaque than the indwelling electrode, and are thus of less significance in the search for non-fermentable sugar substitutes and in the evaluation of the cariogenic potential of foods. If acid formation, measured by an indwelling electrode with telemetric devices, does not reach pH values below 5·7, the substance may be certified as 'safe for teeth' which has gained significance in Switzerland. For public education as well as for the promotion of non-cariogenic foods, such a designation may be very helpful.[82]

3.3. Animal Growth Experiments

A standardized growth experiment, usually with rats, has high informative value for the general assessment of polyols and other bulk sweeteners. Any disturbance of body functions is easily recognizable and allows, together with data on daily food consumption and (at the end of the experiment) on caecal weights, an evaluation of the general wholesomeness of the test substance. An example with α(1 → 5)-linked glucosylfructose, leucrose, is given in Fig. 3.[53a] The evaluation of food and protein utilization efficiencies (g body weight increase per kcal or per g protein consumed) is a first

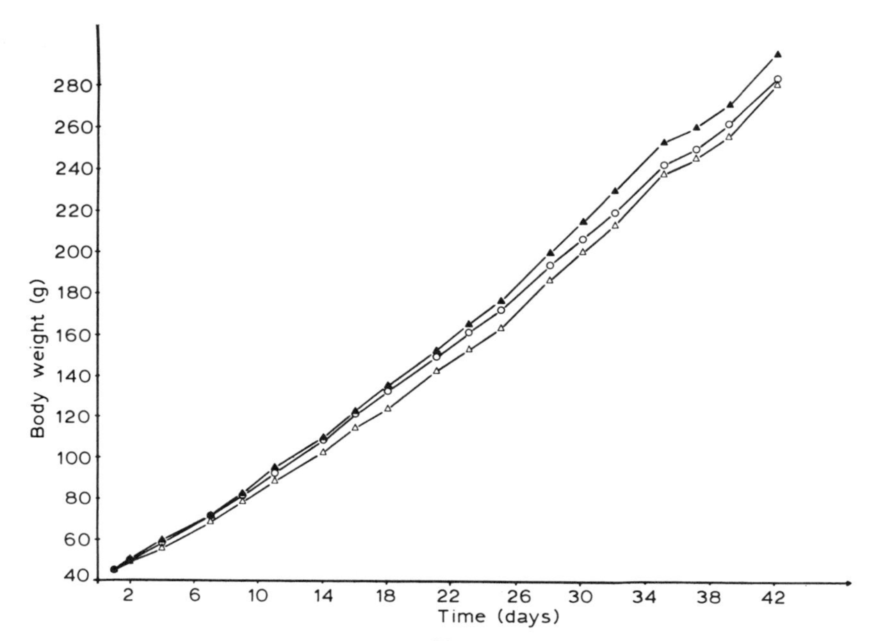

FIG. 3. Growth of rats with leucrose.[54] △, Control (starch); ▲, 20% sucrose; ○, 20% leucrose in diet.

indicator whether a test substance might be termed calorie-reduced or not (see Section 2.4 of this review).

3.4. Animal Studies on Cariogenesis

Because of the high importance of studying the cariogenic potential of foods and food constituents, a highly specialized methodology has been developed. The number of specialized laboratories equipped to adhere to advanced methodology is small. Literature reports therefore require a scrupulous evaluation of their validity. Several books are available[83-87] for the interested reader.

4. SOME PROPERTIES OF POLYOLS AND OTHER BULK SWEETENERS

An overview of polyols and other bulk sweeteners is given in Table 7 and allows a comparison of their properties.

TABLE 7

SOME PROPERTIES OF POLYOLS AND OTHER BULK SWEETENERS

Substance	*Sweetness (sucrose = 1·0)*	*Intermediary metabolism in man*	*Intestinal utilization* in vivo			*Caloric value (4 kcal/g = 100%)*	*Cariogenic potential*	*Suitable in diets for diabetics*	*Intestinal discomfort observed*	*Practical use*
			Small intestine		*Large intestine*					
			Digestion	*Absorption*	*Fermentation*					
D-Glucitol (sorbitol)	0·6	Via glycolysis	—	Incomplete	Yes	100	Low	Yes	+ + +	Yes
D-Mannitol	0·6	Very limited	—	Incomplete	Yes	<100	Low	Yes	+ + +	Yes
D-Xylitol	1·0	Via hexosephosphate pathway	—	Incomplete	Yes	<100	No	Yes	+ + +	Yes
L-Sorbose	0·7	Via glycolytic enzymes, but stereochemical problems	—	Incomplete	Yes	Reduced	Low	Not studied	+ + +	No
L-Glucose	0·7(?)	Not studied	—	Not studied	Not studied	Not studied	Not studied	Not studied	Not studied	No
Leucrose	0·5	Via glycolysis	Complete	Complete	No	100	No	Not studied	No	Proposed
Isomaltulose (Palatinose®)	0·5	Via glycolysis	Almost complete	Almost complete	(Yes)	100	No	Yes	(+)	Yes
Trehalulose	<0·5	Via glycolysis	Not studied *in vivo*	Not studied	Not studied	Not studied	Not studied	Not studied	Not studied	No

Coupling sugar: maltosylsucrose	0·4	Via glycolysis	Not studied	Not studied	Not studied	Not studied	Medium or more	Not studied	Not studied	Proposed
Nystose	0·2	Only microbial fermentation products	No	No	Yes (whole intake)	< 100	Medium	Not studied	+ + + +	Proposed
Lactitol	0·4	Via glycolysis	Incomplete	Incomplete	Yes	50	No	Not studied	+ + + +	Proposed
Maltitol and Malbit®	0·9	Via glycolysis	Incomplete	Incomplete	Yes	60	No	Not studied	+ +	Yes
crystalline	0·9									
liquid	0·6									
Palatinit®	0·5	Via glycolysis	Incomplete	Incomplete	Yes	50	No	Yes	+ +	Yes
Maltodextrins	0	Via glycolysis	Complete	Complete	No	100	Medium or more	No	No	Yes
Glucose and high-fructose syrups	0·6–1·5	Via glycolysis	Complete	Complete	No	100	High	No	No	Yes
Lycasin® 80/55	0·6	Via glycolysis	Incomplete	Incomplete	Yes	Not studied	Low	Not studied	+	Yes

5. SINGLE POLYOLS AND OTHER BULK SWEETENERS

In view of the vast literature, an effort will be made here to cover essential properties of polyols and other bulk sweeteners, concentrating on pure substances or well-defined mixtures. A recent review[87a] concentrates on aspects of dental health.

5.1. Monomeric Compounds (Without Glycoside Bonds)

5.1.1. D-*Glucitol*

D-Glucitol (sorbitol) has been discussed in regard to its metabolism in Sections 2.2 and 2.3 of this review.[11,12,51] When sorbitol is absorbed, it is metabolized after oxidation into either D-fructose or D-glucose, via the glycolytic pathway.[88,89] Thus sorbitol does not pose problems in its fate in the organism apart from its laxative effects.[90,91] Reviews on D-glucitol have appeared recently.[88,91–94] In experimental caries research, D-glucitol serves in plaque-pH telemetry as a yardstick of low acidogenic potential.[83]

D-Glucitol has been used in diets for diabetic patients[95] since 1929,[96] and also for foods with low cariogenic potential.[2,97–99] Some other fields of application belong to food technology and are thus outside the scope of this review.

5.1.2. D-*Mannitol*

D-Mannitol is absorbed incompletely by the human small intestine, with the remainder subject to bacterial digestion in the large intestine.[11] Only a fraction of the absorbed mannitol is metabolized in the human body by funnelling into the glycolytic pathway; the other part is excreted unchanged in the urine.[88,89,94,100] Mannitol is essentially non-cariogenic.[89]

5.1.3. D-*Xylitol*

D-Xylitol is used in foods and oral hygiene products for the purpose of caries prevention. Several reviews have appeared recently[100–102] and cover this substance extensively.[88,89,94,101–104] Its metabolism in man—except for the part which undergoes microbial degradation in the large intestine[51]—consists of an aldolytic cleavage, after oxidation and phosphorylation, in the hexosephosphate pathway[102] and gives glyceraldehyde-3-phosphate (→glycolysis) and glycol aldehyde, eventually leading to glycollic acid.[104]

Many cariological studies with D-xylitol do not report on any other polyol or bulk sweetener and thus make comparison a difficult task.

However, the non-cariogenic nature of D-xylitol is well-established.[101,102] The uses of D-xylitol include parenteral nutrition.[102]

5.1.4. L-*Sorbose*

L-Sorbose has been tried several times as a non-cariogenic bulk sweetener[89,106–110] although its chemical instability[111] certainly limits its wider use. In metabolism, L-sorbose-1-phosphate is formed by the action of fructokinase,[112–114] which has been found to be a very potent inhibitor of hexokinase (k_i a few μmol/litre).[115,116] Aldolytic cleavage yields, besides dihydroxyacetone phosphate, L-glyceraldehyde,[117] the further metabolic fate of which by human tissues depends on a complete inversion at C-2 into D-glyceraldehyde(-3-phosphate), a metabolic problem which has not been solved completely.[111] Besides glycerol kinase from *Candida albicans*, which phosphorylates L-glyceraldehyde and D-glyceraldehyde as well,[118] and alcohol dehydrogenase from yeast of ambiguous stereospecificity, no mechanism has been proposed so far which convincingly explains the inversion of C-2 (the former C-5 of L-sorbose) in glyceraldehyde-3-phosphate.[111]

Some metabolic[119] and side-effects of L-sorbose have been described.[111] In conclusion, this keto sugar has not been used as a sugar substitute in foods.

5.1.5. L-*Glucose*

L-Glucose and other L-sugars have been more frequently put forward in popular media[120] than in the scientific literature as a cure for all the evils of D-glucose. A recently-published article[121] does not give scientifically based concepts for the intended use of L-sugars.

Stereochemical problems, as for L-sorbose, should also apply to L-glucose and other L-sugars. At present little prospect can be foreseen for the wider application of L-glucose as a sugar substitute; *S. mutans* NCTC 10449 ferments it anaerobically.[122]

5.2. Bulk Sweeteners with Glycoside Bonds

As discussed in Section 2.2, the efficiency by which a glycoside bond is hydrolysed enzymically may be a decisive factor in the assessment of polyols and bulk sweeteners for a variety of applications.

5.2.1. Disaccharides

Starting from sucrose, glycosyl transfer by microbial enzymes, in some cases with highly advanced biotechnology, leads to D-glucosyl-D-fructoses with either $\alpha(1\rightarrow1)$[123] or $\alpha(1\rightarrow5)$[124] or $\alpha(1\rightarrow6)$ glycoside bonds.[125]

5.2.1.1. Trehalulose [D-*glycopyranosyl*-α(*1* → *1*)-D-*fructofuranose*]. This is formed as a byproduct in the microbial conversion of sucrose into derivatives such as isomaltulose.[126] Some data on the digestibility of this disaccharide are given in Section 3.1 and ref. 127. It is only very slowly fermented by human dental plaque.[128,129]

5.2.1.2. Leucrose [D-*glycopyranosyl*-α(*1* → *5*)-D-*fructopyranose*]. This occurs in honey[130] as well as in considerable quantities as a byproduct of commercial dextran synthesis from sucrose.[124,131–134] Leucrose has been found to be non-cariogenic in the Cara rat animal model (see Section 3.4).[53,54] This is supported by the non-fermentability of leucrose by several oral micro-organisms and in human dental plaque.[53b]

The absence of any gastrointestinal discomfort in man (100 g doses in solution)[55] as well as the absence of caecal enlargements in rats[53a] (up to 30% leucrose in the diet) indicate its wholesomeness. Laxative effects were never observed. Enzymic data[3,72,73,133,134] are also found in Section 3 of this review; a full publication on leucrose as a non-cariogenic sugar substitute is forthcoming,[53a,53b] including data on intra-oral pH telemetry (see Section 3.2) proving safety for teeth. Leucrose inhibits the fermentation of sucrose by *S. mutans* NCTC 10449,[53a,53b,54,75] does not affect glucosyltransferase,[53a,53b,54,76] but seems also to inhibit the transport of sucrose into *S. mutans* NCTC 10449[135] and thus appears to be a potentially anticariogenic sweetener.

5.2.1.3. Isomaltulose [D-*glucopyranosyl*-α(*1* → *6*)-D-*fructofuranose*] (*Palatinose*®). This has its own significant properties as a sugar substitute, again in non-cariogenic foods, besides being the mother-substance of Palatinit® (see Section 5.2.3.3). Isomaltulose occurs in nature, e.g. in honey.[136] Its digestion in the small intestine is slower than that of sucrose[3,137–140] and may be incomplete in some instances. Naturally, the absorption of glucose and fructose after digestion of isomaltulose does not pose problems, but the increase in blood-sugar is less pronounced and lasts longer than after sucrose, and insulin secretion reacts correspondingly.[138,139] As with many other di- and oligo-saccharides, small amounts of isomaltulose are apparently absorbed uncleaved, since isomaltulose may be found in the urine.[141] So far the wholesomeness of isomaltulose seems to be good.

Provided isomaltulose preparations are pure enough (i.e. no traces of sucrose), acid formation from Palatinose® has not been observed.[54,80] Acid formation, if observed at all, is generally slow enough to keep the pH within safe limits[142–145] and agrees well with pH-telemetric data.

Observations of a practical absence of a cariogenic potential in isomaltulose are in line with the data on acid formation.[146-151] Laboratory observations demonstrate further an interaction of isomaltulose with the glucan-producing enzymes of *Streptococcus mutans*,[76] which may result in a complete suppression of the synthesis of insoluble glucan.[143,152]

5.2.2. Oligosaccharides

Amongst the vast number of oligosaccharides occurring in nature in free form or bound to proteins or lipids, no compound has so far raised great interest as a bulk sweetener. Instead, two groups of oligosaccharides prepared by enzymic biotechnology from sucrose or starch deserve to be mentioned here.

5.2.2.1. Coupling sugars. By the action of cyclodextrin glycosyltransferase, isolated from *Bacillus megaterium*, on a mixture of starch and sucrose a range of oligosaccharides are prepared, called coupling sugars.[153] Glucosyl- and maltosyl-sucroses have been isolated and were subjected, together with the coupling sugars, to a cariological assessment.[154,155] Glucosyl-sucrose and maltosyl-sucrose are substrates for cleavage and fermentation by oral micro-organisms.[156] Their cariogenic potential is therefore higher than that of established non-cariogenic sweeteners. The results reported by two different groups of authors are thus not encouraging for the future use of coupling sugars. Candies prepared with coupling sugars proved acidogenic in plaque pH telemetry.[157] For a more detailed review, see ref. 154.

5.2.2.2. Nystose [D-*glucosyl*-β(*1* → *2*)-D-*fructosyl*-β(*1* → *2*)-D-*fructosyl*-β(*1* → *2*)-D-*fructose*]. This is a prominent constituent of Neosugar, prepared by enzymatic fructosyl transfer onto sucrose.[158] Nystose has been proposed[69] as a non-caloric sweetener because it is not digested at all in the small intestine of man and some animals.[48,54,69] As a consequence, nystose reaches the large intestine unchanged, and gives rise to rapid bacterial fermentation,[48] including laxative effects.[69,159] The organism is thus provided with an as-yet undetermined amount of volatile fatty acids as energy-yielding material. The claim for Neosugar as being non-caloric[69] is therefore untenable.[48]

In addition, nystose is a good substrate for anaerobic fermentation and concomitant acid formation by *S. mutans* NCTC 10449, and by pooled samples of dental plaque[48,54] so that its cariogenic potential will be considerable (some data are presented in Sections 2 and 3 of this review).

5.2.3. Polyols

The compounds to be discussed below all contain a glucose (or galactose) moiety connected via glycoside bonds to a hexitol. Their glycoside bonds resist enzymic cleavage by oral micro-organisms; they are essentially non-cariogenic. A recent review[87a] concentrates on aspects of dental health.

All polyols mentioned below also make less energy available to the organism than the common sugars; they belong to that group of polyols whose glycoside bonds permit only an incomplete digestion in the small intestine (see Sections 2.2 and 3.1 of this review).

5.2.3.1. Lactitol. This polyol [D-galactosyl-β(1→4)-D-glucitol] is prepared by catalytic hydrogenation of lactose,[160] which must be thoroughly purified before reduction. Lactitol is split by small-intestinal carbohydrase preparations at 1–10% of the rate observed with lactose.[161] Its wholesomeness is questionable since unfavourable reactions after the ingestion of lactitol-containing sweets have been observed;[162] furthermore, colonic H_2 production from ingested lactitol has been demonstrated in man.[42,43]

Indirect calorimetry with people who received 50 g lactitol/day for 4 days showed 50% less energy utilization than with sucrose, while body fat in rats on a 16% lactitol-containing diet was reduced by about 30% in comparison with sucrose.[162] The non-cariogenicity of lactitol seems to be established, because lactitol proved to be hypoacidogenic in plaque pH telemetry[81] and *in vitro*.[163] Its usefulness in calorie-reduced foods and for diabetic patients has yet to be established, and might pose some problems[162] not only in daily nutrition but also in its proposed medical use,[164] where a more predictable cathartic effect is described with lactitol than with β-lactulose. A recent study has shown furthermore that galactose metabolism is facilitated in the simultaneous presence of D-glucose, as in the digestion of lactose.[165]

5.2.3.2. Maltitol and Malbit®. Maltitol [D-glucosyl-α(1→4)-D-glucitol] is available from different producers, as essentially pure maltitol with 99% disaccharide alcohol, or as Malbit® with $<90\%$ maltitol and $>5\%$ maltotriitol. Although maltotriitol has effects of its own, e.g. in the inhibition of the enzymic cleavage of maltitol,[8] both preparations are treated together in this review.

Some data on the enzymic cleavage of maltitol are given in Section 2 and Tables 1 and 3.[3,7,8,72,166,167]

The non-cariogenicity seems to be well-established;[168–170] with pure maltitol, practically no acid formation by oral micro-organisms has been observed.[171,172] The intermediary metabolism of maltitol has been investigated in several studies.[43,173–178] Due to the participation of the intestinal microflora (see Section 2) in the degradation of maltitol, its

TABLE 8
COMPOSITION OF MALTITOL-CONTAINING PREPARATIONS (%)

	Pure maltitol	*Malbit®*		*Lycasin® 80/55*
		Crystallized ($n = 12$)	*Liquid* ($n = 5$)	
Glucitol	0	3·8 ± 0·38	1·3 ± 0·07	6·7
Maltitol	98·9	88·4 ± 0·61	79·9 ± 0·76	53·8
Maltotriitol	1·1	5·6 ± 0·42	12·1 ± 0·63	16·3
Total	100·0	97·8	93·3	76·8

caloric value is reduced by about 50%.[179] However, widespread use of maltitol and of Malbit® in foods will finally be determined by its wholesomeness and acceptability.

In general, because of the industrial processes leading from starch to maltitol, there is a graded transition in terms of purity from pure maltitol through Malbit® to Lycasin® 80/55 (see Table 8). With maltitol in these three well-defined preparations, the choice for non-cariogenic, calorie-reduced, technological or other purposes is expanding. Lycasin® 80/55 will be described elsewhere in this review (Section 5.2.4.3) since its properties do not depend on its content of maltitol alone.

5.2.3.3. Palatinit®. This is an equimolar mixture of D-glucosyl-α(1 → 6)-D-glucitol and D-glucosyl-α(1 → 1)-D-mannitol,[180] and is the first polyol with a glycoside bond to undergo a thorough scientific assessment, and has passed a complete toxicological assessment;[180] analytical procedures are available for enzymic methods,[181] gas chromatography[181] and HPLC.[29]

The two glucosyl-hexitols in Palatinit® differ in their conformation, with glucosyl-glucitol being the shorter and bulkier, and glucosyl-mannitol the lengthier molecule.[182,183] Such differences are reflected in the handling of the two molecules by biological systems including disaccharidases,[184] kidney excretion,[29,31,184] inhibitory effects on enzymes,[3] absorption by intestinal mucosa and the like. Despite such fine differentiations, this has not led to any preference for one of the two glucosyl-hexitols over the other in any of the application fields.

Palatinit® has been shown to be non-cariogenic,[81,185,186] about 50% less caloric than the common sugars,[26,27,61–63,66,68,184] and fit for inclusion in the diet of diabetics.[184,187,188]

Furthermore, as partially outlined in Section 2, the metabolic fate of Palatinit® has been studied extensively with reference to digestion in the intestine,[71,184] colonic gas formation,[41,43] bioavailability of glucose,[11] kidney clearance[184] body composition after prolonged feeding,[40,68,184] metabolism in germ-free animals,[27–31] enzymology *in vitro*,[32–34,72] lack of fermentation by oral bacteria,[71,75,81] and its inhibition of sucrose fermentation.[75,80] Doses up to 34·5% in diets for rats ($\cong$28 g/day per kg body weight) are quite safe.[184] Tolerance in humans of 50 g/day has been found to be equal to that of sucrose, with an enhanced tolerance in children.[180]

5.2.4. Starch Hydrolysates

Both oligosaccharides in the strict sense as well as hydrogenated oligosaccharides are dealt with here.

5.2.4.1. Maltodextrins. These are partial hydrolysates of starch, in most cases from maize but also from other sources such as wheat or potato. Their practical use is widespread, and is mainly determined by the technological properties of maltodextrins, and less by nutritional considerations. Depending on the extent of degradation of starch, a range of maltodextrins with different properties can be prepared.

Maltodextrins pose no problem in terms of wholesomeness and provide their full caloric content. In their use in diets for diabetics they resemble starch more than the easily-metabolizable sugars like glucose or maltose. Their cariogenic potential has not been well studied, but by inference from investigations on different types of corn starch, a considerable caries potential is indicated.

5.2.4.2. Glucose and high-fructose syrups. A high yet incomplete degradation of starch leads to products whose designation as glucose syrup indicates their main pattern of use, namely as sweeteners. Upon enzymic conversion of glucose into fructose high-fructose syrups are obtained, and these are used in large amounts as sweeteners in the United States, competing with sucrose.[189] No expectations should be held that such syrups could ever be non-cariogenic, and dental assessment could classify them very close to sucrose.

5.2.4.3. Lycasin® 80/55. This is the modern version of a hydrogenated glucose syrup; its principal composition is given in Table 8 (not including the higher sugar alcohols). Besides technological considerations useful in food manufacturing, it also has a low cariogenic potential.[81]

Metabolically, Lycasin® 80/55 is most probably not of reduced caloric value compared with sugars. Higher homologues of maltotriitol found in Lycasin® 80/55 become gradually closer and closer in properties to the non-hydrogenated oligosaccharides, with no alteration of their metabolic fate.[176] This may explain the fact of some acid formation observed with dental plaque when maltitol and Lycasin® 80/55 are compared.[80]

6. OUTLOOK

The conceptual approach for research on polyols and other bulk sweeteners, as outlined in Sections 2 and 3 above, has demonstrated its usefulness not only in categorizing existing data on these substances but also in the assessment of nystose (Section 5.2.2.2) and leucrose (Section 5.2.1.2) as relative newcomers in the field of sugar substitution. Future developments in polyols and other bulk sweeteners are expected to concentrate less on modified monomeric substances than on modified glycoside bonds (second generation of sugar substitutes). Governed by the intended field of application and by the major role of the gastrointestinal tract, the necessary compromise between non-cariogenicity, wholesomeness and production technology will finally depend on the avoidance of intestinal discomfort.

The availability of safe analytical procedures, not discussed at length here, is a prerequisite in research as well as in food-control laboratories. While polyols and other bulk sweeteners may be analysed by modern techniques like HPLC or gas chromatography, substances like polydextrose® and possibly other bulking agents may pose problems if they lack a defined chemical structure.

D-Fructose, with its high cariogenic potential and its full caloric complement, has been used successfully for diabetics. However, regardless of whether D-fructose is used as such, or in invert sugar or high-fructose syrups, these are all unfit as sugar substitutes due to their cariogenicity.

Reduced-energetic utilization of polyols, the so-called energy-claim, is conceptually quite new and calls for careful nutritional evaluation. Even animal nutritionists, used to dealing with ruminants, have had to adapt their methodology to the fact that polyols and other bulk sweeteners can be assigned two different caloric values, a physical one from complete combustion, and a physiological one in living organisms with their host–microbial interactions in the gut.

Systematic research on polyols and other bulk sweeteners as sugar substitutes will eventually find acceptance amongst scientists, but the transfer of such data into the world of government regulators and into applications by the food industry may require additional efforts by the scientific community before sugar substitutes can exert their full impact to the benefit of the consumer's health. Public acceptance may be positively influenced by the fact that sucrose, starch or lactose is the raw material for most sugar substitutes.

The limited sweetness of most polyols and other bulk sweeteners (listed in Section 4) may call for the addition of small amounts of intense sweeteners. Several of them, e.g. acesulfam-K, cyclamate and saccharin, possess anticariogenic activity,[79,190–192] and thus may be suitable for a wider application in conjunction with non-cariogenic sugar substitutes.

REFERENCES

1. NABORS, L. O'B. and INGLETT, G. E. (1986). In: *Alternative Sweeteners*, L. O'B. Nabors and R. C. Gelardi (Eds), Marcel Dekker, New York, 317–18.
2. BIRKHED, D., SVENSÄTER, G., KALFAS, S. and EDWARDSSON, S. (1987). In: *IV. Würzburger Zuckersymposium*, G. Siebert (Ed.), *Dtsch. Zahnärztl. Z.*, **42** (Supplement), (in press).
3. ZIESENITZ, S. C. (1986). *Z. Ernährungswiss.*, **25,** 253.
4. SEMENZA, G. (1981). In: *Carbohydrate Metabolism and Its Disorders*, Vol. 3, P. J. Randle, D. F. Steiner and W. J. Whelan (Eds), Academic Press, London, 425–79.
5. HAUSER, H. and SEMENZA, G. (1983). *CRC Crit. Rev. Biochem.*, **14**, 319.
6. LEE, L. M. Y., SALVATORE, A. K., FLANAGAN, P. R. and FORSTNER, G. G. (1980). *Biochem. J.*, **187**, 437.
7. YOSHIZAWA, S., MORIUCHI, S. and HOSOYA, N. (1975). *J. Nutr. Sci. Vitaminol*, **21**, 31.
8. WÜRSCH, P. and DEL VEDOVO, S. (1981). *Internat. J. Vit. Nutr. Res.*, **51**, 161.
9. KIMMICH, G. A. (1981). In: *Physiology of the Gastrointestinal Tract*, L. R. Johnson (Ed.), Raven Press, New York, 1035–61.
10. GRAY, G. M. (1981). In: *Physiology of the Gastrointestinal Tract*, L. R. Johnson (Ed.), Raven Press, New York, 1063–72.
11. ZIESENITZ, S. C. (1983). *Z. Ernährungswiss.*, **22**, 185.
12. SCHNELL-DOMPERT, E. and SIEBERT, G. (1980). *Hoppe-Seyler's Z. Physiol. Chem.*, **361**, 1069.
13. GUINAND, P. (1980). Master's Thesis, University of Hohenheim.
14. VAN WEERDEN, E. J., HUISMAN, I. and VAN LEEUWEN, P. (1984). *The Digestion Process of Palatinit® in the Intestinal Tract of the Pig*, ILOB—Instituut voor Lanbouwkundig Onderzoek van Biochemische Producten, Report No. 528a, Wageningen.
15. CUMMINGS, J. H. (1981). *Gut*, **22**, 763.
16. CUMMINGS, J. H. (1983). *Lancet*, **i**, 1206.
17. CUMMINGS, J. H. (1984). *Proc. Nutr. Soc.*, **43**, 35.
18. HUNGATE, R. E. (1984). *Proc. Nutr. Soc.*, **43**, 1.
19. STEPHEN, A. M., HADDAD, A. C. and PHILLIPS, S. F. (1982). *Gastroenterol.*, **82**, 1189.

20. Stephen, A. M., Haddad, A. C. and Phillips, S. F. (1983). *Gastroenterol.*, **85**, 589.
21. Bustos-Fernandez, L. (1986). In: *Proceedings of the XIII International Congress of Nutrition* 1985, T. G. Taylor and N. K. Jenkins (Eds), 1986, John Libbey, London, 224–8.
22. Roediger, W. E. W. (1982). In: *Colon and Nutrition*, H. Kasper and H. Goebell (Eds), MTP Press, Lancaster, 11–25.
23. Argenzio, R. A. and Southworth, M. (1975). *Amer. J. Physiol.*, **228**, 454.
24. McNeil, N. I. (1984). *Amer. J. Clin. Nutr.*, **39**, 338.
25. Demigné, C. and Rémesy, C. (1985). *J. Nutr.*, **115**, 53
26. Siebert, G. and Ziesenitz, S. C. (1985). In: *Die Verwertung der Nahrungsenergie durch Mensch und Tier*, C. Wenk, M. Kronauer, Y. Schutz and H. Bickel (Eds), Wissenschaftliche Verlagsgesellschaft, Stuttgart, 135–7.
27. Ziesenitz, S. C. (1986). In: *Aktuelle Entwicklung und Standard der Künstlichen Ernährung, Infusion Therapy and Clinical Nutrition, Vol. 16*, R. Dölp and D. Löhlein (Eds), 1986, Karger, Basel, 120–32.
28. Ziesenitz, S. C., Vallon, R., Karle, E. J. and Siebert, G., manuscript in preparation.
29. Ziesenitz, S. C. and Steinle, G. Manuscript in preparation.
30. Ziesenitz, S. C. and Steinle, G. Manuscript in preparation.
31. Ziesenitz, S. C. and Siebert, G. Manuscript in preparation.
32. Ziesenitz, S. C., Schorr, H., Karle, E. J. and Siebert, G. Manuscript in preparation.
33. Schorr, H. and Ziesenitz, S. C. Manuscript in preparation.
34. Ziesenitz, S. C. and Friedrich, D. Manuscript in preparation.
35. Drasar, B. S., Shiner, M. and McLeod, G. M. (1969). *Gastroenterol.*, **56**, 71.
36. Moore, W. E. C., Cato, E. P. and Holdeman, L. V. (1978). *Amer. J. Clin. Nutr.*, **31**, 933.
37. Hentges, D. J. (Ed.) (1983). *Human Intestinal Microflora in Health and Disease*, Academic Press, Orlando.
38. Soergel, K. H. (1982). In: *Colon and Nutrition, Falk Symposium: 32*, H. Kasper and H. Goebell (Eds), MTP Press, Lancaster, 27–35.
39. Groha, C. (1979). Personal communication.
40. Kirchgessner, M., Zinner, P. M. and Roth, H.-P. (1983). *Internat. J. Vit. Nutr. Res.*, **53**, 86.
41. Fritz, M., Siebert, G. and Kasper, H. (1985). *Br. J. Nutr.*, **54**, 389.
42. Griessen, M., Bergoz, R., Balant, L. and Loizeau, E. (1986). *Schweiz. Med. Wschr.*, **116**, 469.
43. Würsch, P., and Schweizer, T. (1987). In: *IV. Würzburger Zuckersymposium*, G. Siebert (Ed.), *Dtsch. Zahnärztl. Z.*, **42** (Supplement), (in press).
44. Bond, J. H. and Levitt, M. D. (1972), *J. Clin. Invest.*, **51**, 1219.
45. Cochet, B., Griessen, M., Balant, L., Infante, F., Valloton, M. C. and Bergoz, R. (1981). *Schweiz. Med. Wschr.*, **111**, 192.
46. Pfeffer, M., Ziesenitz, S. C. and Siebert, G. (1985). *Z. Ernährungswiss.*, **24**, 231.
47. Levy, A. G., Benson, J. W., Hewlett, E. L., Herdt, J. D., Doppmann, J. L. and Gordon, R. S. (1976). *Gastroenterol.*, **70**, 157.
48. Ziesenitz, S. C. and Siebert, G. (1987). *J. Nutr.*, **117**, 846.

49. SIEBERT, G. (1987). *Z. Ernährungswiss.*, **26**, 138.
50. ROEDIGER, W. E. W. (1980). *Gut*, **21**, 793.
51. KARIMZADEGAN, E., CLIFFORD, A. J. and HILL, F. W. (1979). *J. Nutr.*, **109**, 2247.
52. CHEN, S. C.-H., TSAI, S. and NESHEIM, M. C. (1980). *J. Nutr.*, **110**, 1023.
53a. ZIESENITZ, S. C., SIEBERT, G. and SCHWENGERS, D. (1987). *J. Nutr.* (*Phila.*), (submitted).
54b. ZIESENITZ, S. C., SIEBERT, G. and IMFELD, T. (1987). *Caries Res.*, (submitted).
54. SIEBERT, G. (1987). In: *IV Würzburger Zuckersymposium*, G. Siebert (Ed.), *Dtsch. Zahnärztl. Z.*, **42** (Supplement), (in press).
55. BENECKE, H. (1986). Personal communication.
56. WENK, C., KRONAUER, M., SCHUTZ, Y. and BICKEL, H. (Eds). (1985). *Die Verwertung der Nahrungsenergie durch Mensch und Tier*, Wissenschaftliche Verlagsgesellschaft, Stuttgart.
57. STOUTHAMMER, A. H. (1978). In: *The Bacteria, A Treatise on Structure and Function*, Vol. VI, J. C. Gunsalus and R. Y. Stanier (Eds), Academic Press, London, 389–462.
58. NÄVEKE, R. and TEPPER, K. P. (1979). *Einführung in die Mikrobiologischen Arbeitsmethoden*, Fischer, Stuttgart.
59. MOORE, W. E. C., CATO, E. P. and HOLDEMAN, L. V. (1978). *Amer. J. Clin. Nutr.*, **31**, 933.
60. RÉRAT, A., FISLEWICZ, M., HERPIN, P., VAUGELADE, P. and DURAND, M. (1985). *C. R. Acad. Sci. Paris*, **300**, série III, 467.
61. ZIESENITZ, S. C., FRITZ, M., PFEFFER, M. and SIEBERT, G. (1985). In: *Die Verwertung der Nahrungsenergie durch Mensch und Tier*, C. Wenk, M. Kronauer, Y. Schutz and H. Bickel (Eds), Wissenschaftliche Verlagsgesellschaft, Stuttgart, 151–4.
62. BERSCHAUER, F. (1985). In: *Die Verwertung der Nahrungsenergie durch Mensch und Tier*, C. Wenk, M. Kronauer, Y. Schutz and H. Bickel (Eds), Wissenschaftliche Verlagsgesellschaft, Stuttgart, 138–9.
63. BERSCHAUER, F. and SPENGLER, M. (1987). In: *IV. Würzburger Zuckersymposium*, G. Siebert (Ed.), *Dtsch. Zahnärztl. Z.*, **42** (Supplement), (in press).
64. THIÉBAUD, D., JACOT, E., SCHMITZ, H., SPENGLER, M. and FELBER, J. P. (1984). *Metabolism*, **33**, 808.
65. GROSSKLAUS, R. (1987). In: *IV. Würzburger Zuckersymposium*, G. Siebert (Ed.), *Dtsch. Zahnärztl. Z.*, **42** (Supplement), (in press).
66. BÄSSLER, K.-H. (1987). In: *IV. Würzburger Zuckersymposium*, G. Siebert (Ed.), *Dtsch. Zahnärztl. Z.*, **42** (Supplement), (in press).
67. VAN WEERDEN, E. J., HUISMAN, J. and VAN LEEUWEN, P. (1984). *Further Studies on the Digestive Process of Palatinit® in the Pig*, ILOB—Institut voor Landbouwkundig Onderzoek van Biochemische Producten, Report No. 530, Wageningen.
68. FÉVRIER, C. R. and PASCAL, G. (1985). *Utilisation Energétique du Palatinit® et du Saccharose chez le Porc en Finition*, INRA, Station de Recherches sur l'Elevage des Porcs.
69. OKU, T., TOKUNAGA, T. and HOSOYA, N. (1984). *J. Nutr.*, **114**, 1574.
70. ZBINDEN, G. (1985). In: *Einsatz von Zuckersubstituten im Kampf gegen Karies*, R. Grossklaus (Ed.), Symposium im Bundesgesundheitsamt 11 November 1985, Berlin, (in press).

71. SIEBERT, G. (1985). In: *Einsatz von Zuckersubstituten im Kampf gegen Karies*, R. Grossklaus (Ed.), Symposium im Bundesgesundheitsamt 11 November 1985, Berlin, (in press).
72. SIEBERT, G. and ZIESENITZ, S. C. (1986). *Z. Ernährungswiss.*, **25**, 242.
73. ZIESENITZ, S. C. (1986). *Z. Ernährungswiss.*, 1986, **25**, 248.
74. LUNSFORD, R. D. and MACRINA, F. L. (1986). *J. Bacteriol.*, **166**, 426.
75. SIEBERT, G., THIM, P. and BRENNER, H. P. (1986). Unpublished data.
76. SIEBERT, G. and FORSTHUBER, F. (1986). Unpublished data.
77. DOYLE, R. J. and CIARDI, J. J. (Eds). (1983). *Glucosyltransferases, Sucrose, Glucans and Dental Caries* (Special Supplement to *Chemical Senses*), Information Retrieval, Washington, DC.
78. ZIESENITZ, S. C. (1987). In: *IV. Würzburger Zuckersymposium*, G. Siebert (Ed.), *Dtsch. Zahnärztl.*, **42** (Supplement), (in press).
79. ZIESENITZ, S. C. and SIEBERT, G. (1986). *Caries Res.*, **20**, 498.
80. SIEBERT, G. and ZIESENITZ, S. C. (1987). In: *IV. Würzburger Zuckersymposium*, G. Siebert (Ed.), *Dtsch. Zahnärztl. Z.*, **42** (Supplement), (in press).
81. IMFELD, T. N. (1983). *Identification of Low Caries Risk Dietary Components*, Monographs in Oral Science, Vol. 11, H. M. Myers (Ed.), Karger, Basel.
82. IMFELD, T. (1987). In: *IV. Würzburger Zuckersymposium*, G. Siebert (Ed.), *Dtsch. Zahnärztl. Z.*, **42** (Supplement), (in press).
83. ANON. (1986). *Scientific Consensus Conference on Methods for Assessment of the Cariogenic Potential of Food. J. Am. Dent. Assoc.*, **112**, 535.
84. NAVIA, J. M. (Ed.) (1977). *Animal Models in Dental Research*, The University of Alabama Press.
85. TANZER, J. M. (Ed.) (1981). *Proceedings*: *Symposium on Animal Models in Cariology*, *Sp. Supp. Microbiology Abstracts*, Information Retrieval, Washington DC.
86. KÖNIG, K. G. (1966). *Möglichkeiten der Kariesprophylaxe beim Menschen und ihre Untersuchung im Kurzfristigen Rattenexperiment*, Verlag Hans Huber, Bern.
87. KÖNIG, K. G., SCHMID, P. and SCHMID, R. (1968). *Arch. oral Biol.*, **13**, 13.
87a. LINKE, H. A. B. (1986). *World Rev. Nutr. Diet*, **47**, 134.
88. WANG, Y.-M. and VAN EYS, J. (1981). *Ann. Rev. Nutr.*, **1**, 437.
89. GRENBY, T. H. (1983). In: *Developments in Sweeteners—2*, T. H. Grenby, K. J. Parker and M. G. Lindley (Eds), Applied Science Publishers, London, 51–88.
90. HYAMS. J. S. (1983). *Gastroenterol.*, **84**, 30.
91. LIFE SCIENCES RESEARCH OFFICE, Federation of American Societies for Experimental Biology (1972). *Evaluation of the Health Aspects of Sorbitol as a Food Ingredient*, SCOGS—9, FDA, Washington.
92. BÄR, A. (1984). In: *Sorbit*, ASPEC (Ed.), c/o ECCO, 19, Rue de l'Orme, Brussels, 15–18.
93. ALLISON, R. G. (Ed.) (1979). *Dietary Sugars in Health and Disease—III. Sorbitol*, Life Sciences Research Office, Federation of American Societies for Experimental Biology, Bethesda, Maryland.
94. SICARD, P. J. (1982). In: *Nutritive Sweeteners*, G. G. Birch and K. J. Parker (Eds), Applied Science Publishers, London, 145–70.
95. FÖRSTER, H. and MEHNERT, H. (1979). *Akt. Ernährung*, **5**, 245.

96. THANNHAUSER, S. J. and MEYER, K. H. (1929). *Münch. med. Wschr.*, **76**, 356.
97. BIRKHED, D., EDWARDSSON, S., KALFAS, S. and SVENSÄTER, G. (1984). *Swed. Dent. J.*, **8**, 147.
98. BÁNÓCZY, J., HADAS, E., ESZTÁRY, I., MAROSI, I. and NEMES, J. (1981). *J. Int. Ass. Dent. Child.*, **12**, 59.
99. GLASS, R. L., (1983). *Caries Res.*, **17**, 365.
100. IBER, F. L. and NASRALLAH, S. M. (1969), *Amer. J. Med. Sci.*, **258**, 80.
101. COUNSELL, J. N. (Ed.) (1978). *Xylitol*, Applied Science Publishers, London.
102. BÄR, A. (1986). In: *Alternative Sweeteners*, L. O'B. Nabors and R. C. Gelardi (Eds), *Food Science and Technology*, Vol. 17, Marcel Dekker, New York, 185–216.
103. HORECKER, B. L., LANG, K. and TAKAGI, Y. (Eds). (1969). *International Symposium on Metabolism, Physiology, and Clinical Use of Pentoses and Pentitols*, Springer, Berlin.
104. DEMETRAKOPOULOS, G. Ev. and AMOS, H. (1978). *Wld Rev. Nutr. Diet.*, **32**, 96.
105. HAUSCHILDT, S., CHALMERS, R. A., LAWSON, A. M. and BRAND, K. (1981). *Z. Ernährungswiss.*, **20**, 69.
106. GRIESHABER, H. (1936). *Z. Klin. Med.*, **129**, 412.
107. DUPAS, C. (1974). *Le Sorbose. Propriétés Physiques et Chimiques. Etudes Toxicologique, Pharmacologique et Pharmacocinétique. Tolérances Clinique et Biologique*, Thése de Pharmacie, Lille.
108. MÜHLEMANN, H. R. (1976). *Schweiz. Mschr. Zahnheilk.*, **86**, 1339.
109. GEHRING, F. and KARLE, E. J. (1978). *Caries Res.*, **12**, 118.
110. ZIMMERMANN, L. (1979). In: *Health and Sugar Substitutes*, B. Guggenheim (Ed.), Proc. ERGOB Conf. (Geneva 1978), Karger, Basel, 145–52.
111. SIEBERT, G., ROMEN, W., SCHNELL-DOMPERT, E. and HANNOVER, R. (1980). *Infusionsther. Klin. Ernähr.*, **7**, 271.
112. HERS, H. G. (1952). *Biochim. Biophys. Acta*, **8**, 416, 424.
113. LEUTHARDT, F. and TESTA, E. (1950). *Helv. Chim. Acta*, **33**, 1919.
114. SANCHEZ, J. J., GONZALES, N. S. and PONTIS, H. G. (1971). *Biochim. Biophys. Acta*, **227**, 67.
115. LARDY, H. A., WIEBELHAUS, V. D. and MANN, K . M. (1950). *J. Biol. Chem.*, **187**, 325.
116. BUEDING, E. and MACKINNON, J. A. (1955). *J. Biol. Chem.*, **215**, 495.
117. TUNG, T.-C., LING, K.-H., BYRNE, W. L. and LARDY, H. A. (1954). *Biochim. Biophys. Acta*, **14**, 488.
118. ASSMANN, U. (1980). *Intermediärprodukte des Sorbosestoffwechsels*, Master's Thesis, University of Hohenheim.
119. WÜRSCH, P., WELSCH, C. and ARNAUD, M. J. (1979). *Nutr. Metab.*, **23**, 145.
120. ANON. (1981). 'A lefty eyes the sweetener game', *Sci. News*, **119**, 276.
121. LEVIN, G. V. (1986). In: *Alternative Sweeteners*, L. O'B. Nabors and R. C. Gelardi (Eds), *Food Science and Technology*, Vol. 17, Marcel Dekker, New York, 155–64.
122. SIEBERT, G. and ZIESENITZ, S. C. (1987). Unpublished data.
123. FUJII, S., KISHIHARA, S., KOMOTO, M. and SHIMIZU, J. (1983). *Nippon Shokuhin Kogyo Gakkaishi*, **30**, 339.
124. STODOLA, F. H., KOEPSELL, H. J. and SHARPE, E. S. (1952). *J. Amer. Chem. Soc.*, **74**, 3202.

125. SCHIWECK, H. (1980). *Alimenta*, **19**, 5.
126. SCHIWECK, H. (1980). Personal communication.
127. YAMADA, K., SHINOHARA, H. and HOSOYA, N. (1985). *Nutr. Rep. Int.*, **32**, 1211.
128. ZIESENITZ, S. D. (1982). In: *Zuckersymposium III, Würzburg*, G. Siebert (Ed.), *Dtsch. Zahnärztl. Z.*, **37**, Spec. Issue 1, 50.
129. ZIESENITZ, S. and SIEBERT, G. (1983). *Caries Res.*, **17**, 163.
130. WATANABE, T. and KIYOSHI, A. (1960). *Tohoku J. Agricult. Res.*, **11**, 109.
131. STODOLA F. H., SHARPE, E. S. and KOEPSELL, H. J. (1956). *J. Amer. Chem. Soc.*, **78**, 2514.
132. BAILEY, R. W. and BOURNE, E. J. (1959). *Nature (London)*, **184**, 904.
133. RUTTLOFF, H., FRIESE, R., TÄUFEL, K. and TOPORSKI, W. (1964). *Die Nahrung*, **8**, 523.
134. RUTTLOFF, H., FRIESE, R. and TÄUFEL, K. (1964). *Naturwiss.*, **51**, 163.
135. ZIESENITZ, S. C. (1986). Unpublished data.
136. SIDDIQUI, I. R. and FURGALA, B. (1967). *J. Apicult. Res.*, **6**, 139.
137. DAHLQVIST, A. (1961). *Acta Chem. Scand.*, **15**, 808.
138. MACDONALD, I. and DANIEL, J. W. (1983). *Nutr. Rep. Int.*, **28**, 1083.
139. KAWAI, K., OKUDA, Y. and YAMASHITA, K. (1985). *Endocrinologica Japonica*, **32**, 933.
140. VAN WEERDEN, E. J., HUISMAN, J. and VAN LEEUWEN, P. (1983). *Digestion Processes of Palatinose® and Saccharose in the Small Intestine and Large Intestine of the Pig*, ILOB—Report No. 520, Wageningen.
141. MENZIES, I. S. (1974). *Biochem. Soc. Trans.*, **2**, 1042.
142. MAKI, Y., OHTA, K., TAKAZOE, I., MATSUKUBO, Y., TAKAESU, Y., TOPITSOGLOU, V. and FROSTELL, G. (1983). *Caries Res.*, **17**, 335.
143. OHTA, K. and TAKAZOE, I. (1983). *Bull. Tokyo Dent. Coll.*, **24**, 1.
144. TOPITSOGLOU, V., SASAKI, N., TAKAZOE, I. and FROSTELL, G. (1984). *Caries Res.*, **18**, 47.
145. GEHRING, F. (1973). *Z. Ernährungswiss.*, Suppl. 15, 16.
146. MÜHLEMANN, H. R. (1985). Personal communication.
147. OOSHIMA, T., IZUMITANI, A., SOBUE, S., OKAHASHI, N. and HAMADA, S. (1983). *Infect. Immun.*, **39**, 43.
148. TAKAZOE, I., FROSTELL, G., OHTA, K., TOPSITSOGLOU, V. and SASAKI, N. (1985). *Swed. Dent. J.*, **9**, 81.
149. SASAKI, N., TOPITSOGLOU, V., TAKAZOE, I. and FROSTELL, G. (1985). *Swed. Dent. J.*, **9**, 149.
150. BIRKHED, D., TAKAZOE, I. and FROSTELL, G. (1987). In: *IV. Würzburger Zuckersymposium*, G. Siebert (Ed.), *Dtsch. Zahnärztl. Z.*, **42** (Supplement), (in press).
151. SIEBERT, G. and ZIESENITZ, S. C. Manuscript in preparation.
152. NAKAJIMA, Y., TOKAOKA, M., OHTA, K. and TAKAZOE, I. (1985). *Seito Gijutsu Kenkyukaishi*, **34**, 58.
153. OKADA, S. and KITAHATA, S. (1975). *J. Jap. Food Ind.*, **22**, 6.
154. IKEDA, T., SHIOTO, T., MCGHEE, J. R., OTAKA, S., MICHALEK, S. M., OCHIAI, K., HIRASAWA, M. and SUGIMOTO, K. (1978). *Infect. Immun.*, **19**, 477.
155. IKEDA, T. (1982). *Int. J. Dent.*, **32**, 33.
156. YAMADA, T., KIMURA, S. and IGARASHI, K. (1980). *Caries Res.*, **14**, 239.

157. IMFELD, T. (1984). In: *Cariology Today*, B. Guggenheim (Ed.), Int. Congr. (Zürich 1983), Karger, Basel, 147–53.
158. HIDAKA, H. (1983). *Kagaku to Seibutsu*, **21**, 291.
159. TOKUNAGA, T., OKU, T. and HOSOYA, N. (1986). *J. Nutr. Sci. Vitaminol.*, **32**, 111.
160. VAN VELTHUIJSEN, A. (1979). *J. Agric. Food Chem.*, **27**, 680.
161. SCHIWECK, H. (1986). Personal communication.
162. GRENBY, T. H. (1986). Personal communication.
163. HAVENAAR, R., HUIS IN'T VELD, J. H. J., BACKER DIRKS, O. and DE STOPPELAAR, J. D. (1979). In: *Health and Sugar Substitutes*, B. Guggenheim (Ed.), Proc. ERGOB Conf. (Geneva 1978), Karger, Basel, 192–8.
164. LANTHIER, P. L., and MORGAN, M. W. (1985). *Gut*, **26**, 415.
165. BARTH, C. A. and KOPRA, N. (1986). *Z. Ernährungswiss.*, **25**, 171.
166. ROSIERS, C., VERWAERDE, F., DUPAS, H. and BOUQUELET, S. (1985). *Ann. Nutr. Metab.*, **29**, 76.
167. ZUNFT, H.-J., SCHULZE, J., GÄRTNER, H. and GRÜTTE, F.-K. (1983). *Ann. Nutr. Metab.*, **27**, 470.
168. IMFELD, T. and LUTZ, F. (1985). *Swiss Dent.*, **6**, 44.
169. FIRESTONE, A. R., SCHMID, R. and MÜHLEMANN, H. R. (1980). *Caries Res.*, **14**, 324.
170. RUNDEGREN, J., KOULOURIDES, T. and ERICSON, T. (1980). *Caries Res.*, **14**, 67.
171. IMFELD, T. (1977). *Helv. Odont. Acta*, **21**, 1.
172. SIEBERT, G. (1986). Unpublished data.
173. LEDERER, P., DELVILLE, P. and CREVECOEUR, E. (1974). *La Sucrerie Belge*, **93**, 311.
174. RENNHARD, H. H. and BIANCHINE, J. R. (1976). *J. Agric. Food Chem.*, **24**, 287.
175. LIAN-LOH, R., BIRCH, G. G. and COATES, M. E. (1982). *Br. J. Nutr.*, **48**, 477.
176. KEARSLEY, M. W., BIRCH, G. G. and LIAN-LOH, R. H. P. (1982). *Starch/Stärke*, **34**, 279.
177. MARANESI, M., GENTILI, P. and CARENINI, G. (1984). *Acta Vitaminol. Enzymol.*, **6**, 3.
178. MARANESI, M., BARZANTI, V., GENTILI, P. and CARENINI, G. (1984). *Acta Vitaminol. Enzymol.*, **6**, 11.
179. FELBER, J. P., TAPPY, L., VOUILLAMOZ, D., RANDIN, J. P., TEMLER, E. and JÉCQUIER, E. (1984). 'Comparative study of maltitol and sucrose by means of continuous indirect calorimetry', presented at the International Symposium *Sugar and Sugar Substitutes in Human Nutrition*, Milan.
180. STRÄTER, P. J. (1986). In: *Alternative Sweeteners*, L. O'Brien Nabors and R. C. Gelardi (Eds), *Food Science and Technology*, Vol. 17, Marcel Dekker, New York, 217–44.
181. GAU, W., KURZ, J., MÜLLER, L., FISCHER, E., STEINLE, G., GRUPP, U. and SIEBERT, G. (1979). *Z. Lebensm. Unters. Forsch.*, **168**, 125.
182. LINDNER, H. J. and LICHTENTHALER, F. W. (1981), *Carbohydrate Res.*, **93**, 135.
183. LICHTENTHALER, F. W. and LINDNER, H. J. (1981). *Liebigs Ann. Chem.*, 2372.
184. GRUPP, U. and SIEBERT, G. (1978). *Res. Exp. Med. (Berl.)*, **173**, 261.
185. GEHRING, F. and KARLE, E. J. (1981). *Z. Ernährungswiss.*, **20**, 96.
186. KARLE, E. J. and GEHRING, F. (1981). *Dtsch. Zahnärztl. Z.*, **36**, 673.
187. SIEBERT, G., GRUPP, U. and HEINKEL, K. (1975). *Nutr. Metabol.*, **18** (Suppl. 1), 191.

188. KIRCHGESSNER, M., ZINNER, P. M. and ROTH, H.-P. (1983), *Internat. J. Vit. Nutr. Res.*, **53**, 86.
189. BUJAKE, J. E. (1986). In: *Alternative Sweeteners*, L. O'Brien Nabors and R. C. Gelardi (Eds), *Food Science and Technology*, Vol. 17, Marcel Dekker, New York, 277–93.
190. SIEBERT, G., ZIESENITZ, S. C. and LOTTER, J. (1987). *Caries Res.*, **21**, 141.
191. ZIESENITZ, S. C. (1987). In: *IV. Würzburger Zuckersymposium*, G. Siebert (Ed.), *Dtsch. Zahnärztl. Z.*, **42** (Supplement), (in press).
192. GRENBY, T. H. and SALDANHA, M. G. (1986). *Caries Res.*, **20**, 7.

Chapter 6

SWEETENERS AND DENTAL HEALTH: THE INFLUENCE OF SUGAR SUBSTITUTES ON ORAL MICROORGANISMS

HARALD A. B. LINKE

Department of Microbiology, New York University Dental Center, New York, USA

SUMMARY

The major dental diseases, dental caries and periodontal disease, are caused by specific Gram-positive and Gram-negative oral microorganisms during interactions with the host's diet. Among the dietary components which are most 'cariogenic' are the carbohydrates. Carbohydrate sweeteners such as sucrose and glucose rank very high on this list, and they are easily converted by oral bacteria into caries-inducive products, e.g. polysaccharides and lactic acid. The other group of carbohydrate sweeteners, the polyols or non-sugar bulk sweeteners, were thought of as being less cariogenic since they are not as readily fermented by oral bacteria. Sorbitol, mannitol, xylitol and other polyols possess dual properties: they are reasonably good sweeteners, and they are considered to be less cariogenic than sucrose. A third group of sweeteners, the intense or artificial sweeteners, are also termed non-caloric because they cannot be utilized as a dietary carbon source. This chapter reviews and discusses the influence of sugar alcohols and intense sweeteners on oral bacteria. Most of the sugar alcohols are metabolized by the oral flora, and subsequently give rise to cariogenic acidic end-products. The mode of action is different in the case of the artificial sweeteners. Saccharin, for example, actively inhibits the growth and acid production of oral bacteria. Therefore some of the intense sweeteners, with their dual properties of being sweet-tasting and 'anti-cariogenic', may have a great potential in caries prevention.

1. INTRODUCTION

Most people in modern western societies enjoy eating sweet-tasting foods. The intake of sucrose-containing foods is often encouraged in our society; especially in children sweet taste sensation is identified with affection and reward. Sweet taste exerts a beta-adrenergic-like stimulatory effect on parotid salivary secretion.[59] This effect contributes to quick dissolution of sweet foods, e.g. candy, in the oral cavity. The interaction of oral bacteria with available sucrose or other carbohydrates leads primarily to dental plaque formation and secondarily to tooth decay, especially in children and young adults. The dental profession, together with nutritionists and responsible food manufacturers, share an interest in finding a dentally safe and palatable sucrose substitute. Several dental related biochemical and clinical studies using polyols as well as artificial sweeteners suggest that some sugar alcohols, e.g. xylitol, besides the non-caloric sweeteners, e.g. saccharin, may be suitable for this purpose. On the contrary, some dietary popular sugar alcohols such as mannitol and sorbitol are easily fermented by oral bacteria, and both have been used in differential media for the selective isolation of *Streptococcus mutans*,[67] the major etiological agent of dental caries.

The chemistry, sources of origin, organoleptic properties and toxicology of the most important sugar alcohols and artificial sweeteners have been described elsewhere[40,52–54,60,88,99,100] and therefore will not be dealt with in this chapter.

This review will discuss pertinent findings on the interaction of oral microorganisms with sugar alcohols and non-caloric artificial sweeteners alone or in combination with the common sugars, e.g. sucrose or glucose. Some of these bacteriological studies indicate that long-term exposure to most sugar alcohols may lead to an adaptation of the oral bacteria to these compounds.[15] The shift in the plaque ecology would then favor bacteria that are able to rapidly ferment most polyols acidogenically. Artificial sweeteners such as saccharin behave like true antibacterial compounds, and do not lead to such adaptation.

2. ROLE OF ORAL MICROORGANISMS IN DENTAL DISEASE

The interaction of oral bacteria with available sucrose or other carbohydrates leads primarily to dental plaque formation. Secondarily, acidic fermentation end-products and/or other extracellular products, such as enzymes and toxins, cause trauma in the host's hard and soft tissues, resulting in the major oral diseases dental caries and periodontal disease.

2.1. Dental Caries

Stephan[110] reported that during rinsing with 10% sucrose solution the pH of dental plaque fell within 2–4 min from about 6·5 to 5·0. Acidic fermentation end-products, e.g. lactic acid, are responsible for this drop in pH, which leads to demineralization of the oral hard tissues and subsequently to dental caries. The association between frequency of sugar consumption and dental caries is well documented and has been reviewed by Sreebny.[108]

Among the oral bacteria that are able to induce dental caries are *Streptococcus mutans*, *Actinomyces viscosus* and *S. sanguis*, together with *S. salivarius*. *S. mutans* produces dental caries preferentially in the presence of sucrose. An extracellular multienzyme complex breaks sucrose down into its two components glucose and fructose with the aid of invertase. The glucosyl transferase moiety of the enzyme complex chains then links glucose units together to form water-insoluble glucan, which eventually forms the basis for the plaque matrix. A third moiety of the enzyme complex, the 'enzyme responsible for adherence',[87] attaches glucan (dextran) to the smooth surface of the oral hard tissue to form the dental plaque matrix. The fructose moiety of sucrose is utilized by *S. mutans* as a carbon and energy source. Fructose is broken down by glycolytic enzymes into acidic fermentation products such as lactic acid, acetic acid, formic acid and other fermentation end-products such as ethanol. Demineralization of the oral hard tissue occurs as soon as the plaque pH drops below the 'critical pH' of 5·5. Furthermore, with the aid of fructosyl transferase, excess of fructose also can be converted into water-soluble fructan (levan). Fructan serves as a storage polysaccharide. Between meals or during the night *S. mutans* is able to convert fructan with the aid of fructanase into fructose, which then also can be utilized for sustained acid production.

Other oral bacteria, e.g. *A. viscosus*, also give rise to dental caries. There is evidence from animal experiments that in the presence of *A. viscosus* glucose seems to be more cariogenic than sucrose, maybe for stereochemical reasons.

Newbrun[88] has written a monograph covering all aspects of dental caries and the relationship between diet and dental caries is discussed by Theilade and Birkhed.[113] The general use of sugar alcohols for the purpose of caries reduction is described elsewhere[11,12,25,44,82,83,98] and has been extensively reviewed by Linke.[74]

2.2. Periodontal Disease

The microbial etiology of periodontal disease has been thoroughly reviewed by van Palenstein Helderman;[121] other reviews have been written on different aspects of the same subject.[86,89,106]

Gram-positive plaque-forming bacteria such as *Streptococcus sanguis*, *Actinomyces viscosus* and *Rothia dentocariosa* are found in sub- and supra-gingival plaque of the healthy gingiva. Under certain circumstances, for example neglected oral hygiene, this plaque may build up to a thickness of 0·4 mm, with a 10- to 20-fold increase in plaque-forming bacteria. Under these circumstances large amounts of soft-tissue-damaging extracellular substances such as neuraminidase are formed, especially by *S. sanguis*, leading to the development of gingivitis, which is characterized by an inflamed gingiva and bleeding of the gums. Other substances produced by *S. sanguis* are a bacteriocin and sialidase, the former contributing to the dominance of *S. sanguis* at the site and the latter aiding build-up of sub- and supra-gingival plaque. Gingivitis is considered to be the mildest form of periodontal disease and can be reversed by therapeutic measures and improved oral hygiene.

Depending on the thickness of plaque at the gingival margin and on the host's salivary oxygen tension, part of the dental plaque environment may become anaerobic. Under these circumstances Gram-negative anaerobic bacteria can colonize the gingival crevice, which is enriched with bacterial degradation products of the host's soft tissue. These new bacteria dwelling in this protein-rich environment contribute, with their extracellular enzymes and toxins, to more severe forms of periodontal disease. The presence of *Bacteroides* species or the so-called BPB (black-pigmented *Bacteroides*) may then promote periodontitis; or the colonization by *Actinobacillus actinomycetemcomitans* may promote the formation of juvenile periodontitis, a specific form of periodontosis. Other forms of periodontal disease are prepubertal periodontitis, adult periodontitis, rapidly progressing periodontitis, acute necrotizing ulcerative gingivitis (ANUG) and Papillon–Lefèvre syndrome. The microbiology of the latter types of periodontal disease is not well understood.

At the outset of periodontal disease, gingivitis is the result of plaque bacteria interacting with common sugars. Therefore, the use of sugar alcohols in the diet may be of some benefit to oral health in suspected or established cases of periodontal disease.[84] The role of sugar alcohols in regard to dental plaque formation is discussed by van Houte.[120]

This review will discuss in chronological order the most pertinent studies *in vitro* and *in vivo* on the influence of polyols, other bulk sweeteners and intense sweeteners on oral microorganisms.

3. BACTERIOLOGICAL STUDIES USING POLYOLS AND OTHER BULK SWEETENERS

3.1. Early Work on Polyol Fermentation

The low fermentability of sugar alcohols by bacteria was observed by Glinka-Tschernorutzky.[36] In a study on the effect of sugar alcohols on growth and nitrogen metabolism of *Bacillus mycoides* it was found that in comparison with glucose and sucrose, growth and incorporation of nitrogen into the cells was less with most of the sugar alcohols tested. Grubb[48] studied the aerobic and anaerobic fermentation of sorbitol, sucrose and glucose at M/10 concentrations by salivary microorganisms (*Lactobacillus acidophilus*, oral streptococci, yeasts, and mixtures of yeasts and *L. acidophilus*). In all cases sorbitol was fermented more slowly than the other sugars. On the basis of these *in vitro* studies, the author suggested that in view of its slow fermentation, sorbitol might be useful in the prevention and control of dental caries. This article also covers the history of the production and microbial fermentation of sorbitol.

3.2. Fermentation of Sorbitol by Oral Bacteria

Shockley *et al.*[101] studied the fermentation of sorbitol by certain acidogenic oral microorganisms (lactobacilli, streptococci, staphylococci). They found that only some bacterial strains of the various species fermented sorbitol. The acid production rate from sorbitol was slow and the quantity of acid was small at optimum growth conditions. In the presence of mixed microflora of saliva, sorbitol selectively promoted the growth of Gram-negative bacteria but not acidogenic Gram-positive rods and cocci. On the basis of these findings the authors concluded that sorbitol would not contribute significantly to acid production and, therefore, would reduce dental caries formation.

Crowly *et al.*[18] investigated the comparative fermentability of sorbitol, glucose and glycerol by common oral microorganisms such as lactobacilli, streptococci, staphylococci and yeasts. They found that sorbitol was not fermented as rapidly as glucose by the oral bacteria tested. Within 24 h all strains produced sufficient acid from glucose to decalcify teeth. This was not observed with sorbitol, whilst glycerol was fermented least of all. The authors indicated that sorbitol could be of importance in limiting the caries process, provided that the fermentable substrate is not retained on or within the tooth structure for more than a week.

3.3. Fermentation of Sorbitol by Dental Plaque

Frostell[23] studied the acid production rates from glucose, sucrose, lactose, fructose, maltose, sorbitol, mannitol, rhamnose and starch by the mixed bacterial population of dental plaque material taken from adult patients. Plaque material (10 mg wet wt) was suspended in Ringer solution and placed under a nitrogen atmosphere. The carbohydrates were then added at a concentration of 0·1 M. The rate of acid production was determined by back titration to initial pH for approximately 1 h. The mean acid production rates from glucose, sucrose, lactose, fructose, maltose and starch ranged between $4{\cdot}50 \times 10^{-9}$ and $10{\cdot}50 \times 10^{-9}$ E/mg wet wt per minute (E = acid production rate). As compared to the other sweeteners the acid production rate from lactose was significantly lower. The rate of acid production from starch was 58% of that from glucose. From sorbitol, mannitol and rhamnose there was little or no acid production. However, this aroused the suspicion that prolonged intake of mannitol and sorbitol by the host may favor the growth of oral bacteria capable of fermenting these sweeteners.

3.4. Synthesis of Polysaccharides from Polyols

Saxton[97] examined the synthesis *in vivo* of both intracellular and extracellular polysaccharides by bacteria in dental plaque from a variety of dietary sugars by electron microscopy. The exposure of plaque *in vivo* to different sugars indicated clearly that sucrose and glucose are rapidly incorporated into intracellular polysaccharide, whereas maltose, fructose, lactose and sorbitol are utilized slowly or not at all by the plaque bacteria. Extracellular polysaccharide was synthesized in quantity only from sucrose, and to a lesser degree from maltose. Synthesis of the polysaccharides seemed to proceed fast. After 30 min the amounts of both intracellular and extracellular polysaccharide exceeded the amount originally in the plaque before exposure to sucrose.

3.5. Lactic Acid Production from Sorbitol and Mannitol

Dallmeier *et al.*[19] found that two strains of plaque streptococci, No. 249 and No. 275, isolated from caries-susceptible patients, fermented glucose, sorbitol and mannitol almost exclusively via the Embden–Meyerhof pathway. Utilization of glucose by both strains produced mainly lactic acid (88·0 and 85·5 mol%, respectively). Fermentation of sorbitol and mannitol produced less lactic acid: 36·5 and 16·4 mol% from sorbitol, and 19·9 and 33·8 mol% from mannitol. As compared with sorbitol, mannitol fermentation produced more formic acid and also some ethanol. Since formic acid, unlike lactic acid, is not a chelating agent, the authors concluded that the cariogenicity of sorbitol and mannitol must be lower than that of hexoses such as glucose.

3.6. Lactic Acid Production from Xylitol and Lycasin®

Gehring[31] studied the fermentation of sucrose, mannitol, sorbitol, xylitol, Lycasin® and calcium–sucrose–phosphate by oral and other bacteria. After 24 h strain S.227 from a group of cariogenic streptococci produced the most acid from sucrose and sorbitol, little acid from calcium-sucrose-phosphate, and after an adaptation period of three days, a large amount of acid from Lycasin®, but a very small amount of acid, if any, from xylitol. Two reference strains of the Lancefield group *Q* produced acid from xylitol. Repeated passage of cariogenic streptococci in Lycasin® and xylitol media led to adaptation to metabolize these sugar alcohols. *S. sanguis* strains in the presence of 5% sucrose and 1–2% calcium–sucrose–phosphate produced such large amounts of polysaccharide that the medium became a gel.

Bramstedt and Trautner[13] examined the utilization of Lycasin® by plaque streptococci. They found a much lower acid and intracellular polysacchardie production by streptococci utilizing Lycasin® rather than glucose. Cells pregrown on Lycasin® underwent a change in their pool of enzymes responsible for synthesis of intracellular polysaccharides; reincubation of these cells with glucose resulted in a reduced ability to synthesize glycogen.

Stegmeier *et al.*[109] studied the metabolism of mannitol and sorbitol by plaque streptococci. Both are metabolized via the Embden–Meyerhof pathway by phosphorylation to mannitol-1-phosphate or sorbitol-6-phosphate which are then converted enzymically to fructose-6-phosphate. The major end-products of this pathway are lactic acid, ethanol and formic acid.

Brown and Wittenberger[14] confirmed the results obtained by Stegmeier *et al.*[109] They found that *Streptococcus mutans* NCTC 10449 adapted to mannitol or sorbitol produced high levels of mannitol-1-phosphate dehydrogenase and sorbitol-6-phosphate dehydrogenase activity. Neither of these enzymes was present at significant levels in cells grown on glucose or other carbon sources. Their presence indicated that strains of *S. mutans* ferment both mannitol and sorbitol by a pathway that involves phosphorylation of the substrates prior to their oxidation by distinct enzymes of the Embden–Meyerhof pathway via fructose-6-phosphate.

Gehring and Patz[33] found in a Warburg test under aerobic conditions that in comparison with glucose, only a small amount of sorbitol was metabolized by *S. mutans* NCTC 10449. However, under anaerobic conditions much more glucose was utilized, while the rate of sorbitol utilization remained unchanged.

Platt and Werrin[93] studied acid production from sucrose, glucose, mannitol, sorbitol and xylitol by oral streptococci (*S. salivarius*, *S. sanguis*, *S. mitis*, *S. mutans* NCTC 10449 and *S. mutans* BHT). Only *S. mutans* BHT was able to produce lactic acid from all the carbohydrates, with the exception of xylitol. Xylitol was not fermented by the oral streptococci tested in this study.

3.7. Fermentation of Polyols in the Mouth of Rodents

Gehring[32] isolated from the oral cavity of rats and hamsters microorganisms capable of metabolizing mannitol, sorbitol and xylitol. Certain streptococci of the Lancefield group *Q*, enterococci-like streptococci, *Proteus morganii* and other unknown Gram-negative rods were able to ferment all three sugar alcohols, with the exception of the two latter groups which did not ferment sorbitol, and the *P. morganii* group which did not ferment mannitol.

3.8. Growth Inhibition of *S. mutans* by Xylitol

Knuuttila and Mäkinen[61] found that 2·5% xylitol in the medium retarded the growth of *S. mutans* (Ingbritt) in the presence of 0·25% glucose, but the extracellular polysaccharide-forming activity of the cells was increased up to 80% at xylitol concentrations of up to 55 mg/ml growth medium. Xylitol had no effect on the uptake of glucose by the cells. In contrast to these studies Assev *et al.*[1a] found that xylitol exhibited a dose-related inhibition of growth of *S. mutans* OMZ-176 in brain–heart infusion medium, and they suggested that the mechanism involved may be an effect of the translocation of glucose across the bacterial cell membrane. Sorbitol produced no similar effect but caused a delay in the formation of the stationary phase of the growth curve.

3.9. 'Turku Sugar Studies'

Mäkinen *et al*[81] discovered during the 'Turku sugar studies' that persons on a restricted sucrose, fructose or xylitol diet showed significant differences in their salivary peroxidase activity. In contrast to the sucrose and fructose groups, xylitol consumption increased considerably the salivary lactoperoxidase activity, suggesting that persons habitually consuming a xylitol-containing diet may avoid certain oral inflammatory infections because of the antibacterial properties of the enzyme. According to Clem and Klebanoff,[16] peroxidase plays an important role in the inhibition of oral lactobacilli and streptococci. The lactoperoxidase–thiocyanate–hydrogen peroxide system in saliva also inhibits the net transport of glutamate into the cell.

3.10. Mannitol Production by Oral Bacteria

Kornman and Loesche[62] reported on the excretion of mannitol by *S. mutans* and *Lactobacillus casei*. *S. mutans* GS-5 or 3720 was grown in the presence of fructose, glucose, or sucrose, and then inoculum from the log-phase was transferred to phosphate buffer with 0·5–2·5% sucrose. After set intervals the supernatants of these cultures were assayed by gas–liquid chromatography for the presence of mannitol. Sixty-minute cultures of *S. mutans*, adapted to fructose, exhibited peaks for fructose, mannitol, and glucose, plus other unidentified peaks. Mannitol was not found in cultures with sucrose concentrations below 1·5%. Mannitol peaks were not detected in cultures of *S. sanguis* but were present in those of *L. casei*. No mannitol was detected in resting cell cultures grown with glucose or sucrose. The authors believed that the excretion of mannitol by *S. mutans* in the presence of high concentrations of sucrose could involve the conversion of fructose to mannitol, which prevents the build-up of fructose and allows polysaccharide synthesis to proceed.

Loesche and Kornman[75] found that resting cell suspensions of *S. mutans* 6715 in the presence of up to 2·5% sucrose or 1% glucose and under anaerobic conditions (85% N_2, 10% H_2 and 5% CO_2) produced mannitol as a metabolic end-product, determined with the aid of gas–liquid chromatographic analysis. They suggested that mannitol may be formed by a reduction of fructose-6-phosphate via mannitol-1-phosphate. This pathway would permit the regeneration of NAD under anaerobic conditions, thereby giving maximal catabolism of hexoses via the glycolytic pathway in *S. mutans*. This hypothesis might explain why some mutants of *S. mutans* which are weak mannitol fermenters have a reduced ability to use sucrose. Furthermore, the results indicate that the mannitol catabolism as described by Brown and Wittenberger[14] can be reversed by *S. mutans* under certain growth conditions.

3.11. Acid Production from Maltitol and Lycasin®

Edwardsson *et al.*[22] compared the acid production from maltitol with the acid production from Lycasin®, sorbitol and xylitol by several oral streptococci and lactobacilli. The polyols were added at a final concentration of 1·0% to two different basal media. The final pH was recorded after a seven-day incubation period at 37°C. While maltitol was fermented by only two-thirds of the lactobacilli, Lycasin® was fermented by all strains of *S. faecalis*, 90% of the lactobacilli, 50% of the *S. sanguis* strains, approximately 30% of the *S. mutans* strains and by a few other streptococci. Acid production from sorbitol was observed by 80% of the *S. mutans* strains, all *S. faecalis* strains and most of the lactobacilli. Only *L. salivarius* and *S. avium* were able to ferment xylitol.

3.12. Acid Production from Sorbitol by Dental Plaque Suspensions

Birkhed *et al.*[8] studied the acid production from glucose and sorbitol in dental plaque suspensions (10 mg wet wt/ml) and the pH changes in dental plaque *in vivo* after mouth rinses (for 30 s) with 10% solutions of glucose and sorbitol before and after 4–6 weeks of frequent daily mouth rinses with sorbitol in 18 volunteers. The mean acid production from sorbitol (as a percentage of that from glucose) increased about 21% ($P < 0{\cdot}001$) and the mean initial/resting plaque pH values were approximately 0·2 units higher ($P < 0{\cdot}01$) after the sorbitol adaptation period. The pH decreases from sorbitol were more pronounced after the six-week adaptation period ($P < 0{\cdot}01$). The acid production from sorbitol and the pH decreases after mouth rinses with sorbitol were much smaller than the corresponding values found with glucose before as well as after the adaptation period.

3.13. Amylase and Lycasin® Metabolism in Dental Plaque

In a study *in vitro* on amylase in starch and Lycasin® metabolism in human dental plaque, Birkhed and Skude[10] tested the acid production activity from glucose, boiled soluble starch and Lycasin® at a concentration of 3·0% (w/v) in 0·1 ml samples of plaque suspension, with the aid of an automatic titration method. The acid production activity (APA) from soluble starch as well as from Lycasin® was significantly lower ($P < 0{\cdot}001$) than that from glucose. Lycasin® was metabolized much more slowly ($P < 0{\cdot}01$) than soluble starch. The APA from soluble starch and Lycasin® was optimal at a substrate concentration of 0·03–6%. At a lower concentration (0·003%) the APA was about 70% and at a higher concentration (12%) 25% lower compared with the optimal substrate concentration range. Furthermore, the amylase (α-1,4-glucan-4-glucanohydrolase, EC 3.2.1.1) activity in saliva was about 15 times higher than in dental plaque; 50% of the total plaque amylase activity was found after centrifugation in resuspended plaque pellets. This indicates that the microbial amylase activity in plaque is low compared with that of saliva. There was a correlation between total plaque amylase activity and APA from soluble starch ($r = 0{\cdot}94$). No correlation was found between amylase activity in saliva and APA from soluble starch or between amylase activity in saliva or plaque and APA from Lycasin®.

3.14. Tracer Studies Using ^{14}C-Labeled Xylitol

Hayes and Roberts[51] studied the metabolism of glucose, xylitol, arabinitol, mannitol, ribitol and sorbitol by human dental plaque bacteria suspended in Ringer solution with centrifuged and heat-treated saliva. Most sugar alcohols produced a reduction in pH after aerobic and anaerobic incubation, with the exception of xylitol, which even after serial transfers in saliva–xylitol media, did not produce any significant change in final pH. $^{14}CO_2$ production from U-^{14}C-glucose or -xylitol during fermentation at pH 7·5 indicated that carbon dioxide release from xylitol amounted to 7–27% of that released from glucose. Therefore xylitol had no inhibitory effect on plaque sugar metabolism, but was itself metabolized, although at a slower rate than glucose.

3.15. Automatic Titration of Acids in Dental Plaque Suspensions

Birkhed and Edwardsson[7] studied the acid production of several carbohydrate sweeteners in human dental plaque suspensions *in vitro* using an automatic titration method. The acid production rates expressed as percentages of the rate from glucose were: from mannitol and sorbitol 0%; maltitol and sorbitol, 10–30%; French Lycasin® (80/55), 20–40%; lactose, 40–60%; Swedish Lycasin®, 50–70%; fructose, 80–100%; glucose syrups (DE 40 and 60) and invert sugar and sucrose, 100%. The authors indicated that these data do not give complete information about the cariogenicity of the sweeteners tested, but their fermentability by dental plaque would probably be one of the most important factors to take into consideration.

3.16. Acid Production from Swedish and French Lycasins®

Frostell and Birkhed[24] determined the acid production from Swedish Lycasin® (candy quality) and French Lycasin® (80/55) at a final concentration of 3% (w/v) in suspensions of human dental plaque using an automatic titration method. Compared with the glucose control, both were significantly less fermentable. Acid production fron Lycasin® 80/55 was about 57% lower than from Swedish Lycasin®.

3.17. Fermentation of Polyols by Lactobacilli

Havenaar *et al.*[50] reported that no predominant oral bacteria were found capable of fermenting xylitol and L-sorbose, whilst sorbitol, lactitol, maltitol and Lycasin® were fermented to some degree by strains of *Streptococcus*, *Actinomyces* and *Lactobacillus*. *S. mutans* was able to ferment sorbitol, lactitol, maltitol and Lycasin®, whereas *L. casei* utilized L-sorbose as well as Lycasin®. Adaptation of *S. mutans* by frequent subculturing in sorbitol, lactitol, maltitol and Lycasin® resulted in a sharp increase in fermentation (production of acid and some intracellular polysaccharide) of these sugars; however, when the adapted strain was subcultured subsequently once in glucose, most of its adaptability was lost. No adaptation to xylitol and L-sorbose by *S. mutans* C67-1 was observed.

3.18. Acid Production from Licorice and Hydrogenated Potato Starch

Toors and Herczog[115] studied the acid production from licorice and different sugar substitutes in *S. mutans* 6715 (serotype *d*) monoculture and pooled plaque–saliva mixtures. The fermentability of potato starch, sorbitol, glycerol, regular licorice (35 vol. % sucrose), and xylitol by *S. mutans*, expressed as a percentage of the glucose control, was 3, 40, 3, 76, and 3% respectively. With the pooled plaque–saliva mixture the fermentability of potato starch, sorbitol, experimental licorice, xylitol, and hydrogenated potato starch was 82, 12, 68, 5, and 60% respectively. A very similar pattern was observed in both systems.

3.19. Acid Production from Palatinit®

Karle and Gehring[57] examined the cariogenic properties of Palatinit®, an equimolecular mixture of the disaccharide alcohols α-D-glucopyranosyl-1,6-sorbitol and α-D-glucopyranosyl-1,6-mannitol, *in vitro* and *in vivo*. They exposed five strains of *S. mutans* to this bulk sweetener as well as to sucrose. After seven days of fermentation the final pH of the Palatinit® medium was still above 6·5, whereas the sucrose control fell to a pH below 4·5. In addition, no extracellular polysaccharides were produced by the *S. mutans* strains in the presence of Palatinit®.

The ability of strains of *S. mutans*, *S. sanguis*, *S. mitis*, *Actinomyces viscosus*, *A. odontolyticus*, *A. naeslundii*, *A. israelii* and the genus *Bacterionema* to ferment Palatinit® was also studied by van der Hoeven.[119] All the strains of *S. mutans*, *A. viscosus* and *A. israelii* were found to produce acid from Palatinit® to some degree. Since no Palatinit®-splitting activity was found in cell-free extracts of *S. mutans* C67-1 and OMZ 176, it was suggested that this bulk sweetener is transported unchanged into the cell rather than hydrolyzed extracellularly. In addition, the oral flora in Palatinit®-fed rats showed no adaptation, and Palatinit®-fermenting bacteria did not accumulate. However, it was observed that the rats on a Palatinit® diet developed massive amounts of dental plaque, possibly due to the presence of *A. viscosus* which is a strong plaque producer even in the absence of added sugars in the diet.

3.20. Long-term Studies using Lycasin®

In a study on the effects of three months' frequent consumption of Lycasin®, maltitol, sorbitol and xylitol on human dental plaque, Birkhed *et al.*[9] found that the acid production in dental plaque suspensions from Lycasin®, maltitol and sorbitol, expressed as a percentage of that from glucose, was approximately the same before and after the test period. However, no acid production from xylitol was observed either before or after that test period. The amount of plaque (wet wt) in the Lycasin® group was higher after than before the three-month test period. In the maltitol, sorbitol and xylitol groups, however, it was approximately the same before and after the test period. No difference in the relative numbers of facultative streptococci, e.g. *S. mutans*, or facultative anaerobic lactobacilli in the plaque either at the beginning or at the end of the test period was observed.

3.21. Fermentation of L-Sorbose

With the aid of the Warburg technique, the ability of saliva and plaque to utilize L-sorbose was investigated by Lutz and Gülzow.[77] Compared with sucrose, L-sorbose produced only traces of carbon dioxide; therefore the authors assumed that L-sorbose must have an extremely low cariogenic effect.

3.22. Tracer Studies Using ^{14}C-Labeled Glycerol

The glycerol uptake by strains of *S. mutans* was studied by Kral and Daneo-Moore.[63] They found that only a few strains of *S. mutans* were able to incorporate the non-fermentable substrate, ^{14}C-glycerol; the amount incorporated ranged from 0·15 to 0·43% of the cellular dry weight and followed simple saturation kinetics, i.e. the amount of incorporated glycerol depended solely on the concentration of glycerol in the growth medium.

3.23. pH Suppression by Trichlorosucrose and Polyols

Drucker and Verran[21] compared the effects of glucose, sorbitol, sucrose, trichlorosucrose, xylitol, trichlorosucrose plus sucrose, and xylitol plus sucrose on lowering of pH by *S. mutans* NCTC 10832 and *S. mutans* NCTC

10440 (both strains serotype *c*) grown in four different media *in vitro*. Sucrose was used as the positive control and plain bacterial suspension as the negative control. The lowest pH levels were usually recorded in the presence of sucrose and glucose. Higher pH values were generally attained with sorbitol, trichlorosucrose, and xylitol. Acid production from sucrose was more often diminished by the presence of trichlorosucrose rather than by the presence of xylitol.

3.24. Growth of *S. mutans* in Synthetic Media Containing Polyols

Soyer and Frank[107] compared the growth of *S. mutans* ATCC 25175 in complex and synthetic culture media supplemented with sorbitol, mannitol, Lycasin® 80/55, sorbose, xylitol, sucrose, glucose and fructose. A fall in pH was related to the microbial growth, and especially to the production of lactic acid. The different sugars influenced only slightly the bacterial growth in the complex medium, whereas in the synthetic medium there was no bacterial growth in the presence of sorbose, xylitol or Lycasin® 80/55. An adaptation period was necessary to induce bacterial growth in the synthetic medium supplemented with sorbitol and mannitol.

3.25. Fermentation of Stereoisomeric Components of Palatinit®

Gehring and Karle[35] studied the acid production from Palatinit® by *S. mutans*, which was compared with the two stereoisomeric components α-D-glucopyranosyl-1,6-sorbitol (GPS) and α-D-glucopyranosyl-1,6-mannitol (GPM) of Palatinit®, sorbitol, mannitol and sucrose. The final pH of the medium (DIFCO phenol red broth base) after 48 h of incubation in the presence of 1% carbohydrate was found to be in the order GPM > control (no carbohydrates added) > Palatinit® > GPS > 5·5 > sorbitol > mannitol > sucrose, indicating that almost no acid was formed from Palatinit® or its stereoisomeric components GPS and GPM. However, a *L. casei* strain was able to degrade Palatinit® slowly and form appreciable amounts of acid from it.

3.26. Acid Production from Polyols and its Effects on Hydroxyapatite

Grenby and Saldanha[39] compared boiled sweets made from a Lycasin® with conventional sucrose sweets in their ability to serve as a substrate for acid production by oral microorganisms (measured by pH and titration). The action of the acid on hydroxyapatite (demineralization) was quantified by spectroscopic calcium and phosphate microanalysis, and the synthesis of polysaccharide by oral microorganisms was also investigated.

Consistently less acid was produced from Lycasin® than from the sucrose-containing sweets. This was also reflected in the reduced demineralization of hydroxyapatite. In addition, significantly less polysaccharide was synthesized by the oral microorganisms growing on Lycasin® rather than on the sucrose-based sweet media. In summary, major differences were detected in the microbial metabolism of Lycasin®- and sucrose-containing media.

The fermentation of L-sorbose by microorganisms of human dental plaque was studied by Lohmann *et al.*,[76] who were able to isolate *Lactobacillus casei* as a major L-sorbose-degrading microorganism from dental plaque. L-Sorbose consumption and acid production were both analyzed with the aid of high-pressure liquid chromatography (HPLC). After 48 h incubation *L. casei* produced more lactic acid from L-sorbose than *S. mutans* did from sucrose under comparable conditions.

3.27. Acid Production from Xylitol in Sorbitol-Conditioned Plaque

Rölla *et al.*[96] found that 2·5% xylitol added to BHI medium inhibited the growth of *S. mutans* OMZ 176 and *S. mutans* GS5. Sorbitol as well as glucose promoted the growth of both strains under the same conditions. Tests were also carried out on three-day-old sucrose-induced plaque *in vivo*: application of 15% sucrose, 16% sorbitol, 4% xylitol and 16% sorbitol + 4% xylitol solutions yielded plaque pH values of 4·40, 5·65, 6·33 and 6·20 respectively, in one of the subjects. These data indicate that a combination of sorbitol and small amounts of xylitol is more effective in reducing the acid production of plaque than a similar amount of sorbitol alone.

3.28. Genetic Transfer of the Ability to Ferment Sorbitol

Westergren *et al.*[124] found that the ability to ferment sorbitol was transferred genetically to non-sorbitol-fermenting (Srl^-) strains of *S. sanguis* by exposing them to naked DNA extracted from sorbitol-fermenting (Srl^+) strains of *S. sanguis* and *S. mitior*. DNA from *S. mutans*, *S. salivarius* and strains of *Lactobacillus* did not transform these Srl^- to Srl^+ strains. The transformant strain produced the same amount of acid from sorbitol as the donor strain, i.e. 40% of that from glucose. No acid was produced from sorbitol by a glucose-grown donor or transformant strains.

3.29. Effect of Lycasin® on Polysaccharide Synthesis

Gehring[34] reported that the microbial conversion of Lycasin® 80/55 by mixed flora from plaque and saliva, as well as bacteria important in the etiology of dental caries, occurs slowly if at all. In addition, only slight acid production and no extracellular polysaccharide synthesis was observed.

3.30. Using Manometric Warburg Technique in Palatinit® Metabolism

Gülzow[49] utilized the manometric Warburg technique to compare the anaerobic fermentation of Palatinit® by undefined bacterial suspensions from saliva and three-day-old plaque with that of sucrose, xylitol and sorbitol. The metabolism of Palatinit® was similar to that of sorbitol. In comparison with the rapid fermentation of sucrose, Palatinit® showed initially a lag phase followed by a steady and continued fermentation. Xylitol was not fermented within a time span of 6 h. After 6 h of incubation an average yield of carbon dioxide (in μl) was obtained with the saliva sample (or the plaque sample) in the test groups: control (H_2O) 280·43 (19·64), M/100 xylitol 295·44 (21·12), M/100 sorbitol 442·07 (51·63), M/100 Palatinit® 418·47 (48·57), M/100 sucrose 617·49 (306·31). Carbon dioxide was a measure of the organic acids produced during fermentation of the various carbohydrate sweeteners.

3.31. Acid Production from Dextrin H

Würsch and Koellreutter[125] studied acid production from various carbohydrates in a suspension of human dental plaque material. The following carbohydrates were added to the suspension at a final concentration of 1·2%: glucose, starch, dextrin H, Lycasin® 80/55, maltitol and maltotriitol. After incubation of the suspensions, starch produced the most acid, even more than glucose. Lower and seemingly concentration-dependent acid production was seen with dextrin H. Addition of maltitol and particularly maltotriitol significantly inhibited acid production from dextrin H. Acid was produced at a lower rate from Lycasin® 80/55, and least acid was produced from maltitol, followed by maltotriitol. The assay was repeated with plaque suspended in whole saliva which had considerable α-amylase activity. Maltotriitol not only inhibited α-amylase, but also affected the utilization of fermentable carbohydrates such as glucose.

3.32. Variation in Acid Production due to Variation of Oral Microflora

Maki *et al.*[79] studied acid production from sucrose, isomaltulose, sorbitol and xylitol in suspensions of human dental plaque. In comparison with the endogenous acid production from sucrose (=100%), significantly less acid was produced from isomaltulose and even less acid from sorbitol. No acid was produced by the dental plaque flora from xylitol. The variation in acid production among individuals is due to differences in the plaque microflora. The authors suggest that isomaltulose would be a promising substitute for sucrose to reduce the incidence of dental caries, provided it is nontoxic to the host and is well tolerated.

3.33. Metabolism of Sorbitol and Xylitol in *S. mutans*

Slee and Tanzer[105] indicated that sorbitol utilization by *S. mutans* NCTC 10449 was signalled by the induction of sorbitol-specific phosphoenolpyruvate-dependent phosphotransferase and sorbitol-6-phosphate dehydrogenase activity which persisted throughout the growth cycle in the presence of sorbitol. In the presence of low glucose levels sorbitol transport and catabolism were rapidly repressed, and did not start again until all glucose had been utilized. Dills and Seno[20] found that in mutant strains of *S. mutans* lacking phosphoenolpyruvate:glucose phosphotransferase activity, 2-deoxy-D-glucose did not inhibit sugar alcohol uptake. While some sugar alcohols, e.g. mannitol and sorbitol, are easily fermented by cariogenic bacteria, others, like xylitol, are thought not to be metabolized by oral bacteria. However, Karle and Gehring[58] observed that some strains of streptococci in Lancefield's serotype *Q* can utilize xylitol to a certain extent *in vitro*. Mäkinen *et al.*[80] studied the growth of three strains of *Candida albicans* in 1% peptone broth supplemented with 5% xylitol in the presence and absence of 0·2% glucose. Cells of *C. albicans* were not able to utilize xylitol to support growth after a five-month maintenance period, but the presence of 0·2% glucose enhanced growth. On the contrary, Gallagher and Fussell[26] reported that oral microorganisms isolated from saliva and dental plaque were able to utilize the pentose sugar alcohols xylitol, L-arabinitol and ribitol. Among the isolated strains they found lactobacilli, propionibacteria and *C. albicans*. All of these strains were able to ferment xylitol to acid under anaerobic as well as aerobic conditions.

3.34. Xylitol Metabolism in *C. albicans*

In *C. albicans* xylitol is metabolized into D-xylulose with the aid of the enzyme NAD–polyol dehydrogenase,[122] which requires NAD. In the presence of ATP, D-xylulose is converted to D-xylulose-5-phosphate, which then may enter the Embden–Meyerhof pathway via D-glyceraldehyde-3-phosphate.

3.35. Polyol Metabolism by Intestinal Microflora

Siebert and Grupp[102] deduced from rat experiments that a symbiosis between host and the intestinal microflora might be possible. A fraction of ingested sugar alcohols such as sorbitol and Palatinit® is metabolized by these bacteria to acetate, which then can be taken up by the host.

3.36. Ultrastructure of Xylitol-Grown Cells

Tuompo *et al.*[117] examined the effect of xylitol and other carbon sources on the cell wall of *S. mutans* ATCC 27351. Transfer of actively growing cells into a xylitol-containing medium caused distinct alterations in bacterial ultrastructure without notable effects on their viability. With the aid of electron microscopy, some degradation of these xylitol-grown cells, as well as autolysis, intracellular vacuoles and lamellated formations in the cytoplasmic membrane, were frequently seen, independently of the xylitol concentration. Incubation of cells of this strain in media containing either glucose, fructose, sucrose, lactose, sorbitol or mannitol as the primary carbon source did not affect their cellular ultrastructure.

3.37. Fermentation of Xylitol by *Lactobacillus casei*

Vadeboncoeur *et al.*[118] studied the effect of xylitol on growth and glycolysis of various acidogenic oral bacteria (*S. mutans*, *S. salivarius*, *S. sanguis*, *Actinomyces* spp. and *Lactobacillus casei*). Most *S. mutans* strains were inhibited by 0·5% xylitol in the presence of 0·2% glucose; the sensitivity to xylitol of these strains was interpreted as a delay in reaching the stationary phase of the growth curve. The growth pattern of all sensitive strains was identical, whether xylitol was added at the beginning of the experiment or during growth. With the exception of *S. sanguis* 10556, this effect was not observed with the other acidogenic oral bacteria tested. The sensitive strains exhibited in the presence of xylitol a decreased capacity to produce acid, ranging from 38 to 69%. The rate of acid production from glucose by *S. mutans* GS5-2 was not affected by xylitol.

3.38. Fermentation of Polyols by *Propionibacterium avidum* and *S. mutans*

Gallagher and Pearce[27] found that the plaque bacterium *Propionibacterium avidum* fermented sorbitol and xylitol as rapidly as it did sucrose, into the main end-products, acetic acid and propionic acid, as well as the minor end-products, formic acid, isovaleric acid and succinic acid. The drop in pH was higher with sorbitol than with sucrose or xylitol. Therefore neither sugar alcohol can be regarded as universally inert with respect to dental caries since many individuals carry significant numbers of propionibacteria as well as other polyol fermenters in their dental plaque. Gallagher *et al.*[28] produced artificial carious lesions on bovine dental enamel slabs incubated with acidogenic *P. avidum* in the presence of xylitol.

The resulting lesions were graded more severe than those produced by *S. mutans* in the presence of mannitol, but not as severe as those produced by *S. mutans* in the presence of glucose.

Gauthier *et al.*[30] studied the effect of xylitol in concentrations of 0·5, 1·0 and 2·0% on the growth of *S. mutans* LG-1 (serotype *c*) in the presence of 0·2% sucrose, glucose, fructose, mannose, lactose, mannitol, or sorbitol. Xylitol increased the time usually needed by the cells to reach the stationary phase in the presence of most of the carbohydrates, but no effect was noted in the presence of sucrose or fructose. This xylitol-mediated growth inhibition was modified neither by temperature nor by variation of pH or by the presence or absence of oxygen. Repeated growth of *S. mutans* LG-1 in the presence of xylitol with one of the other carbohydrates led to improved xylitol tolerance. However, the cells were still unable to grow at the expense of xylitol. The authors concluded that this insensitivity of the cells to xylitol resulted from mutation. The increase in the mean generation time observed when the insensitive cells were cultured in the presence of fructose suggested an alteration in the fructose transport system. In *Escherichia coli* it had been shown by Reiner[94] that fructose-specific components of the phosphoenolpyruvate:sugar phosphotransferase system were involved in xylitol toxicity. This would explain why xylitol did not inhibit the growth of the cells in the presence of fructose, since fructose would prevent the uptake of xylitol. Evidence for the presence of a xylitol phosphotransferase system in *S. mutans* OMZ 176 was also presented by Assev and Rölla.[2] This *S. mutans* strain was found to take up xylitol by a phosphotransferase system in spite of lack of ability to catabolize this pentitol. Resting cells grown on U-^{14}C-xylitol produced a metabolite with an R_f value in thin-layer chromatography similar to that of xylitol-5-phosphate. Treatment of this metabolite with phosphatase yielded xylitol. Apparently xylitol is transported and phosphorylated via the fructose phosphotransferase system, and then xylitol phosphate (presumably xylitol-5-phosphate) accumulates inside the cell.

3.39. Accumulation of Xylitol Inside the Bacterial Cell

Wåler *et al.*[123] presented evidence that *S. mitior* JC43 and *S. mutans* OMZ-176 take up xylitol and transfer it to xylitol phosphate. Using ^{14}C-xylitol and autoradiography they could demonstrate that the metabolite formed by both resting cell cultures and growing cells could be xylitol-5-phosphate. Dental plaque also takes up xylitol; a xylitol/protein complex is formed in addition to xylitol phosphate and some unidentified labelled components. The authors suggest that accumulation of xylitol phosphate inside the cells may 'poison' the bacteria and therefore may possibly explain the alleged caries-therapeutic effect of xylitol.

3.40. Phosphorylation of Xylitol by *S. mutans*

Trahan *et al.*[116] conducted a further study on transport and phosphorylation of xylitol by a fructose phosphotransferase system in *S. mutans* ATCC 27352 (serotype *g*) and LG1 (serotype *c*). Added U-^{14}C-xylitol was taken up by growing cells at the expense of glucose. Alkaline phosphatase treatment followed by enzymatic analysis and thin-layer chromatography revealed that the metabolite accumulating was xylitol phosphate. Xylitol was phosphorylated at the expense of phosphoenolpyruvate by toluenized cells of strain LG1. This process was dependent on phosphoenolpyruvate and required the presence of both soluble and cellular membrane fractions in the reaction mixture. The phosphoenolpyruvate-dependent phosphorylation by isolated membranes of strain LG1 in the presence of the soluble fraction was inhibited by fructose but not by glucose, mannose or galactose. It was concluded that xylitol was transported and phosphorylated by a phosphoenolpyruvate:fructose phosphotransferase system. The data also suggested that xylitol toxicity in *S. mutans* is caused by the intracellular accumulation of xylitol phosphate.

3.41. Fermentation of Glycerol by Immature Dental Plaque

Gallagher *et al.*[29] examined the flora of immature dental plaque capable of fermenting glycerol to acid. The proportion of cultures fermenting glycerol into acid (pH 5·0–5·5) of the total viable count in the dental plaque of six young adults ranged from 19 to 63%; this did not change after four days of intense mouth rinsing with a glycerol- and urea-containing plaque-mineralizing solution, indicating no short-term adaptation of the oral flora. Significant calcium deposition in the plaque of most subjects suggested that base production from urea may overwhelm acid production from glycerol.

3.42. Acid Production from Isomaltulose and Other Polyols

Bibby and Fu[6] used a new method *in vitro* to compare the ability of natural dental plaque to convert sucrose, several sugar alcohols and two artificial sweeteners into acids. The tests were carried out at concentrations of 0·1, 1·0 and 10·0%. Isomaltulose gave the lowest pH, followed in order by Lycasin®, mannitol, Palatinit®, sorbitol, sorbose, and then, with negligible acid production, xylitol, besides the two artificial sweeteners saccharin and aspartame. Xylitol and saccharin also gave smaller pH depressions in higher concentrations.

3.43. Effect of Carbohydrate Sweetener Interaction on Acid Production

Linke and Jarymowycz[72] found an interesting interaction of carbohydrate sweeteners. They studied acid and glucan production *in vitro* as well as the adherence to smooth surfaces of *S. mutans* (serotypes *c* and *d*) in the presence of 2% sucrose medium supplemented with 1% solutions of glucose, fructose, maltose or soluble starch. The results were striking and unexpected. During sucrose fermentation, a larger amount of combined acids (lactic plus acetic acid) was produced by serotype *c* than by serotype *d*. Whilst the former serotype produced more lactic acid, the latter produced more acetic acid. Fermentation of sucrose supplemented with glucose produced up to 22% less combined acids whereas supplementation with fructose yielded up to 32% less combined acids. Similar quantitative relations were seen during glucan production and in the adherence experiment. The results indicated that fermentation of sucrose in combination with other carbohydrate sweeteners by *S. mutans* produced less cariogenic acidic end-products than sucrose alone.

3.44. Effect of Oxygen on Polyol Metabolism

Kalfas and Edwardsson[56] used three agar media that indicate acid production from sorbitol by oral microorganisms, especially in an anaerobic or microaerophilic atmosphere containing 5–6% carbon dioxide. A total of 16 strains (out of 64 pure cultures of bacteria and yeasts evaluated) within the genera *Actinomyces*, *Lactobacillus*, *Rothia* and *Streptococcus* fermented sorbitol.

Svensäter *et al.*[111] studied the anaerobic and aerobic metabolism of sorbitol in *S. sanguis* and *S. mitior*. Washed cell suspensions of both strains were incubated with sorbitol, and after 20 min the production of lactate, formate, ethanol and acetate was analyzed by quantitative enzymic tests. Under anaerobic conditions both strains produced lactate, formate, ethanol and acetate, but the amounts of these were somewhat reduced when these cells were exposed to air. When grown under aerobic conditions the cells produced only lactate and acetate. Cells grown under strictly anaerobic conditions produced lactate dehydrogenase and pyruvate formate-lyase activities. However, the level of pyruvate formate-lyase was significantly reduced when the cells were exposed to air, whereas the level of lactate dehydrogenase remained almost unchanged under both conditions. In summary, the results indicated that *S. sanguis* and *S. mitior* both metabolize sorbitol differently under anaerobic and aerobic conditions. This difference may depend on the oxygen-sensitivity of the pyruvate formate-lyase of these oral bacteria.

3.45. Fermentation of Neosugar® by *Bifidobacterium*

Tokunaga *et al.*[114] mention in their study on the influence of chronic intake of Neosugar®, a fructo-oligosaccharide, on growth and gastrointestinal function of the rat, that this new sweetener stimulates the proliferation of *Bifidobacterium* and suppresses the growth of *Escherichia* and *Clostridium*.

3.46. Inhibition of Glucan Synthesis

Nisizawa *et al.*[90] found that there is a difference in mode of inhibition between α-D-xylosyl-β-D-fructoside (XF) and α-isomaltosyl-β-D-fructoside (IF) in glucan synthesis by *S. mutans* 6715 D-glucosyltransferase. In the presence of IF a large proportion of low-molecular-weight glucan (LMWG) and a series of non-reducing oligosaccharides are synthesized. However, in the presence of XF the production of reducing sugars and the sucrose consumption was significantly suppressed, and no LMWG or oligosaccharides were produced. The authors concluded that IF acted as an alternative acceptor for the D-glucosyl and/or D-glucanosyl transfer reactions of the enzyme, and served to reduce the formation of insoluble and soluble D-glucan. On the other hand, XF inhibited competitively the sucrose-splitting activity of the enzyme, acting as an analog to sucrose, thereby diminishing the synthesis of D-glucan.

3.47. Aerobic and Anaerobic Metabolism of Sorbitol in Dental Plaque

Kalfas and Birkhed[55] studied acid production from sorbitol and glucose under aerobic as well as anaerobic conditions in suspensions of two-day-old dental plaque and oral streptococci (*S. mitior*, *S. sanguis* and *S. mutans* strains). Both were incubated with sorbitol or glucose (final concentration 90 mM) at 37°C for 30–80 min. The pH of the solutions was monitored, and the organic acids produced (lactate, acetate, formate) and ethanol were determined enzymically. The differences in concentration of the four fermentation products in the presence and absence of air were small, with lactic acid being the lowest. The amounts of formate and ethanol were greater in the suspensions incubated with sorbitol than with glucose. The pH drop was smaller in the aerobic than in the anaerobic suspensions. The pH decrease from glucose was the same under aerobic as under anaerobic conditions.

3.48. Delayed Uptake of Sorbitol in Xylitol-Pretreated Cells

Assev and Rölla[4] found a delayed uptake of sorbitol by *S. mutans* when the cells were pretreated with xylitol. Further metabolism of sorbitol did not take place in these cells even though sorbitol was found intracellularly. In untreated *S. mutans* cells, sorbitol was taken up and then further metabolized, providing a carbon and energy source. Xylitol metabolism was unchanged in the presence of sorbitol. It appears that xylitol changes the pathway of sorbitol uptake via an inducible permease. The exact mechanism by which the accumulated intracellular sorbitol enhances the xylitol inhibition is still unclear.

3.49. Xylitol Inhibition of Glycolysis in *S. mutans*

In a further study on the growth inhibition of *S. mutans* OMZ 176 by xylitol, Assev and Rölla[3] found that thin-layer chromatography of extracts of cells which had been exposed to U-^{14}C-xylitol indicated that xylulose phosphate was produced by the cells in addition to xylitol phosphate. Cells pretreated with xylitol and then exposed to U-^{14}C-glucose accumulated ^{14}C-hexose-6-phosphates, indicating that a xylitol metabolite, or metabolites, were competing with fructose-6-phosphate for the phosphofructokinase, since glycolysis is inhibited at this step. After accumulation of xylitol-5-phosphate, the bacteria were able to expel xylitol, presumably after intracellular dephosphorylation of the xylitol-5-phosphate. In summary, there is a 'futile cycle' in these cells, whereby xylitol is taken up and phosphorylated, and later dephosphorylated and expelled. It is believed that three different mechanisms may be involved in the inhibition of *S. mutans* by xylitol: (a) use of PEP during phosphorylation of xylitol; (b) use of NAD^+ during dehydrogenation of xylitol-5-phosphate; and (c) competitive inhibition of glycolysis at the phosphofructokinase level by either xylitol-5-phosphate and xylulose-5-phosphate or by both.

3.50. Effect of Lactitol on Demineralization

Grenby and Phillips[47] compared the dental properties of lactitol, a polyol derived from lactose, with glucose, sucrose, sorbitol, mannitol and xylitol *in vitro*. The tests were carried out using an otherwise carbohydrate-free medium inoculated with mixed human dental plaque bacteria. The acid production after 24 h of incubation and the demineralizing action of the acid on standardized samples of hydroxyapatite, powdered dental enamel and intact enamel surfaces were determined. Glucose and sucrose as substrates gave rise to by far the most acid and demineralization. Less acid production and demineralization were observed in the tests with sorbitol and mannitol as substrates. Fermentation of lactitol and xylitol was only very slight, resulting in a final pH of 6·6–6·9, and producing extremely low enamel demineralization figures as well, significantly below all the others.

3.51. Acidogenesis of Leucrose and Polyglucose PL-3

Siebert and Ziesenitz[103] studied the cariogenic properties of two novel sugar substitutes, leucrose, the α-(1 $\rightarrow$ 5)-glycoside bond analog of sucrose, and polyglucose PL-3, a linear dextran-like glucan of 18·5 average chain length. Acidogenesis by oral bacteria revealed that virtually no acids were produced from either substrate. However, leucrose, though not split by invertase, inhibits this enzyme. The authors believed that leucrose, possessing a certain sweetness and giving bulk to a foodstuff, and polyglucose PL-3, a tasteless substance for bulking purposes and possibly a carrier for intense sweeteners, may have promising properties as non-cariogenic sugar substitutes for the future.

4. BACTERIOLOGICAL STUDIES USING INTENSE SWEETENERS

4.1. Inhibition of Streptococcal Growth and Acid Production by Intense Sweeteners

Linke and Chang[64] studied the interaction of the artificial sweeteners, cyclamate, saccharin, naringin and neohesperidin dihydrochalcone, and aspartyl-phenylalanine methyl ester (aspartame) on the growth pattern and acid producion of seven glucose-grown *Streptococcus mutans* strains representing the serotypes *a* to *e*. Most sweeteners were tested at a concentration range from 0·02 to 20·00 mg/ml. Whilst the latter three sweeteners had no significant effect on growth and acid production of the glucose-grown *S. mutans* strains tested, cyclamate reduced growth of all *S. mutans* strains tested, especially at the highest concentration used. Saccharin significantly inhibited the growth of all *S. mutans* strains increasingly with rising concentrations and acid formation was therefore reduced. It was suggested that saccharin may have some potential in reducing the incidence of dental caries.[65]

An extension of this study produced evidence[66] that saccharin inhibited the growth of most oral streptococci, including *S. mutans* (serotype *a* to *e*), *S. mitis*, *S. salivarius*, *S. sanguis*, *S. faecalis*, *S. faecalis* var. *liquefaciens*, *S. pyogenes*, *S. durans*, *S. agalactiae* and *S. lactis*. A significant decrease of acid production was observed with increasing saccharin concentrations. Compared with the control, the pH increased at the highest saccharin concentration (20·00 mg/ml) from 0·90 units (*S. faecalis* var. *liquefaciens*) up to 2·00 units (*S. mutans* B-14); the average pH increase for all *S. mutans* strains was found to be 1·65 units. The final pH of the glucose medium after

24 h of incubation at the highest saccharin concentration (20·00 mg/ml) ranged from 5·25 (*S. durans*) to 6·45 (*S. pyogenes*). These pH levels were significantly above the 'critical pH' of 5·50, suggested as necessary for demineralization of the teeth and subsequent caries development. Indeed, it has been shown that saccharin, when added to a cariogenic diet (e.g. 56% sucrose) at moderate concentrations, can reduce the caries score in hamsters[68] and rats.[42,95,112] Since the growth-inhibitory effect of saccharin on oral microorganisms can be recognized below its threshold level of sweetness and even in the presence of carbohydrate sweeteners such as sucrose, a method for treating teeth to reduce dental caries was provided wherein the teeth are in contact with saccharin material, optionally in the presence of sugar, for example in a chewing gum composition, the saccharin material being present in an amount sufficient to inhibit growth of *S. mutans* in the oral cavity or on the teeth.[78]

4.2. Effect of Intense Sweeteners on Adherence of *S. mutans*

The effect of aspartame as well as saccharin and cyclamate on the adherence of *S. mutans* 6715 to Nichrome wire *in vitro* was studied by Olson.[91] *S. mutans* 6715 was grown for 24 h at 37°C in the presence of 4·5% sucrose with and without artificial sweeteners at concentrations of 0·45 and 0·95 mg/ml. The wires were transferred into freshly inoculated tubes daily for five days, and the plaque dry weight was recorded. At the 0·90 mg/ml concentration aspartame, saccharin and cyclamate produced 8·1, 10·2 and 14·4 mg plaque (dry wt), respectively. The author concluded that the reduction with aspartame indicates that peptides of aspartic acid and phenylalanine might be useful in the reduction of adherent plaque.

4.3. Influence of Aspartame on Acid Production in Whole Saliva

Mishiro and Kaneko[85] studied the effect of aspartame alone and in mixture with amino acids and glucose, on lactic acid production in human whole saliva. While lactate production from glucose was apparently promoted by aspartame, the expected drop in pH did not occur. These effects of aspartame could not be produced by its L-aspartic acid and L-phenylalanine moieties. The results of this investigation indicate a dual effect of aspartame: (a) a lactate-production-promoting factor (like streptogenin); and (b) a limitation of the pH fall in the presence of supplemental sugar.

4.4. Effect of Saccharin on Plaque Bacteria

The behavior of mixed cultures of bacteria from dental plaque was examined by Grenby and Bull[38] in the presence of 0, 0·25, 1·00 and

2·00 mg/ml sodium saccharin. The bacteria were cultured on agar and in liquid bacterial growth media at 37°C. The bacterial numbers were determined by colony counts and turbidity measurement, acid production by pH measurement and NaOH titration; total carbohydrate in the bacterial cell suspension, and the protein content of the suspension were also determined. Whilst 0·25 mg/ml saccharin only slightly suppressed the rate of growth and metabolism, 2·00 mg/ml saccharin was most effective in reducing bacterial growth, acid production, complex carbohydrate as well as protein production. The investigators pointed out that such levels of saccharin can easily be reached in the mouth when foods and drinks sweetened with saccharin instead of sugar are ingested.

4.5. Effect of Saccharin and Acesulfame-K on Adherence of *S. mutans*

The adherence of *S. mutans* NCTC 10449 (serotype *c*) to smooth glass surfaces was studied by Linke[69] in 1% sucrose medium in the presence of saccharin (Na salt), acesulfame-K and aspartame in concentrations from 0·02 to 20·00 mg/ml. All the artificial sweeteners reduced overall growth (adherent + suspended cells), but particularly suspended cells. Compared with the control, adherence was optimal with 2·00 mg/ml sodium saccharin, 2·00–20·00 mg/ml acesulfame-K and 4·00 mg/ml aspartame. At higher concentrations, the order of adherence of the *S. mutans* NCTC 10449 cells to the smooth glass surface was: sodium saccharin $>$ acesulfame-K $>$ aspartame. These findings are in agreement with an earlier clinical study by Grenby[37] on the effects of a low-calorie sweetener composed of glucose and saccharin, on plaque formation in 24 dental student volunteers. After a three-day test period the amount of plaque formed in the treated group was significantly lower than in the control group, but the protein content of the plaque (both water-soluble and insoluble components) of the treated group was significantly higher than in the control. This elevated protein content could probably be due to an increase in the number of bacterial cells relative to the extracellular polysaccharide matrix of the plaque.

4.6. Aspartame as Nitrogen Source for Oral Bacteria

Grenby and Saldanha[41] studied the utilization of aspartame as a nitrogen source by oral microorganisms in comparison with ammonium sulfate, L-aspartic acid, L-phenylalanine alone and in a mixture, other amino acids and bovine albumin. All compounds were tested in a basal nitrogen-free liquid growth medium, and after inoculation with a standard culture of mixed saliva and incubation at 37°C for 24 h, turbidity, cell protein, pH and titratable acid were measured. The data from 25 individual experiments indicated that microbial growth and acid production were at a maximum when the medium contained ammonium sulfate, L-aspartic acid or mixtures of amino acids. Aspartame as well as L-phenylalanine or bovine albumin were poorly utilized as a nitrogen source by acidogenic oral microorganisms. Grenby[43] pointed out that the particular role of proteins that carbohydrates cannot fulfil is to act as a source of nitrogen, which is necessary for the growth of microorganisms. Since aspartame seems to be poorly utilized as a nitrogen source for oral bacteria, as reflected in poor growth and acid production of these bacteria, the better are the prospects of such a compound for dental health. On the other hand, Coulombe and Sharma[17] voiced concern regarding the widespread use of aspartame because of the possibility that its consumption may produce neurobiochemical and behavioral effects in humans, particularly in children and susceptible individuals, suggesting that there is a need for additional research on the safety of this sweetener.

4.7. Mechanism of Saccharin Inhibition in *S. mutans*

The mechanism of the saccharin inhibition of acid production from glucose by oral streptococci has been studied by Linke and Kohn.[70] Saccharin apparently inhibits vital enzymes of the glycolytic scheme. In cell-free extracts of *S. mutans* NCTC 10449, 10^{-2}M-sodium saccharin inhibited the enzymes hexokinase, glyceraldehyde-3-phosphate dehydrogenase, phosphoglycerate mutase, pyruvate kinase, 3-phosphoglycerate kinase and phosphofructokinase 48, 45, ~16, 12, 3 and 1·4%, respectively. Sodium chloride in the concentration range from 10^{-3}–10^{-2}M produced little, if any, inhibition of the glycolytic enzymes tested under the same conditions.

4.8. Uptake of U-^{14}C-Glucose by *S. mutans* in Presence of Saccharin

The uptake of U-^{14}C-glucose by resting cells of *S. mutans* OMZ-176 was studied by Linke and Jarymowycz[71] in the presence of saccharin (Na salt) in concentrations of 0·02, 0·20, and 2·00 mg/ml. The incorporation of the radioactive-labeled sugar was observed for 150 min. Compared with the control (no saccharin added) and the sodium-chloride-treated control (no saccharin added), saccharin at the highest concentration more than doubled the accumulation of radioactive-labeled carbon within the cells. This suggests that the glucose metabolism of *S. mutans* is curtailed by saccharin, e.g. glucose is metabolized at a reduced rate via the Embden–Meyerhof pathway, and this leads to an accumulation of ^{14}C within the microbial cell. This finding is in agreement with the previously reported inhibition of glycolytic enzymes by saccharin.[70]

4.9. Inhibition by Saccharin of Caecal Bacteria

Pfeffer *et al.*[92] found that acesulfame-K, cyclamate and saccharin (sodium salts of the latter two compounds) effectively inhibited anaerobic acid production from glucose by the caecal bacterial flora of standard-chow-fed rats. They also indicated that caecal bacteria require six to eight times higher concentrations of the individual sweeteners than oral microorganisms such as *S. mutans* NCTC 10449 before they show the same inhibitory effect.[126]

4.10. Inhibition of Gram-Positive and Gram-Negative Bacteria by Saccharin

The inhibitory effect of saccharin on growth is not restricted to *Streptococcus* spp. Growth and acid production of glucose-grown Gram-positive and Gram-negative rods as well as cocci from the human oral cavity were studied by Linke and Doyle[73] in the presence of saccharin (Na salt) in concentrations of 0·02, 0·20, 2·00 and 20·00 mg/ml. All the Gram-positive rods tested, i.e. *Actinomyces viscosus*, *Lactobacillus acidophilus*, *Bacillus subtilis* and *Corynebacterium diphtheriae*, and Gram-positive cocci, i.e. *Streptococcus* spp., *Staphylococcus aureus* and *Micrococcus luteus*, were inhibited significantly by saccharin, especially at the higher concentrations (Fig. 1(a) and (c)). While Gram-negative cocci, i.e. *Veillonella* sp. and *Neisseria sicca*, were strongly inhibited by all the saccharin concentrations, Gram-negative rods, i.e. the enterics and *Acinetobacter* sp., exhibited little, if any, restraint (Fig. 1(b) and (d)). Saccharin caused a significant reduction in fermentative acid production along with curtailed growth.

In an earlier study on the effect of sodium saccharin in the diet on caecal microflora of rats Anderson and Kirkland[1] found that saccharin could reduce the amount of ammonia produced from urea by *Proteus vulgaris*; this inhibition of urease activity by saccharin appeared not to be a direct effect on the enzyme but to be due to an inhibition of microbial growth. In another study on the effect of sodium saccharin upon bacteria in small intestinal contents of rats Naim *et al.*[87a] found that saccharin inhibited growth and acid production of *L. acidophilus*, *L. jensenii*, *L. fermentum* and several *Escherichia coli* strains.

4.11. New Method to Determine Acid Production from Sweeteners *in vitro*

Bibby and Fu[6] used a new method *in vitro* to examine the ability of natural dental plaque to convert several sweeteners into acids. The tests were carried out at concentrations of 0·1, 1·0 and 10·0%, or, in the case of aspartame and saccharin, at similar sucrose sweetness equivalents. Whilst some sugar alcohols produced low pH values, aspartame, saccharin and xylitol were converted to negligible amounts of acid. Xylitol and saccharin also gave smaller pH depressions in higher concentrations of test solutions.

4.12. Effect of Talin® and Other Intense Sweeteners on Metabolism of Plaque Bacteria

The effects of aspartame, saccharin (Na salt), acesulfame-K, cyclamate and Talin® (thaumatin) on the growth and metabolism of oral bacteria in a pooled plaque sample as well as *S. mutans* NCTC 10449 were studied by Grenby and Saldanha[45,46] in a peptone liquid medium, an agar medium by the zone inhibition method, and a tryptone soya medium by a modification of the minimum inhibitory concentration (MIC) method. Whilst aspartame was more effective in the mixed plaque cultures, saccharin was more effective against the *S. mutans* strain tested. For the MIC test Talin® was excluded because it coagulated in the liquid tryptone soya medium after sterilization in an autoclave. The remaining four artificial sweeteners were tested at concentrations of 0·005, 0·020, and 0·050 g/ml. Acesulfame-K and cyclamate had no significant effect on microbial growth and acid production. However, both aspartame and saccharin exerted an inhibitory action, the latter having a greater effect than the former. Whilst the highest concentration of aspartame had a small influence in elevating the final pH, saccharin at all three concentrations kept the final pH higher, with a greater effect as the concentration was increased.

4.13. Effect of Thaumatin on Acid Production by Oral Bacteria

Au and Berry[5] studied the effect of thaumatin on growth and acid production of *S. mutans*, *S. sanguis*, *A. naeslundii* and *A. viscosus* in presence of glucose, sucrose, fructose, mannitol and sorbitol. The fermentability of thaumatin was also compared with that of the five carbohydrates or polyols. All tests were carried out using a basal medium composed of thioglycollate broth without dextrose or indicator. The results indicated no significant difference in growth and acid production by each strain metabolizing either the five carbohydrates/polyols alone or in the presence of thaumatin. However, the authors observed a significant decrease in growth and acid production of each strain metabolizing thaumatin alone.

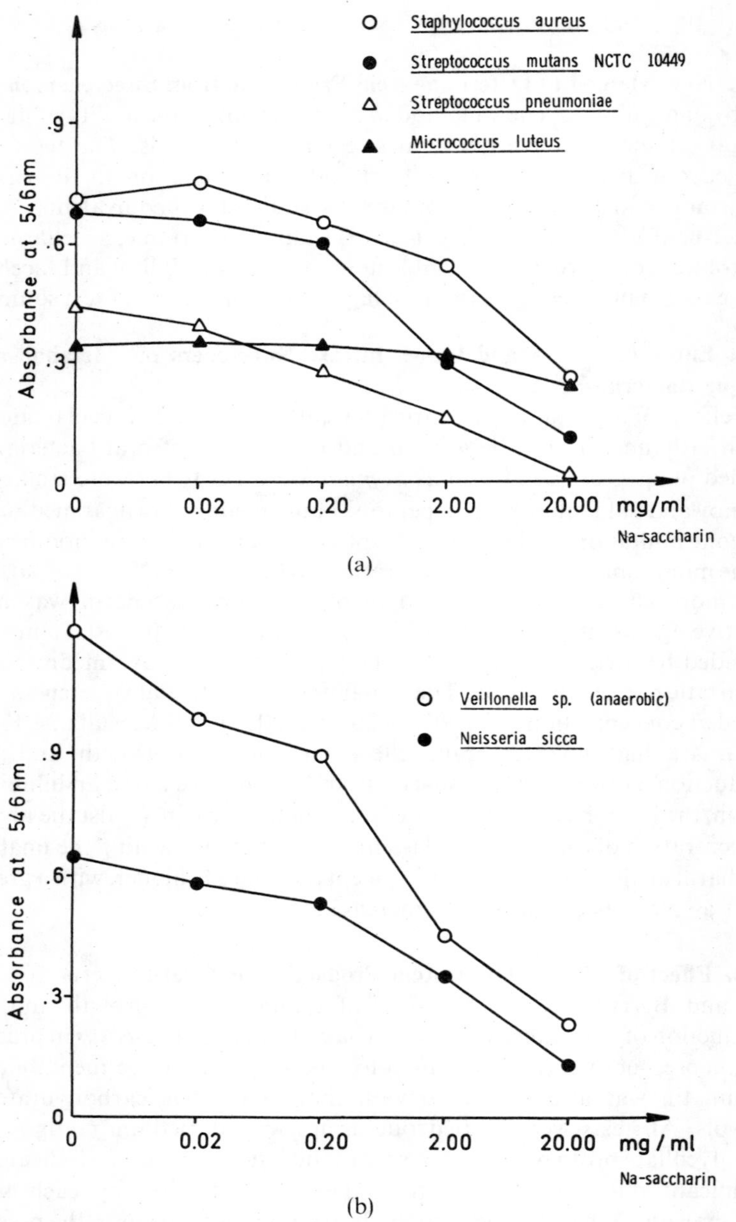

FIG. 1. Effect of saccharin (sodium salt) on growth (absorbance) of pathogenic and oral bacteria after 24 h of incubation: (a) Gram-positive cocci; (b) Gram-negative cocci.

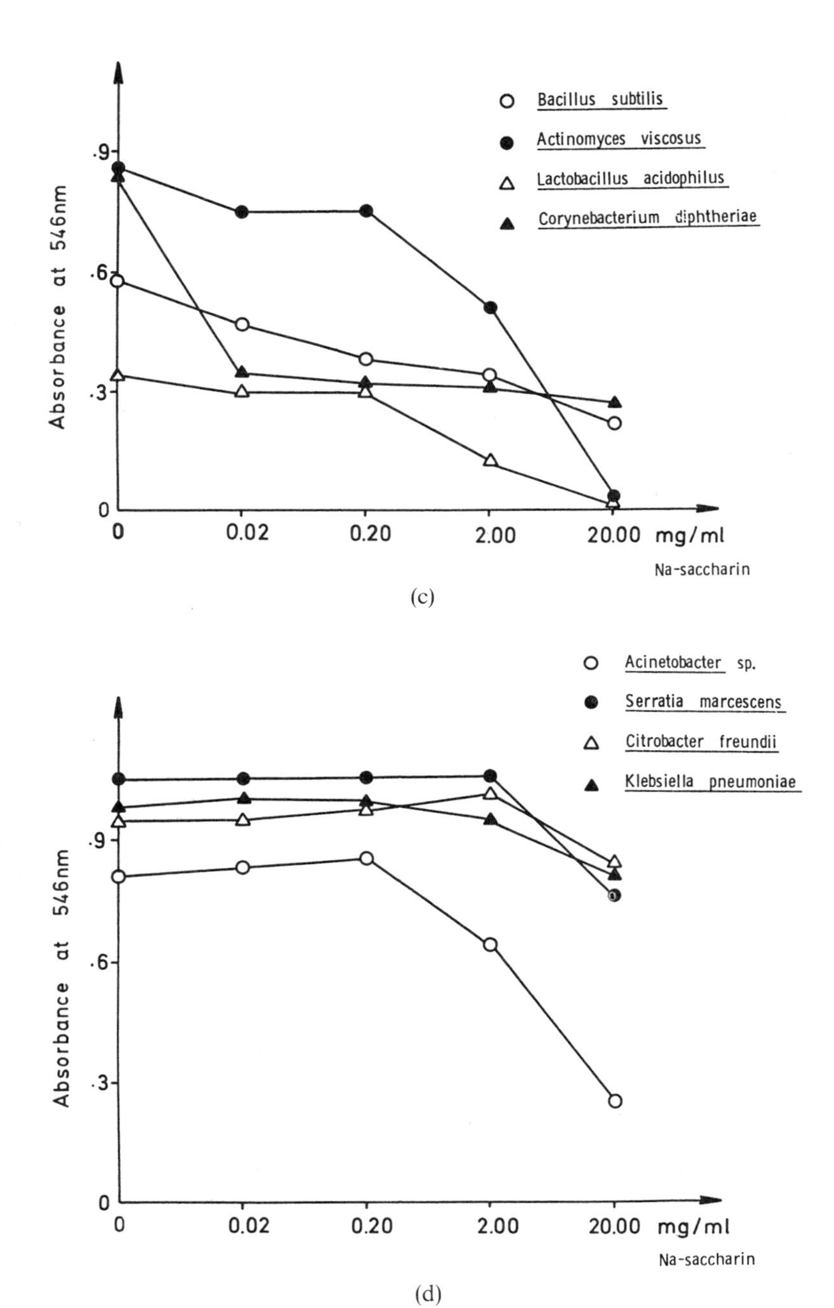

FIG. 1.—*contd.* (c) Gram-positive rods; (d) Gram-negative rods.

Thaumatin was only minimally fermented by the four bacterial strains and did not exert an inhibitory effect on growth and acid production of the carbohydrates.

4.14. Effect of a 'Sweetener Mix' on Anaerobic Acidogenesis by Oral Bacteria

Ziesenitz and Siebert[128] reported that anaerobic acidogenesis from sucrose by *S. mutans* NCTC 10449, *Lactobacillus casei* LSB 132 and *A. viscosus* Ny 1 was inhibited by acesulfame-K, cyclamate and saccharin but not by aspartame. Moreover, a combination of the effective sweeteners ('sweetener mix') resulted in a strong synergism of their inhibitory effects; added fluoride (50 μm) or xylitol (10 mM) or both further enhanced this inhibitory effect. The ED_{50} values in millimoles for the inhibitory effect of acesulfame K, cyclamate and saccharin on the acidogenesis from sucrose by *S. mutans* NCTC 10449 alone (and in mix; molar ratios 1·00:1·67:0·10, respectively) were 22 (12), 75 (34) and 4·3 (1·5), respectively. In comparison with the control, the inhibition of acidogenesis from sucrose by *S. mutans* NCTC 10449 was 37% with the sweetener mix, 49% with the sweetener mix + fluoride, and 61% with the sweetener mix + fluoride + xylitol. Acidogenesis by all three strains is inhibited by the sweetener mix regardless of whether sucrose, D-glucose or D-fructose serves as fermentable sugar.[127]

4.15. Competitive Inhibition of NADH by Saccharin in *S. mutans*

Brown and Best[14a] proposed a mechanism for the influence of saccharin on glucose metabolism by *S. mutans* NCTC 10449. During a study of the inhibition of lactate dehydrogenase, the enzyme which catalyzes the terminal, regulatory step in the Embden–Meyerhof–Parnas pathway, they found that saccharin competitively inhibited the binding of its reduced coenzyme, NADH, to the enzyme. This competitive interaction of saccharin with NADH came as no surprise, given the structure of this sweetener. Saccharin is composed of a bicyclic ring system with one heterocyclic moiety containing a nitrogen and a sulfur atom as well as a keto group at the 1-position. NADH contains two structural and functional heterocyclic ring systems, adenine and pyridine, which are neither structurally nor spatially dissimilar to one or both ring components of saccharin. Saccharin also inhibited two other NAD-dependent dehydrogenases from *S. mutans*, sorbitol-6-phosphate dehydrogenase and mannitol-1-phosphate dehydrogenase. Furthermore, saccharin inhibited the NADPH-dependent glutamate dehydrogenase from *S.*

mutans, thus possibly serving to restrict carbon flow into certain anabolic pathways. In addition, saccharin also had a negative effect upon the NADP-dependent glyceraldehyde-3-phosphate dehydrogenase from *S. mutans*, thus limiting the production of the NADPH required for many reductive reactions within biosynthetic pathways.

5. CONCLUSIONS

Most of the bacteriological studies cited on sugar alcohols indicate that these compounds are metabolized in a similar manner to the common sugars, e.g. sucrose or glucose. The majority of polyols are taken up by the bacterial cell, are phosphorylated, enter the glycolytic pathway, and finally are metabolized into acidic end-products such as lactic acid (Fig. 2). The delay in utilization of some sugar alcohols by *S. mutans* is either due to competitive inhibition, for example by minute amounts of glucose, or to the fact that some anabolic enzymes are not present in the cell as constituents, and have to be induced after contact with the substrate to be metabolized. It

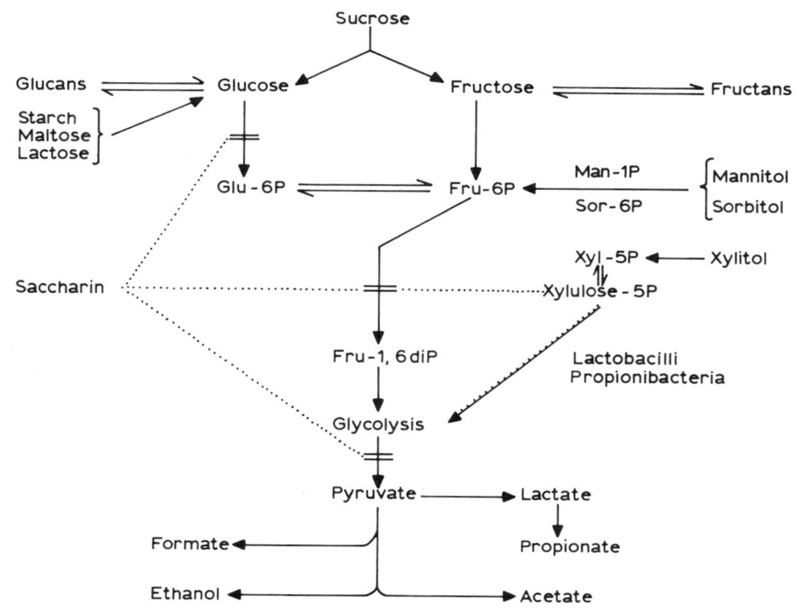

FIG. 2. The principal metabolism of carbohydrate sweeteners by *Streptococcus mutans* (··· ǂ, inhibition; ⟶, different pathway).

is also important to note that the oral microflora of man in modern Western societies is biochemically geared to the almost exclusive use of the common dietary carbohydrate sweeteners sucrose, glucose and fructose. However, increased usage of dietary sugar alcohols—with exceptions—will lead within a few years to an adaptation of the oral microflora to these compounds and would render the sugar alcohols as acidogenic as the common sugars. Such an adaptive shift of the oral ecology in the dominating presence of sugar alcohols has been observed in long-term (years) clinical studies which were recently reviewed by Linke.[74]

The intense sweeteners, whether natural or synthetic in origin, interact in a completely different manner with oral bacteria. In the case of saccharin, there is a true inhibition of bacterial growth and acidogenesis, due to its influence on vital glycolytic enzymes. Saccharin therefore has a potential in caries prevention. Even more powerful in caries reduction are mixtures of intense sweeteners.[104] During the last ten years some research has been devoted to the interaction of intense sweeteners with oral bacteria, but there is much more refined research needed in this challenging area, especially in regard to the properties of the intense sweeteners *in vivo*.

REFERENCES

1. Anderson, R. L. and Kirkland, J. J. (1979). *Fd Cosmet. Toxicol.*, **18**, 353.
1a. Assev, S., Vegarud, G. and Rölla, G. (1980). *Act. Path. Microbiol. Scand., Sect. B*, **88**, 61.
2. Assev, S. and Rölla, G. (1984). *Act. Path. Microbiol. Immunol. Scand., Sect. B*, **92**, 89.
3. Assev, S. and Rölla, G. (1986). *Act. Path. Microbiol. Immunol. Scand., Sect. B*, **94**, 97.
4. Assev, S. and Rölla, G. (1986). *J. Dent. Res.*, **65**, 290, Abstr. 1089.
5. Au, G. S. and Berry, C. W. (1986). *J. Dent. Res.*, **65**, 329, Abstr. 1441.
6. Bibby, B. G. and Fu, J. (1985). *J. Dent. Res.*, **64**, 1130.
7. Birkhed, D. and Edwardsson, S. (1978). In: *Health and Sugar Substitutes. Proc. ERGOB Conf. Geneva*, B. Guggenheim (Ed.), Karger, Basel, 211.
8. Birkhed, D., Edwardsson, S., Svensson, B., Moskovitz, F. and Frostell, G. (1978). *Archs. Oral Biol.*, **23**, 971.
9. Birkhed, D., Edwardsson, S., Ahlden, M.-L. and Frostell, G. (1979). *Act. Odont. Scand.*, **37**, 103.
10. Birkhed, D. and Skude, G. (1978). *Scand. J. Dent. Res.*, **86**, 248.
11. Birkhed, D., Edwardsson, S., Kalfas, S. and Svensäter, G. (1984). *Swed. Dent. J.*, **8**, 147.
12. Birkhed, D., Kalfas, S., Svensäter, G. and Edwardsson, S. (1985). *Internatl Dent. J.*, **35**, 9.

13. BRAMSTEDT, F. and TRAUTNER, K. (1971). *Dtsch. Zahnärztl. Z.*, **26**, 1135.
14. BROWN, A. T. and WITTENBERGER, C. L. (1973). *Archs. Oral Biol.*, **18**, 117.
14a. BROWN, A. T. and BEST, G. M. (1986). *Caries Res.*, **20**, 406.
15. CARLSSON, J. (1978). In: *Health and Sugar Substitutes. Proc. ERGOB Conf. Geneva*, B. Guggenheim (Ed.), Karger, Basel. 205.
16. CLEM, W. H. and KLEBANOFF, S. J. (1966). *J. Bacteriol.*, **91**, 1848.
17. COULOMBE, R. A., JR and SHARMA, R. P. (1986). *Toxicol. Appl. Pharmacol.*, **83**, 79.
18. CROWLEY, M. C., HARNER, V., BENNETT, A. S. and JAY, P. (1956). *J. Amer. Dent. Assoc.*, **52**, 148.
19. DALLMEIER, E., BESTMANN, H.-J. and KRÖNCKE, A. (1970). *Dtsch. Zahnärztl. Z.*, **25**, 887.
20. DILLS, S. S. and SENO, S. (1983). *J. Bacteriol.*, **153**, 861.
21. DRUCKER, D. B. and VERRAN, J. (1980). *Archs. Oral Biol.*, **24**, 965.
22. EDWARDSSON, S., BIRKHED, D. and MEJARE, B. 1977). *Act. Odont. Scand.*, **35**, 257.
23. FROSTELL, G. (1964). *Act. Odont. Scand.*, **22**, 457.
24. FROSTELL, G. and BIRKHED, D. (1978). *Caries Res.*, **12**, 256.
25. GALLAGHER, I. H. C. and PEARCE, E. I. F. (1977). *New Zealand Dent. J.*, **73**, 200.
26. GALLAGHER, I. H. C. and FUSSELL, S. J. (1979). *Archs. Oral Biol.*, **24**, 673.
27. GALLAGHER, I. H. C. and PEARCE, E. I. F. (1983). *New Zealand Dent. J.*, **79**, 75.
28. GALLAGHER, I. H. C., PEARCE, E. I. F. and CUTRESS, T. W. (1983). *Archs. Oral Biol.*, **28**, 317.
29. GALLAGHER, I. H. C., PEARCE, E.I.F. and HANCOCK, E. M. (1985). *New Zealand Dent. J.*, **81**, 20.
30. GAUTHIER, L., VADEBONCOEUR, C. and MAYRAND, D. (1984). *Caries Res.*, **18**, 289.
31. GEHRING, F. (1971). *Dtsch. Zahnärztl. Z.*, **26**, 1162.
32. GEHRING, F. (1974). *Dtsch. Zahnärztl. Z.*, **29**, 769.
33. GEHRING, F. and PATZ, J. (1974). *Dtsch. Zahnärztl. Z.*, **29**, 1026.
34. GEHRING, F. (1981). *Kariesprophylaxe*, **3**, 1.
35. GEHRING, F. and KARLE, E. J. (1981). *Z. Ernährungswiss.*, **20**, 96.
36. GLINKA-TSCHERNORUTZKY, H. (1930). *Biochem. Zeitschr.*, **221**, 125.
37. GRENBY, T. H. (1975). *Br. Dent. J.*, **139**, 129.
38. GRENBY, T. H. and BULL, J. M. (1979). *Caries Res.*, **13**, 16, 89.
39. GRENBY, T. H. and SALDANHA, M. G. (1981). *J. Dent. Res.*, **60**, 1192, Abstr. 211.
40. GRENBY, T. H. (1983). In: *Development in Sweeteners—2*, T. H. Grenby, K. J. Parker and M.G. Lindley (Eds), Applied Science Publishers, London, 51.
41. GRENBY, T. H. and SALDANHA, M. G. (1983). *J. Dent. Res.*, **62**, 685.
42. GRENBY, T. H. (1984). *Caries Res.*, **18**, 178, Abstr. 71
43. GRENBY, T. H. (1984). *Dental Health*, **23**, 8.
44. GRENBY, T. H. (1984). *J. Roy. Soc. Health*, **104**, 27.
45. GRENBY, T. H. and SALDANHA, M. G. (1986). *Caries Res.*, **20**, 7.
46. GRENBY, T. H. and SALDANHA, M. G. (1986). *Caries Res.*, **20**, 181, Abstr. 92.
47. GRENBY, T. H. and PHILLIPS, A. (1987). *Caries Res.*, **21**, 170, Abstr. 33.
48. GRUBB, T. C. (1945). *J. Dent. Res.*, **24**, 31.
49. GÜLZOW, H.-J. (1982). *Dtsch. Zahnärztl. Z.*, **37**, 669.

50. Havenaar, R., Huis in't Veld, J. H. J., Backer Dirks, O. and de Stoppelaar, J. D. (1978). In: *Health and Sugar Substitutes, Proc. ERGOB Conf. Geneva*, B. Guggenheim (Ed.), Karger, Basel. 192.
51. Hayes, M. L. and Roberts, K. R. (1978). *Archs. Oral Biol.*, **23**, 445.
52. Higginbotham, J. D. (1983). In: *Development in Sweeteners—2*, T. H. Grenby, K. J. Parker and M. G. Lindley (Eds), Applied Science Publishers, London, 119.
53. Imfeld, T. (1983). In: *Monographs in Oral Science*, Vol. 11, H. M. Myers (Ed.), Karger, Basel, 117.
54. Inglett, G. E. (1978). In: *Health and Sugar Substitutes, Proc. ERGOB Conf. Geneva*, B. Guggenheim (Ed.), Karger, Basel. 184.
55. Kalfas, S. and Birkhed, D. (1986). *Caries Res.*, **20**, 237.
56. Kalfas, S. and Edwardsson, S. (1985). *J. Clin. Microbiol.*, **22**, 959.
57. Karle, E. J. and Gehring, F. (1978). *Dtsch. Zahnärztl. Z.*, **33**, 189.
58. Karle, E. J. and Gehring, F. (1981). *Dtsch. Zahnärztl. Z.*, **36**, 673.
59. Kemmer, T. and Malfertheiner, P. (1983). *Res. Exp. Med.*, **183**, 35.
60. Kinghorn, A. D. and Compadre, C. M. (1985). *Pharm. Internatl*, **6**, 201.
61. Knuuttila, M. L. E. and Mäkinen, K. K. (1975). *Caries Res.*, **9**, 177.
62. Kornman, K. S. and Loesche, W. J. (1975). *J. Dent. Res.*, **54** (Spec. Iss. A), 204, Abstr. 634.
63. Kral, T. A. and Daneo-Moore, L. (1980). *Infect. Immun.*, **30**, 759.
64. Linke, H. A. B. and Chang, C. A. (1976). *Z. Naturforsch.*, **31c**, 245.
65. Linke, H. A. B. (1976). *Abstr. Ann. Meet. Amer. Soc. Microbiol.*, D15, 54.
66. Linke, H. A. B. (1977). *Z. Naturforsch.*, **32c**, 839.
67. Linke, H. A. B. (1977). *J. Clin. Microbiol.*, **5**, 604.
68. Linke, H. A. B. (1980). *Ann. Dent.*, **39**, 71.
69. Linke, H. A. B. (1983). *Microbios*, **36**, 41.
70. Linke, H. A. B. and Kohn, J. S. (1984). *Caries Res.*, **18**, 12.
71. Linke, H. A. B. and Jarymowycz, M. O. (1984). *Microbios*, **40**, 41.
72. Linke, H. A. B. and Jarymowycz, M. O. (1985). *Caries Res.*, **19**, 189, Abstr. 105.
73. Linke, H. A. B. and Doyle, G. A. (1985). *Microbios*, **42**, 163.
74. Linke, H. A. B. (1986). *Wld Rev. Nutr. Diet*, **47**, 134.
75. Loesche, W. J. and Kornman, K. S. (1976). *Archs. Oral Biol.*, **21**, 551.
76. Lohmann, D., Gehring, F. and Karle, E. J. (1981). *Caries Res.*, **15**, 263.
77. Lutz, D. and Gülzow, H.-J. (1979). *Dtsch. Zahnärztl. Z.*, **34**, 168.
78. Mackay, D. A. M., Witzel, F. and Linke, H. A. B. (1981). US Patent 4 291 045.
79. Maki, Y., Ohta, K., Takazoe, I., Matsukubo, Y., Takaesu, Y., Topitsoglou, V. and Frostell, G. (1983). *Caries Res.*, **17**, 335.
80. Mäkinen, K. K., Ojanotko, A. and Vidgren, H. (1975). *J. Dent. Res.*, **54**, 1239.
81. Mäkinen, K. K., Tenovuo, J. and Scheinin, A. (1976). *J. Dent. Res.*, **55**, 652.
82. Mäkinen, K. K. (1978). *Experienta Suppl.*, **30**, 1.
83. Mäkinen, K. K. and Scheinin, A. (1982). *Ann. Rev. Nutr.*, **2**, 133.
84. Mäkinen, K. K. (1985). *Int. Dent. J., London*, **35**, 23.
85. Mishiro, Y. and Kaneko, H. (1977). *J. Dent. Res.*, **56**, 1427.
86. Moore, W. E. C., Holdeman, L. V., Cato, E. P., Smibert, R. M., Burmeister, J. A., Palcanis, K. G. and Ranney, R. R. (1985). *Infect. Immun.*, **48**, 507.
87. Mukasa, H. and Slade, H. D. (1974). *Infect. Immun.*, **10**, 1135.

87a. NAIM, M., ZECHMAN, J. M., BRAND, J. G., KARE, M. R. and SANDOVSKY, V. (1985). *Proc. Soc. Exp. Biol. Med.*, **178**, 392.
88. NEWBRUN, E. (1983). *Cariology*, 2nd edn, Williams & Wilkins, Baltimore.
89. NEWMAN, M. G. (1985). *J. Periodontol.*, **56**, 734.
90. NISIZAWA, T., TAKEUCHI, K., IMAI, S., KITAHATA, S. and OKADA, S. (1986). *Carbohydr. Res.*, **147**, 135.
91. OLSON, B. L. (1977). *J. Dent. Res.*, **56**, 1426.
92. PFEFFER, M., ZIESENITZ, S. C. and SIEBERT, G. (1985). *Z. Ernährungswiss.*, **24**, 231.
93. PLATT, D. and WERRIN, S. R. (1979). *J. Dent. Res.*, **58**, 1733.
94. REINER, A. M. (1977). *J. Bacteriol.*, **132**, 166.
95. REUSSNER, G. and GALIMIDI, A. (1981). *J. Dent. Res.*, **60** (Suppl. A), 315. Abstr. 17.
96. RÖLLA, G., OPPERMANN, R. V., WÅLER, S. M. and ASSEV, S. (1981). *Scand. J. Dent. Res.*, **89**, 247.
97. SAXTON, C. A. (1969). *Archs. Oral Biol.*, **14**, 1275.
98. SCHEININ, A. (1979). *Int. Dent. J., London*, **29**, 237.
99. SCHIFFMAN, S. S., CROFTON, V. A. and BEEKER, T. G. (1985). *Physiol. Behav.*, **34**, 369.
100. SHAW, J. H. and ROUSSOS, G. G. (1978). *Sweeteners and Dental Caries, Spec. Suppl. Feeding, Weight & Obesity Abstracts*, Information Retrieval, Washington, DC.
101. SHOCKLEY, T. E., RANDLES, C. I. and DODD, M. C. (1956). *J. Dent. Res.*, **35**, 233.
102. SIEBERT, G. and GRUPP, U. (1977). *Dtsch. Zahnärztl. Z.*, **32** (Suppl. 1), S36.
103. SIEBERT, G. and ZIESENITZ, S. C. (1987). *Caries Res.*, **21**, 169, Abstr. 31.
104. SIEBERT, G., ZIESENITZ, S. C. and LOTTER, J. (1987). *Caries Res.*, **21**, 141.
105. SLEE, A. M. and TANZER, J. M. (1983). *Archs. Oral Biol.*, **28**, 839.
106. SLOTS, J. and DAHLEN, G. (1985). *Scand. J. Dent. Res.*, **93**, 119.
107. SOYER, C. and FRANK, R. M. (1979). *J. Biol. Buccale*, **7**, 295.
108. SREEBNY, L. M. (1982). *Wld Rev. Nutr. Diet.*, **40**, 19.
109. STEGMEIER, K., DALLMEIER, E., BESTMANN, H.-J. and KRÖNCKE, A. (1971). *Dtsch. Zahnärztl. Z.*, **26**, 1129.
110. STEPHAN, R. M. (1940). *J. Amer. Dent. Assoc.*, **27**, 718.
111. SVENSÄTER,G., TAKAHASHI-ABBE, S., ABBE, K., BIRKHED, D., YAMADA, T. and EDWARDSSON, S. (1985). *J. Dent. Res.*, **64**, 1286.
112. TANZER, J. M. and SLEE, A. M. (1983). *J. Amer. Dent. Assoc.*, **106**, 331.
113. THEILADE, E. and BIRKHED, D. (1986). In: *Textbook of Cariology*, A. Thylstrup and O. Fejerskov (Eds), Munksgaard, Copenhagen, Chapter 8.
114. TOKUNAGA, T., OKU, T. and HOSOYA, H. (1986). *J. Nutr. Sci. Vitaminol.*, **32**, 111.
115. TOORS, F. A. and HERCZOG, J. I. B. (1978). *Caries Res.*, **12**, 60.
116. TRAHAN, L., BAREIL, M., GAUTHIER, L. and VADEBONCOEUR, C. (1985). *Caries Res.*, **19**, 53.
117. TUOMPO, H., MEURMAN, J. H., LOUNATMAA, K. and LINKOLA, J. (1983). *Scand. J. Dent. Res.*, **91**, 17.
118. VADEBONCOEUR, C., TRAHAN, L., MOUTON, C. and MAYRAND, D. (1983). *J. Dent. Res.*, **62**, 882.
119. VAN DER HOEVEN, J. S. (1979). *Caries Res.*, **13**, 301.

120. van Houte, J. (1978). In: *Health and Sugar Substitutes, Proc. ERGOB Conf. Geneva*, B. Guggenheim (Ed.), Karger, Basel. 199.
121. van Palenstein Helderman, W. H. (1981). *J. Clin. Periodontol.*, **8**, 261.
122. Veiga, L. A. (1968). *J. Gen. Appl. Microbiol.*, **14**, 79.
123. Wåler, S. M., Rölla, G., Assev, S. and Ciardi, J. E. (1984). *Swed. Dent. J.*, **8**, 155.
124. Westergren, G., Krasse, B., Birkhed, D. and Edwardsson, S. (1981). *Archs. Oral Biol.*, **26**, 403.
125. Würsch, P. and Koellreutter, B. (1982). *Caries Res.*, **16**, 90.
126. Ziesenitz, S. C. and Gross, C. (1985). *J. Dent. Res.*, **64**, 300.
127. Ziesenitz, S. C. and Siebert, G. (1987). *Caries Res.*, **21**, 170 Abstr. 32.
128. Ziesenitz, S. C. and Siebert, G. (1987). *Caries Res.*, **20**, 498.

Chapter 7

DENTAL ADVANTAGES OF SOME BULK SWEETENERS IN LABORATORY ANIMAL TRIALS

R. HAVENAAR

Department of General and Industrial Microbiology, CIVO/TNO Institutes, Zeist, The Netherlands

SUMMARY

Animal experiments in caries research have proved their value for the human situation if they can be carried out under controlled conditions. Recommendations on experimental design, standardization, health control and cariogenic diets are discussed. The potential cariogenicity of sorbitol, xylitol, Lycasin®, maltitol, lactitol, Palatinit®, palatinose, L-sorbose and coupling sugars is reviewed, along with the anti-cariogenic and remineralizing properties of some bulk sweeteners in animal trials under ad libitum *and controlled feeding conditions and after prolonged consumption. Special attention is paid to secondary effects of sugar substitutes on plaque flora, general health and eating patterns of experimental animals. Based on controlled animal studies, bulk sweeteners may be classified as moderate/low-cariogenic (sorbitol), non-cariogenic (Lycasin 80/55) and anti-cariogenic (xylitol). Other alternative sweeteners and combinations of sweeteners are promising but need further research. Up to now, permanent adaptation of plaque flora to metabolize the different sweeteners has not been found.*

It can be assumed that the bulk sweeteners have dental benefits for humans.

1. INTRODUCTION

The crucial test to prove that a sugar substitute is non-cariogenic must be based on clinical trials. Prior to studies in humans knowledge obtained from animal experiments is indispensable. Animal experiments with sugars and sugar substitutes are mostly carried out in rats. Before discussing animal trials in relation to the potential cariogenicity, the non-cariogenicity or even the anti-cariogenicity of sugars and sugar substitutes, it is important to know how relevant animal experiments are to the human situation.

The 'Animal Caries Working Group' of the Scientific Consensus Conference on Methods for the Assessment of the Cariogenic Potential of Foods (San Antonio, USA, 1985) achieved consensus on the statement that the rat is an appropriate animal model to use in studying caries and the cariogenicity of foods and to determine the relative cariogenic potential of foods. This was based on studies showing that the pathogenesis of dental caries is essentially the same in rats and in humans, and on the observation that foods shown to be cariogenic or non-cariogenic in humans have a similar effect in rats.

2. RECOMMENDATIONS

Although it is concluded that rat experiments can result in valuable information which can be extrapolated to a great extent to the human situation, this cannot be said for all rat trials on sugar substitutes which have been published. Experiments with animals other than rats, such as hamsters and monkeys, can also have a high scientific value, but the scientific value of animal experiments and the extent to which the findings can be extrapolated to the human situation depend on the way the experiments are carried out.

The Animal Caries Working Group therefore made a list of recommendations on rat experiments to study the cariogenic potential of foods. Because of their importance for animal trials on sugar substitutes, they are summarized below:

—every experiment must have a detailed protocol;
—all tests must be carried out to the highest standards of animal husbandry (see ref. 67);
—the animals must be free from *Streptococcus mutans* (at the start of the experiments) and from *Strep. bovis* as well as from *Mycoplasma*, parasites and most viruses, which should be checked;
—sex of the animals is not important in short-term experiments;
—the animals must be randomized within experimental groups;
—the animals should have an infection by a culture in the log-phase of

growth of *Strep. mutans*, with or without additional organisms such as *Actinomyces* and *Lactobacillus*, before day 22 of age;
—the infection should be monitored at least twice during the experiment;
—it is recommended to use a gel diet with or without sucrose to sustain *Strep. mutans* in the dental plaque;
—the test diet should be prepared in small, uniform particle-size and if possible provided by a programmed feeder as described by König *et al.*[45] 17 times daily at one-hour intervals;
—sucrose, containing no more than 3% cornstarch, should be used as positive control and uncooked cornstarch as negative control;
—all caries lesions should be scored using the Keyes method,[40] although only the severity of the sulcal lesions would be included in the calculation of the Cariogenic Potential Index (CPI);[3]
—the initial and final weights of the animals should be recorded, and littermates identified to aid in statistical analysis of the data, whilst any multiple comparison technique may be used to analyze the data.

Although these recommendations are specially directed towards rat experiments to determine the CPI of foods, most of them are also very useful for animal studies on sugar substitutes. However, in my opinion some qualifications should be made. The first is on the caries scoring. It is suggested that the method described by Keyes[40] should be used. However, the sectioning and scoring technique described by Dalderup and Jansen[7] and modified by König *et al.*[43] has proved its value in many experiments. This method is probably even more sensitive for detecting small fissure lesions, especially when they are not located in the central part of the fissures. Comparison of the two techniques was made by Van Reen and Cotton[70] and Green.[16] There is no reason to suggest that animal experiments in which the scoring procedures according to König *et al.* are used should be considered as less valuable.

A second point is the gel diet described by Navia.[57] The purpose of this wet gel diet is to avoid contact between the nutrients in the main diet and the molars of the rats. Another way of doing this is by tube feeding.[3,63] However, the gastric intubation of young rats twice or more a day is not only very time-consuming, but, according to Shaw,[63] can also lead to 'higher than desirable mortalities'. This was the main reason for choosing the gel diet in the above-mentioned recommendations. However, the exclusive consumption of meals containing the test food, such as sugar substitutes, is not comparable with the human situation. If dental plaque comes in contact only with the test product and not with sugar-containing main meals, the accumulation of plaque and the activity of the microflora could be reduced in a way that would never happen in the human situation. Although tube feeding and gel diets have scientific advantages, these methods are removed from normal circumstances, and may lead to results which are more optimistic than can be proved in humans.

To revert to the programmed feeding of the rats, the advantage of determining the CPI of foods using a programmed feeder has been demonstrated in collaborative studies[63], but it is not always necessary to test alternative sweeteners mixed up with the food. The meal frequency is extremely important and easily influenced by experimental variables.[44] Meal frequency should be checked properly on animals fed *ad libitum* the test and control diets, for example with (electronically modified) registration cages.[45,65] A König/Hofer programmed feeder is indeed essential if it is doubtful that the meal frequency is equal in the test and control groups, and if alternate feeding of sugars (e.g. in main meals) and sugar substitutes (in additional meals), more or less conforming to the human situation, is wanted.

3. STANDARDIZATION

International recommendations or guidelines, for example as described above, can lead to better standardized animal experiments, and make a contribution toward experiments which are comparable with each other. This could lead to fewer experiments because there will be fewer contradictory results. Animal experiments on sugar substitutes are sometimes very hard to compare with each other because of differences in procedure. Conflicting results can be caused by differences in the experimental set-up. Although international approved recommendations have great advantages and should be applied as far as possible, one should be prepared to deviate from them if there are valid reasons.

4. SUGAR SUBSTITUTES AND HEALTH OF EXPERIMENTAL ANIMALS

Since the introduction of the cariogenic diet 580 by Stephan and Harris[66] and diet 2000 by Keyes and Jordan,[41] these diets have been used in many caries experiments in hamsters, rats, gerbils[9] and mice.[58] Certainly, they have proved their value, especially in short-term caries experiments, but they are not entirely suitable for studies on sugar alcohols, the majority of caloric sugar substitutes. As a consequence of the substitution of sucrose by sugar alcohols in concentrations above 5–10%, the increase in body weight is greatly retarded,[1,46,62] the animals suffer from severe diarrhoea and often do not survive the experimental period.[5,17,21,52,53,64] These pathological findings, due to diets 580 and 2000 in combination with sugar substitutes, could be caused partly by the lactose in the skim milk powder and the low concentrations of proteins and fat.[25] Because of the absence of lactase in (adult) rats, lactose cannot be metabolized and this results in osmotic diarrhoea, which is intensified by slowly absorbed sugar alcohols like mannitol, sorbitol and xylitol. It is hardly necessary to state that results obtained from experiments in which the animals suffer poor general health are at least questionable, quite apart from the ethical aspects.

To overcome these complications it is advisable to use diets which satisfy the nutrient requirements of the animals and are free from lactose. The concentrations of sugar substitutes in the diet must be increased gradually so that the animals can get used to them. On the whole the maximum concentration of sugar alcohols should not rise above about 20%. A purified powder diet for rats to test sugar substitutes without detrimental effects is described by Havenaar *et al.*[25] This diet (SSP), containing all nutrients required by rats but free from lactose, is recommended for testing polyols because of the following properties: within reasonable limits sugar alcohols can be added without influencing food uptake, body weight gains and general health; within the carbohydrate fraction (50% of the diet) all desired carbohydrate combinations can be achieved; *Strep. mutans* colonizes the mouth very well and in a reproducible manner. Testing xylitol in diet SSP, there was no influence on the absolute and relative weights of liver, spleen and kidneys of rats, but a reversible enlargement of the caecum was found.[25,28] Enlargement of the caecum, accompanied by histological alterations of the mucous membranes due to sugar alcohols in a purified diet, was also described by Karle.[35] Virtanen and Mäkinen[71] found no histological and histochemical differences in the palatinal mucosa, submandibular glands and the duodenum wall in rats fed on xylitol compared with glucose. Not only xylitol, but also sorbitol, Lycasin and (to a lesser extent) L-sorbose induce an enlargement of the caecum.[26,51] It is however unlikely that this has an influence on caries development.

Some authors have tried to get around these difficulties by applying the sugar substitutes topically.[8,15,30] In combination with controlled feeding or *ad libitum* feeding of diet 2000 (containing 20–56% sucrose), there was no diarrhoea and the animals remained in good health. The sugar substitutes did not affect body weight gains significantly, except for 60% sorbitol and 60% xylitol applied in the study reported by Gey and Kinkel.[15]

In general, alternative bulk sweeteners which are not hydrogenated to sugar alcohols cause less undesirable gastrointestinal disturbances, because they are better and faster absorbed. Rats fed *ad libitum* on diet 2000f containing 30% L-sorbose gained less body weight compared with rats fed on 30% sucrose, but all animals finished the caries test in apparently good health.[51] In two studies described by Ooshima *et al.*[59,60] the experimental rats did not suffer from diarrhoea when they were fed on a modified diet 2000 containing 56% palatinose. The absence of laxative properties is also claimed for 'coupling sugar' preparations.[32,33]

5. MODERATE-CARIOGENIC BASIC DIETS

Apart from the general health effects, there is another complication in the use of highly cariogenic diets like diet 2000.[25] Based on the finding that the caries-reducing effect of xylitol in rats was most pronounced in the upper molars, in which the caries progression was slower than in the lower molars, the authors pointed out that diets with a high cariogenic potential are less suitable to detect relatively slight differences in cariogenicity. As mentioned above, diet SSP was recommended in rat trials for health reasons, but a further advantage is that this diet induces reproducible and moderate numbers of carious lesions, suitable for detecting relatively slight differences in potential cariogenicity between diets with different food additives. In several rat and mice experiments under *ad libitum* as well as under programmed feeding conditions this diet demonstrated its benefits.[26,31,34] Because of its moderate cariogenicity the rats develop hardly any smooth-surface lesions when low concentrations (20%) of sucrose are used for 8–9 weeks. In some experiments this could be a disadvantage, but it should be noted that in humans most carious lesions are found in the fissures of the teeth. In combination with sucrose- or glucose-containing drinking water, this diet gave high numbers of fissure and approximal lesions and also moderate numbers of smooth-surface lesions (Havenaar, unpublished work).

6. THE POTENTIAL CARIOGENICITY OF SOME BULK SWEETENERS

6.1. Sorbitol

One of the first substitutes for easily soluble and fermentable sugars to be tested for cariogenicity in an animal model was sorbitol. Kleiner and

Kinkel[42] reported that the replacement of (part of) the carbohydrate in the diet by sorbitol did not reduce the incidence of caries in rats. However, Shaw and Griffiths[64] showed that sorbitol had little cariogenicity compared with sucrose. This was more or less supported by a large number of other studies in hamsters and rats.[13,20,21,36,37,46,52,53]

Most of these results must be interpreted very cautiously, because the intake of sorbitol-containing diets was often lower than that of the sucrose-containing controls, and the general health of the animals fed on sorbitol was very poor in most of the experiments: the animals suffered from diarrhoea, showed (dramatically) decreased weight gains and/or did not survive the experimental period. It is very likely that in these animal trials the meal frequency in the sorbitol groups was quite different from that in the control groups. In fact animal studies which do not satisfy the minimal conditions of a valid experimental set-up should not be used to provide final conclusions on the dental advantages of a particular sugar substitute. The main question on sorbitol is: under which circumstances is sorbitol less harmful to the teeth than sucrose or glucose? Further, is sorbitol better or worse in comparison with other alternative bulk sweeteners?

Karle and Gehring[37] reported the results of a programmed feeding experiment on rats with sorbitol-containing chocolate. The growth curves of the rats indicated similar increases in body weight (except for the last week) on sucrose, sorbitol and the basic control diet. The health status of the animals is not mentioned, but it may be inferred that it was at least acceptable, in which case the results on caries incidence can be useful. Sorbitol-containing chocolate was less cariogenic than normal sucrose chocolate, although not statistically significantly ($P < 0{\cdot}10$). Sorbitol chocolate was significantly less cariogenic in comparison with diet 2000S ($P < 0{\cdot}005$). Other rat experiments under programmed feeding conditions are described by Havenaar *et al.*[26] In two separate studies a purified basic diet (SSP) was used. This did not induce diarrhoea or diminish the general health of the rats when starch was substituted stepwise by sorbitol up to levels of 20% or 25%. The results of these two experiments showed that the average incidence of fissure caries on sorbitol was significantly lower than on sucrose ($P < 0{\cdot}05$). However, the uncooked starch groups had significantly fewer fissure lesions than the sorbitol groups ($P < 0{\cdot}01$). The average numbers and percentages of *Strep. mutans* in the sorbitol, sucrose and starch groups were similar. This indicates that the caries scores were not influenced by differences in the colonization of *Strep. mutans*.

The cariogenicity of sorbitol in relation to that of sucrose can be expressed as the Cariogenic Potential Index (CPI), according to Bowen *et al.*[3] If in these three animal trials the CPI of sucrose is set at 100, then the CPI of sorbitol is between 53 and 63. It can be concluded from these studies that in rat experiments under controlled conditions sorbitol can be classified as a moderate-cariogenic sweetener.

6.2. Xylitol

Narkates *et al.*[56] reported in an abstract that xylitol at a 5% level in the diet of rats resulted in more smooth-surface lesions than a cornstarch diet. Xylitol at 10% and 20% resulted in caries scores identical with those in the starch control group. These findings are not supported by previous or later animal trials. Under *ad libitum*[2,21,37,52,53] as well as programmed feeding conditions[26,35,37] the low cariogenicity of the basic diet was either not changed or made even lower by the addition of xylitol. In rats of which the salivary glands had been removed[36,37] xylitol failed to raise the number of carious lesions. In some of the above publications it was reported that the animals suffered from severe diarrhoea and/or died before the end of the experiment. These health problems undoubtedly had their influence on the results of the experiments. But also (or especially) in animal trials in which the rats were in good general health the results showed clearly that xylitol is not cariogenic.

6.3. Lycasins

Lycasins are hydrogenated starch hydrolysates. The composition differs from one product to another depending on its manufacturing process, which can lead to variations in properties, e.g. in fermentability by dental plaque.[12] This means that each time a Lycasin® is tested, the type of product must be specified. The first Lycasin was manufactured by Lyckeby Stärkelseförädling AB in Sweden as 'product 6563'. Caries studies on animals showed a lower potential cariogenicity of this and other types of Swedish Lycasins as compared with sucrose.[1,11,13,36,46] However, Swedish types of Lycasin are no longer to be had. Nowadays different types of Lycasin are made by Roquette Frères, Lille, France. The most extensively studied type is obtainable commercially as Lycasin® 80/55, a clear, colourless syrup. For animal studies it is also available as a powder, containing 6–8% sorbitol, 50–55% maltitol and about 40% higher hydrogenated saccharides.

The first publication about animal caries experiments on Lycasin 80/55 was by Firestone *et al.*[8] In their experiments Lycasin was applied topically to rats' teeth (five times daily, 200 μl of a 50% solution) in combination with a diet containing 20% sucrose. The Lycasin group had similar numbers of carious lesions to the groups which had sucrose or water applied topically. Rat experiments described by Grenby and Saldanha[20] showed that Lycasin 80/55 fed *ad libitum* was significantly less cariogenic than sucrose ($P < 0{\cdot}05$). The total number of lesions (mean caries score 2·1 versus 7·5) and the average size of the lesions (1·2 versus 1·4) were both smaller. In this experiment Lycasin was more cariogenic than sorbitol ($P < 0{\cdot}05$; mean caries score 2·1 versus 0·8), but the food intake (n.s.) and body weight gains ($P < 0{\cdot}001$) of the rats in the sorbitol group were lower than in the Lycasin and sucrose groups. The animals were not inoculated with a cariogenic *Strep. mutans* strain and the microorganisms responsible for the carious lesions were not investigated. Another series of five experiments, reported by Grenby,[18] showed that two different types of Lycasin were less cariogenic than sucrose and similar to maize-starch in producing very low levels of caries. Only one type produced more lesions than sorbitol.

In two rat caries experiments described by Havenaar *et al.*,[26] Lycasin 80/55 at 20–25% in purified diet SSP was tested under programmed feeding conditions. In the first experiment only the test or control diets were available for the rats (18 meals per day). In the second trial the test or control diets (14 meals per day) were alternated with meals containing sucrose (20%) and glucose (10%) (four times per day). This arrangement was chosen to get similar colonization patterns of *Strep. mutans* in the test and control groups. In both experiments the average incidence of fissure caries, taking into account the severity of the lesions, was significantly lower in the Lycasin groups than in the sucrose and the sorbitol groups ($P < 0{\cdot}01$). The levels of dental caries on the Lycasin regime were closely similar to those on xylitol, L-sorbose and uncooked starch. The rats were in good health. The body weight gains of the rats in the Lycasin groups were slightly lower than (experiment 1) or similar to (experiment 2) those of the rats in the sucrose groups. The consistency of the faeces was not influenced by the high amount of sugar alcohols. In another rat experiment[27] the animals were inoculated with *Strep. mutans* serotype *c* or *d*, sometimes in combination with *Actinomyces viscosus*. The diet contained 20% Lycasin 80/55 and was given *ad libitum*. The average number of meals per day, the meal size and the eating time in the Lycasin and sucrose groups were similar. However, the percentages and numbers of *Strep. mutans* and *Act. viscosus* were lower in the Lycasin groups than in the sucrose controls. The caries scores showed that Lycasin was not cariogenic at all under these experimental conditions. A disadvantage of the types of Lycasin that have a high grade of maltitol is that the product becomes hygroscopic. Lycasin 80/55 as powder was rather hygroscopic, even when mixed into the diet of the animals. In the above rat experiments the Lycasin-containing diets were stored in desiccators which were opened only once a day. Nevertheless the diet lost its powdery appearance and became more sticky. It cannot be said whether this had an influence on the potential cariogenicity of Lycasin. Possibly it prolonged the oral clearance time.

Under the experimental conditions described Lycasin 80/55 is virtually non-cariogenic: in the two controlled feeding experiments the CPIs of Lycasin and uncooked starch were between 18 and 22 and between 23 and 25, respectively. In the *ad libitum* study the CPI of Lycasin was even lower: between 5 and 8.

6.4. Maltitol

Maltitol is a hydrogenated disaccharide (α-1,4-glucose-sorbitol) produced by the reduction of maltose, and one of the main components of Lycasin 80/55. Although it is widely used in Japan in chewing gum, candies, bakery products and jam, there are only a few relevant publications. A caries trial in rats inoculated with *Strep. mutans* was carried out with maltitol in the diet by Mukasa,[54] in which it was shown to be non-cariogenic. Maltitol has one disadvantage in common with Lycasin 80/55: it is very hygroscopic. Firestone *et al.*[8] described a rat experiment in which maltitol was applied topically (five times a day, 200 μl of 50% solution) to the molars in combination with a cariogenic diet (containing 20% sucrose). *Strep. mutans* and *Act. viscosus* were introduced in the mouths of the animals. The results showed that the numbers of dentinal fissure lesions in the animals treated with maltitol were closely similar to those in animals treated with water or sucrose.

The potential cariogenicity of maltitol is most probably comparable with that of Lycasin 80/55.

6.5. Lactitol

Another reduced disaccharide (β-1,4-galactose-sorbitol) is lactitol, derived from lactose. Although it has been put forward as a less cariogenic sugar substitute no evidence is found in the literature, at least not as far as animal trials are concerned.

6.6. Palatinit®

After hydrogenation of isomaltulose an equimolecular mixture of two disaccharide alcohols are formed: glucopyranosyl-sorbitol (= isomaltitol = α-1,6-glucose-sorbitol) and glucopyranosyl-mannitol (= α-1,6-glucose-mannitol). So far the cariogenic properties of Palatinit mixed into a basic diet have been tested in two animal studies: one programmed feeding experiment with rats inoculated with two different strains of *Strep. mutans* described by Karle and Gehring,[38] and one study with rats fed *ad libitum*, reported by Van der Hoeven.[69] In the programmed feeding study (18 meals per 24 h) Palatinit (increased stepwise in the diet up to 30%) was significantly less cariogenic than sucrose ($P < 0{\cdot}001$), but significantly more cariogenic than xylitol ($P < 0{\cdot}05$). According to the authors the animals never showed any symptoms of diarrhoea during the experiment. In spite of controlled feeding the body weight gains of the rats in the Palatinit and xylitol groups were somewhat lower than in the sucrose and basic diet groups. The other study consisted of two separate experiments in which the rats were fed *ad libitum* on diets containing 16% Palatinit or 16% sucrose. In the first experiment 8-week-old non-inoculated rats were used. The numbers of dentinal fissure lesions, scored every fortnight, were very low in the Palatinit group in comparison with the sucrose group. In the second experiment 28-day-old rats were divided into non-inoculated and those inoculated with *Strep. mutans* C67-1 and *Act. viscosus* Ny 294, since these organisms fermented Palatinit to final pH 4·9 and 5·1 respectively.[68] The caries incidence in both was slightly higher on the sucrose than on the Palatinit regimes ($P < 0{\cdot}01$). Both in the inoculated and the non-inoculated groups the CPI of Palatinit was about 60. In the controlled feeding study the CPIs of Palatinit and the basic control diet were 38 and 15 respectively.

Topical application of Palatinit (five times daily, 150 or 200 μl, 50% solution) in combination with a cariogenic diet, showed inconsistent results.[8] In one out of two experiments the numbers of dentinal carious lesions in the Palatinit group were significantly lower than in the groups treated with water ($P < 0{\cdot}05$) or sucrose ($P < 0{\cdot}001$). In the other experiments the B-type dentinal lesions and the smooth-surface lesions in the Palatinit group were lower compared with the water group ($P < 0{\cdot}05$), but not compared with the sucrose group.

In conclusion, these few animal trials make it seem likely that Palatinit belongs to the category of moderate-cariogenic bulk sweeteners.

6.7. Palatinose

Palatinit is formed by the reduction (hydrogenation) of isomaltulose (= α-1,6-glucose-fructose), a structural isomer of sucrose. Another name for isomaltulose is palatinose, which has been found in honey and sugar cane extract. Palatinose can be derived from sucrose by the use of a glucosyltransferase from *Protaminobacter rubrum*.

Palatinose was tested for cariogenicity in rat experiments by Ooshima *et al.*[59] Rats 18 days old were infected with *Strep. mutans* strain 6715 and fed *ad libitum* on diet 2000 containing 56% pulverized palatinose, sucrose, glucose, fructose or cornstarch for 55 days. The caries scores of the rats in the palatinose and cornstarch groups were very low (CPI = 8 and 4 respectively) in comparison with the sucrose group ($P < 0{\cdot}001$). The feeding of palatinose or cornstarch as the only carbohydrate (apart from the lactose in skim milk) resulted in the recovery of very low numbers of *Strep. mutans* 6715, serotype *d*/*g* (= *Strep. sobrinus*) from the molar surfaces of the rats, compared with the sucrose, glucose or fructose groups. This suppressed colonization of *Strep. mutans* could have had an influence on the caries scores, which was more or less supported by the results.[59] Some other groups of rats were fed on diets containing combinations of sugars. The recoveries of *Strep. mutans* from rats fed on sucrose + cornstarch (28/28%), sucrose + palatinose (28/28%) or sucrose alone (56%) were similar. The caries scores in the sucrose + palatinose group (CPI = 38) were lower than the sucrose group ($P < 0{\cdot}01$), but were now higher than that of the sucrose + cornstarch group (CPI = 19). The authors suggested prematurely from this study that the partial replacement of sucrose by palatinose had a caries-inhibiting effect. Fortunately they rectified this in a later publication,[60] concluding that the lower caries scores in the sucrose/palatinose (28/28%) group compared with the sucrose (56%) group were most probably caused by the lower concentration of sucrose in the diet. This second rat caries study had the same experimental set-up as described above, except that the animals were inoculated with *Strep. mutans* serotype *c*, strain MT8148R. Unlike strain 6715, which was used in the preceding experiment, this strain can utilize palatinose, especially after subculturing in a 1% palatinose-containing broth medium. *Strep. mutans* MT8148R easily colonized the mouth of the rats fed on palatinose and other sugars, irrespective of the presence of sucrose. The caries scores for palatinose (56%) and starch were both higher than in the first study (CPI = 18), but again significantly lower than for sucrose ($P < 0{\cdot}005$), glucose and fructose. The average caries scores of the rats fed on sucrose + palatinose (28/28%) and sucrose + starch (28/28%) were similar (CPI = 67 and 55 respectively) and both lower than on 56% sucrose ($P < 0{\cdot}01$). The suggestion that the low cariogenicity of palatinose was attributable to the lack of colonization of *Strep. mutans* serotype *d*/*g* was not completely supported by this study, in which serotype *c* was used.

Only a tentative conclusion can be drawn: in animal trials palatinose seems to be a low-cariogenic bulk sweetener but further investigations are needed before making a final judgement.

6.8. L-Sorbose

L-Sorbose is a reducing sugar, belonging to the L-series of the ketohexose family ($C_6H_{12}O_6$), produced by fermentation of D-sorbitol by the 'sorbose bacteria'. A rat experiment under *ad libitum* feeding conditions is reported by Mühlemann.[51] Diet 2000f (64% flour) with replacements of flour by 30% sucrose or 30% L-sorbose was given for 35 days to Osborne–Mendel rats infected with *Strep. mutans* OMZ 176 and *Act. viscosus* Ny-1. Smooth-surface as well as dentinal fissure lesions were absent from the L-sorbose group, in contrast to the sucrose group ($P < 0{\cdot}001$). In a separate experiment it was ascertained that this was not caused by differences in meal frequency between the L-sorbose and sucrose groups. Although according to the author the animals appeared healthy and showed no diarrhoea, the body weight gains of the rats in the L-sorbose group were much lower compared with the starch and sucrose control groups.

The finding that L-sorbose is non-cariogenic is more or less in agreement with programmed feeding studies. Karle and Gehring[39] reported a study with xerostomized and non-xerostomized rats infected with *Strep. mutans* R1 and R2. The animals were fed for six weeks on stepwise increased concentrations of L-sorbose, xylitol or sucrose up to 30% (18 meals per day). The L-sorbose groups showed significantly lower numbers of fissure lesions than the sucrose control groups ($P < 0{\cdot}001$; CPI = 38 and 28 for the non-xerostomized and xerostomized rats, respectively), but slightly higher scores compared with the xylitol- and starch-containing control groups (CPI = 20–21 and 15–19 respectively). In a controlled feeding study by Havenaar *et al.*,[26] Osborne–Mendel rats were inoculated with *Strep. mutans* serotype *c* and fed on stepwise increased concentrations of L-sorbose, xylitol, sorbitol, Lycasin, starch or sucrose up to 20% (experiment I) or 25% (experiment II) for 71–72 days (for further information see under Lycasin). In both experiments L-sorbose as well as xylitol, Lycasin and starch induced hardly any fissure caries lesions, in contrast to sucrose and sorbitol ($P < 0{\cdot}01$). The CPIs for L-sorbose in experiments I and II were 18 and 10 respectively, and were even lower than for uncooked starch (CPI = 23 and 25, respectively).

Topical applications of L-sorbose (150 or 200 μl of a 50% solution, five times daily) in combination with a cariogenic diet resulted, in one out of the four experiments, in lower fissure caries scores compared with the application of water as well as sucrose ($P < 0{\cdot}001$).[8]

Thus in these rat experiments L-sorbose appears to be a very low- or non-cariogenic alternative for sucrose and glucose.

6.9. Coupling Sugar

In Japan coupling sugar preparations can be used freely in food products, because they are regarded as natural sweeteners containing a mixture of monosaccharides, glucosyl-sucrose, oligosaccharides and oligosaccharides terminated at the reducing end by sucrose. Coupling sugar is manufactured from a mixture of starch and sucrose using cyclodextrin glucosyltransferase from *Bacillus megaterium*.[32]

One of these coupling sugars, designated as CSSF, was tested in an animal trial by Ikeda *et al.*[33] Gnotobiotic and conventional rats were infected with *Strep. mutans* strain 6715, which can ferment CSSF. Diet 305, containing 5% CSSF, 5% sucrose or a combination of 5% CSSF and 5% sucrose (gnotobiotic rats only) was given *ad libitum* for 25 days. The average caries scores for buccal, sulcal and proximal surfaces were significantly lower in the CSSF groups compared with the sucrose groups ($P < 0{\cdot}01$). This effect was somewhat more pronounced in the gnotobiotic rats (CPI for CSSF = 35) than in the conventional rats (CPI = 49; relative cariogenicity of CSSF calculated from the mean buccal + sulcal + proximal caries scores). Gnotobiotic rats fed on sugar-free diet 300 showed nearly similar numbers of lesions (CPI = 29) to gnotobiotic rats fed on diet 305 with 5% CSSF (CPI = 35). The animals in test and control groups were healthy, free from diarrhoea and made similar gains in body weight. The number of animal trials with coupling sugar is too small to reach a firm conclusion, but it is probably a moderate- or low-cariogenic bulk sweetener.

7. ADAPTATION OF DENTAL PLAQUE TO BULK SWEETENERS

It has been suggested by Fostell[10] that the prolonged consumption of sorbitol could result in the selection of a plaque flora with an increased ability to ferment sorbitol. Fermentation experiments *in vitro* have demonstrated that repeated cultivation of several oral bacteria, in particular *Strep. mutans*, in media containing sugar substitutes resulted in adaptation. This adaptation, which found expression in a faster and lower pH fall, has been ascertained for sorbitol, lactitol, maltitol and Lycasin[14,24] and for palatinose.[59] However, it is partly lost after subculturing in a glucose medium.[24] Since in man the consumption of sugar substitutes will always be alternated with the consumption of glucose or sucrose, it seems unlikely that adaptation will ever be complete *in vivo*. Although this assumption is supported by Carlsson[4] it requires confirmation *in vivo*.

One of the first studies in animals to investigate the capacity of micro-organisms in dental plaque to adapt to sugar substitutes was done on monkeys by Cornick and Bowen.[6] Two groups of four monkeys (*Macaca fascicularis*) were fed on a sucrose-free diet providing about 1 g/day sorbitol. One group was infected with plaque containing *Strep. mutans* from a caries-active monkey. The second group already had an established population of *Strep. mutans*. Over a two-year period several parameters were monitored to determine whether adaptive processes took place. The results showed that the numbers of sorbitol-fermenting bacteria (determined at weekly intervals) as well as the acid-producing capacity of plaque from sorbitol (measured intra-orally) were not altered during the experiment. The results suggest that long-term consumption of sorbitol does not induce an adaptation of the plaque flora.

In five successive experiments (I–V) with rats fed *ad libitum* on diet SSP containing Lycasin 80/55 (20%) or sucrose (20%), the adaptation of the dental plaque flora was studied by Havenaar *et al.*[27] In experiment I the rats were infected with washed cultures of *Strep. mutans* serotype *c* (strain C67-1) and *Act. viscosus* (strain Ny-1). In three successive transmission experiments (II–IV) the rats were infected with plaque suspensions derived from the animals of the Lycasin or sucrose groups of the preceding experiment. The rats in experiment V were inoculated with washed cultures of the original strains or with plaque suspensions harvested from experiment IV. At the end of each experiment the numbers and percentages of *Strep. mutans* and *Act. viscosus* in both the sucrose and Lycasin groups were determined in plaque from the upper and lower molars. Caries was assessed in the lower molars of experiments I and V. In this long-term study Lycasin 80/55 was non-cariogenic in contrast to sucrose ($P < 0{\cdot}001$), irrespective of whether the rats were infected with the original strains or with *Strep. mutans* and *Act. viscosus* strains which had survived several rat experiments (I–IV) on either a sucrose or a Lycasin diet. The colonization (average numbers) of *Strep. mutans* and *Act. viscosus* on the molars was lower in the Lycasin groups compared with the sucrose groups. However, the ratios of these numbers between the sucrose and the corresponding Lycasin groups were not changed during the five successive experiments. This indicates that long-term exposure of the plaque flora to Lycasin 80/55 did not alter the composition of the microflora under the set conditions. Fermentation of Lycasin *in vitro* by the plaque flora of the rats fed on Lycasin was slightly faster than by the plaque flora from rats fed on sucrose. Probably some extra enzyme activity had been induced during this prolonged exposure to Lycasin.

From the very few animal trials on the long-term consumption of alternative bulk sweeteners, the only conclusion that can be drawn is a provisional one, that neither a selection nor an adaptation of the plaque flora is induced by sorbitol and Lycasin.

8. ANTI-CARIOGENIC PROPERTIES OF BULK SWEETENERS

8.1. Xylitol

Xylitol has been tested for its cariogenicity in animal experiments (see Section 6.2) as well as in clinical trials reported by Scheinin and Mäkinen.[61] The results showed that it is non-cariogenic, and moreover, based on the clinical studies, it was suggested that xylitol possesses anti-cariogenic or therapeutic (remineralizing) properties.

The caries-inhibiting effect of xylitol in the rat model was described for the first time by Mühlemann *et al.*[53] The rats were fed *ad libitum* on diet 2000f in which equal amounts of wheat flour were substituted by sucrose, different sugar alcohols or combinations of sucrose + sugar alcohols. The oral microflora was enlarged by superinoculations with *Strep. mutans* and *Act. viscosus.* The combination of sucrose/xylitol (15/15%) resulted in significantly lower numbers of dentinal fissure lesions compared with sucrose (15%) alone ($P < 0{\cdot}05$). Neither the sucrose/xylitol combination (25/25%) nor sucrose/sorbitol (25/25%) decreased the cariogenicity of sucrose (25%). Despite the addition of an antidiarrhoeic product, the animals suffered from severe diarrhoea which decreased body weight gains and increased morbidity and mortality so that the results were questioned by the authors themselves. Mundorff and Bibby[55] reported in an abstract that under programmed feeding conditions sucrose/xylitol combinations in different ratios (5, 10 or 20% xylitol) resulted in caries reductions of between 10 and 20% compared with rats fed on 67% sucrose. Similar findings were reported by Childers *et al.*,[5] but they suggested that the observed caries inhibition of 30% in the xylitol/sucrose (17/17%) group, compared with the sucrose (17%) group, may be related, at least in part, to profound diarrhoea and reduced weight gains of the rats. In their study a linear relationship between survival rates and caries indices in the xylitol groups existed. Short-term experiments (21 days) with low quantities of xylitol (3–6%) in a modified Stephan 580 diet fed *ad libitum* to rats also gave far lower body weights,[48] but the authors could not detect differences in the eating patterns between test and control groups. They found that the addition of 3% and 6% xylitol to a high-cariogenic diet produced less caries than the same diet without the addition of xylitol (3%: n.s.; 6%: $P < 0{\cdot}02$). Also feeding *ad libitum* of sucrose meals and starch + 3% or 6% xylitol meals on alternate days resulted in lower caries scores than unsupplemented starch meals ($P < 0{\cdot}001$).

The partial substitution of uncooked starch by xylitol (5%) in a sucrose (20%) and glucose (5%) diet for rats (diet SSP 20/5) resulted in significantly lower numbers of dentinal fissure lesions ($P < 0{\cdot}02$) compared with the same diet without xylitol.[25] This was found in rats inoculated with both *Strep. mutans* serotype *c* and serotype *d*. These results, demonstrating caries-inhibiting properties of xylitol, were confirmed in a later study.[28] In a separate study,[29] rats fed *ad libitum* on diet SSP 20/5, with or without 5% xylitol, had exactly the same average meal frequency, meal size and eating time. This implies that the findings were not caused by differences in eating patterns. An anti-cariogenic effect was also demonstrated for xylitol-containing drinking water in combination with a moderate cariogenic diet.[23] The incidence of fissure caries in rats which received 2% or 4% xylitol in their drinking water was lower than in the tap-water control group ($P < 0{\cdot}02$). The caries reduction in the 4% xylitol drinking water group was similar to that in the 5% xylitol diet group. However, in comparison with fluoridation of the drinking water (NaF; 10–40 ppm F^-), the caries reduction due to xylitol was relatively small.

Under programmed feeding conditions the partial replacement of starch by xylitol (5%) in diet SSP 20/5 as well as the alternate feeding of sucrose-containing main meals (12 times per day) and 8–33% xylitol-containing additional meals (3–6 times per day) showed identical anti-cariogenic effects of this sugar substitute.[22] If equal amounts of xylitol per 24 h were consumed by the rats, this effect seemed to be independent of the concentration of xylitol in the additional meals, but positively correlated with the frequency with which xylitol was consumed. In neither of these above-mentioned experiments was the colonization of *Strep. mutans* influenced by xylitol. This is in accordance with results of clinical trials.[47]

These results are not in agreement with the data of Grenby and Colley.[19] They did not find an anti-cariogenic effect of xylitol in rats with a non-specified plaque flora. These conflicting results can not simply be explained by possible differences in meal frequency between test and control groups or by the fact that the rats were not superinfected with *Strep. mutans*. To what extent the absence or presence of *Strep. mutans* in plaque played an essential role in the caries-inhibiting effect of xylitol is not clear. Grenby and Colley used a highly cariogenic, lactose-containing diet not designed for studying sensitive de- and re-mineralization processes, for as long as eight weeks. In some experiments this caused very high and wide-ranged caries scores. Their findings about poor general health of the rats are in agreement with other reports (see Section 4). The relatively moderate caries-inhibiting effect of xylitol would be harder to detect when highly cariogenic diets are used and/or when the experimental animals are not in good health, which could influence their eating habits and/or normal natural resistance.

Except for the experiments of Grenby and Colley,[19] most other studies suggest that xylitol possesses an anti-cariogenic effect. The mechanism behind this property cannot be ascertained from these animal trials.

8.2. L-Sorbose

Mühlemann[51] reported that in rats superinfected with *Strep. mutans* and *Act. viscosus*, the cariogenic effect of 30% sucrose was significantly depressed by the addition of 30% L-sorbose to the diet ($P < 0{\cdot}001$). Although these animals had extremely retarded body weight gains (67% less than the sucrose group) had a 20% reduction of the average eating time (n.s.), they had the same meal frequency compared with the control group. *Ad libitum* and programmed intermittent feeding experiments with rats inoculated with *Strep. mutans* serotype *d* tested the anti-cariogenic properties of L-sorbose.[22] The partial substitution of starch by L-sorbose (5%) in diet SSP 20/5 resulted in two separate experiments in a reduction of the numbers of dentinal fissure lesions (40–70% reduction) in comparison with the same diet without L-sorbose ($P < 0{\cdot}01$). The feeding of diet SSP 20/5 (12 times per day), alternated by 33% L-sorbose-containing additional meals (six times per day) also resulted in a significant caries reduction (35%) compared with starch-containing additional meals ($P < 0{\cdot}02$).

It is probably premature to come to a final conclusion that L-sorbose possesses anti-cariogenic properties like xylitol, but the results of these animal trials are very promising.

8.3. Other Bulk Sweeteners

The other bulk sweeteners have not been tested for their anti-cariogenic properties as extensively as xylitol and L-sorbose. Local applications of a variety of sugar substitutes showed inconsistent results, on which conclusions cannot be based.[8,15,30] Feeding *ad libitum* of diet SSP 20/5 in which 5% starch was substituted by sorbitol, Palatinit or Lycasin 80/55 did not give lower numbers of fissure lesions compared with an unsupplemented diet.[22] Neither intermittent feeding of sucrose + glucose (20/5%) in main meals (12 times per day) nor 33% sorbitol-containing additional meals (six times per day) resulted in lower numbers of fissure lesions than sucrose + glucose main meals and starch-containing additional meals.

As far as is known, these animal trials indicate that sorbitol, Lycasin and Palatinit do not inhibit caries development. Other bulk sweeteners have not yet been tested for anti-cariogenic effects. It should not be necessary to test moderate- or low-cariogenic sweeteners for this property.

8.4. Combinations of Bulk Sweeteners

In a programmed feeding study with rats, combinations of Lycasin 80/55 + xylitol and Lycasin 80/55 + L-sorbose were tested for their anti-cariogenic effects.[22] The rats were all fed on diet SSP, containing 20% sucrose and 10% glucose (16 times per day). Besides, they received additional meals (eight times per day) containing, in groups I–VI respectively: sucrose, Lycasin (25%), sucrose + xylitol, sucrose + L-sorbose, Lycasin + xylitol or Lycasin + L-sorbose (25/10%). The results demonstrated that xylitol (group III) inhibited the cariogenic potential of sucrose (group I; $P < 0{\cdot}01$) and changed the non-cariogenic potential of Lycasin 80/55 (group II) to an anti-cariogenic combination of Lycasin + xylitol (group V; $P < 0{\cdot}02$). Lycasin + L-sorbose had similar but smaller effects.

9. REMINERALIZING PROPERTIES OF SOME BULK SWEETENERS

One of the first remineralizing studies in animals in relation to sugar substitutes was reported by Karle.[35] Gnotobiotic rats, mono-associated with *Strep. mutans*, were fed on a sucrose diet for three weeks, followed by a xylitol diet for five weeks. The average number of fissure lesions in this group was lower than in the group receiving the sucrose diet for three weeks only. Leach and Green[48] demonstrated in several rat experiments a remineralizing effect of xylitol. Caries-active rats with a non-specified plaque flora were fed a high-cariogenic diet (days 1–5). Thereafter they received a starch diet, with or without 6% xylitol supplementation (days 6–15). A number of dentinal fissure lesions, produced by the sucrose diet, were reversed by subsequent exposure to the xylitol-supplemented starch diet, in contrast to the unsupplemented starch diet ($P < 0{\cdot}05$). However, this effect was not consistent in all their experiments, e.g. sucrose (days 1–4) followed by starch + xylitol or starch alone (days 5–12), resulted in similar numbers of lesions. Prolonged feeding of starch + 6% xylitol (days 6–11,

11–16 and 16–21) showed a progressive regression of the fissure lesions induced by sucrose (days 1–5), in contrast to unsupplemented starch ($P < 0{\cdot}01$). These results suggest a remineralizing effect of xylitol in the absence of easily fermentable carbohydrates.

A study by Havenaar *et al.*[28] had basically the same set-up, although it used a moderate-cariogenic diet, containing sucrose and glucose (diet S), for 54 or 75 days. This was followed by a three-week period in which the rats were fed on the same diet supplemented with 5% xylitol (diet SX). The rats which changed from diet S to diet SX on day 54 or 75 had significantly lower numbers of initial dentinal fissure lesions in comparison with rats fed on diet S only for 54 or 75 days ($P < 0{\cdot}05$). But the numbers of advanced lesions were not definitely altered, while the total of very advanced lesions was increased. This result indicates that xylitol induced remineralization of the initial lesions, notwithstanding the high sugar content of the diet. Most of the larger lesions were probably too far advanced to be influenced by xylitol and consequently progressed further.

It is suggested by Leach *et al.*[50] that sweetening agents stimulate the salivary flow and raise the pH and calcium concentrations in saliva. This would promote a remineralizing environment in dental plaque if the sweetener is not fermented by the plaque flora. According to the authors, this explains why not only xylitol, but also mannitol and sorbitol[50] and Lycasin 80/55[49] supplementing a starch diet resulted in reversals of dentinal fissure caries in rats, produced by prior exposure to sucrose. This hypothesis could be right if no easily-fermentable substrates are available, but must be rejected if the sugar substitute is fed simultaneously with sucrose and/or glucose, as was done in some of the xylitol studies. Apparently, the anti-cariogenic and remineralizing effect of xylitol (and probably also L-sorbose) is based on properties which sorbitol, Lycasin 80/55 and Palatinit® most probably do not have.

10. CONCLUSIONS

Animal trials on the dental properties of alternative bulk sweeteners, especially sugar alcohols, should be carried out with special diets, so that no detrimental effects arise. In general, more attention must be paid to general health and eating patterns of the animals, and programmed feeding experiments, whether or not with alternating meals, are to be preferred. A measure of international standardization on these points is desirable.

Based on a number of well-controlled animal trials with bulk sweeteners on the prevention of dental caries, these sugar substitutes can be classified as: (a) moderate/low-cariogenic, (b) non-cariogenic or (c) anti-cariogenic. From the results above it can be concluded that under the experimental conditions, we must regard (a) sorbitol and provisionally also Palatinit, Palatinose and coupling sugars, as moderate- or low-cariogenic; (b) Lycasin 80/55 and probably also maltitol as non-cariogenic, and (c) xylitol and probably also L-sorbose as anti-cariogenic bulk sweeteners. Further, there is some evidence that no adaptation of practical importance develops in the dental plaque flora to sorbitol and Lycasin 80/55. Combinations of xylitol and Lycasin 80/55 seem promising. Such combinations can be valuable and need further research.

Because of the many similarities between humans and rats in factors which play a role in the caries process, there is no evident reason to suppose that the above results and conclusions may not be extrapolated to humans. In fact it can be assumed that different bulk sweeteners, having proved to be low- or non-cariogenic or even anti-cariogenic in animal trials, will have dental benefits to humans.

REFERENCES

1. Birkhed, D. and Frostell, G. (1978). *Caries Res.*, **12**, 250.
2. Bowen, W. H. and Amsbaugh, S. M. (1978). Unpublished data. Some experimental data are mentioned in: Amsbaugh, S. M. and Bowen, W. H. (1981). The influence od carbohydrate on the cariogenicity of test diets in rats. In: *Proc. Animal Models in Cariology*, J. M. Tanzer (Ed.), Information Retrieval, Washington, DC, 309–17.
3. Bowen, W. H., Amsbaugh, S. M., Monell-Torrens, S., Brunelle, J., Kuzmiak-Jones, H. and Cole, M. F. (1980). *J. Amer. Dent. Ass.*, **100**, 677.
4. Carlsson, J. (1979). In: *Health and Sugar Substitutes, Proc. ERGOB Conf. Geneva*, B. Guggenheim (Ed.), Karger, Basel, 205–10.
5. Childers, N. K., Kreitzman, S. N. and Yagiela, J. A. (1978). *IADR Abstract No.* 149; *J. Dent. Res.*, **57** (Spec. Issue A), 112.
6. Cornick, D. E. R. and Bowen, W. H. (1972). *Archs. Oral Biol.*, **17**, 1637.
7. Dalderup, L .M. and Jansen, B. C. P. (1955). *Int. Z. Vitam. Forsch.*, **26**, 235.
8. Firestone, A. R., Schmid, R. and Mühlemann, H. R. (1980). *Caries Res.* **14**, 324.
9. Fitzgerald, D. B. and Fitzgerald, R. J. (1965). *Archs. Oral Biol.*, **11**, 139.
10. Frostell, G. (1965). In: *Nutrition and Caries Prevention*, Almquist & Wiskell, Stockholm, 60–6.
11. Frostell, G. (1971). *Dtsch. Zahnärztl. Z.*, **26**, 1181.
12. Frostell, G. and Birkhed, D. (1978). *Caries Res.*, **12**, 256.
13. Frostell, G., Keyes, P. H. and Larson, R. H. (1967). *J. Nutr.*, **93**, 65.

14. Gehring, F. (1971). *Dtsch. Zahnärztl. Z.*, **26**, 1162.
15. Gey, F. and Kinkel, H. T. (1978). *J. Dent. Res.*, **57** (Spec. Issue A), Abstract 147.
16. Green, R. M. (1981). In: *Proc. Symposium on Animal Models in Cariology*, J. M. Tanzer (Ed.), Information Retrieval, Washington, DC, 189–93.
17. Green, R. M., Leach, S. A. and Hartles, R. L. (1976). *J. Dent. Res.*, **55D**, 143.
18. Grenby, T. H. (1982). *J. Dent. Res.*, **61**, 557.
19. Grenby, T. H. and Colley, J. (1983). *Archs. Oral Biol.*, **28**, 745.
20. Grenby, T. H. and Saldanha, M. G. (1983). *Proc. Nutr. Soc.*, **42**, 78a.
21. Grunberg, E., Beskin, G. and Brin, M. (1973). *Int. J. Vitam. Nutr. Res.*, **43**, 227.
22. Havenaar, R. (1984). *Sugar Substitutes and Dental Caries. Animal Experiments and Microbiological Aspects.* Doctoral thesis, University of Utrecht, The Netherlands.
23. Havenaar, R. (1984). *J. Dent. Res.*, **63**, 120.
24. Havenaar, R., Huis in't Veld, J. H. J., Backer Dirks, O. and Stoppelaar, J. D. (1979). In: *Health and Sugar Substitutes, Proc. ERGOB Conf. Geneva*, B. Guggenheim (Ed.), Karger, Basel, 192–8.
25. Havenaar, R., Huis in 't Veld, J. H. J., de Stoppelaar, J. D. and Backer Dirks, O. (1983). *Caries Res.*, **17**, 340.
26. Havenaar, R., Drost, J. S., de Stoppelaar, J.D., Huis in 't Veld, J. H. J. and Backer Dirks, O. (1984). *Caries Res.*, **18**, 375.
27. Havenaar, R., Drost, J. S., Huis in't Veld, J. H. J., Backer Dirks, O. and de Stoppelaar, J. D. (1984). *Archs. Oral Biol.*, **29**, 993.
28. Havenaar, R., Huis in 't Veld, J. H. J., de Stoppelaar, J. D. and Backer Dirks, O. (1984). *Caries Res.*, **18**, 269.
29. Havenaar, R. and Gerats-Boom, A. (1984). *Caries Res.*, **18**, 536.
30. Hefti, A. (1980). *Caries Res.*, **14**, 136.
31. Huis in 't Veld, J. H. J., de Boer, J. S. and Havenaar, R. (1982). *J. Dent. Res.*, **61**, 1199.
32. Ikeda, T. (1982). *Int. Dent. J.*, **32**, 33.
33. Ikeda, T., Shiota, T., McGhee, J. R., Otake, S., Michalek, S. M., Ochiai, K., Hirasawa, M. and Sugimoto, K. (1978). *Infect. Immun.*, **19**, 477.
34. Kamp, E. M., Huis in 't Veld, J. H. J., Havenaar, R. and Backer Dirks, O. (1981). In: *Animal Models and Cariology*, J. M. Tanzer (Ed.), Information Retrieval, Washington, DC, 121–30.
35. Karle, E. J. (1977). *Dtsch. Zahnärztl. Z.*, **32** (Suppl. I), 89.
36. Karle, E. and Büttner, W. (1971). *Dtsch. Zahnärztl. Z.*, **26**, 1097.
37. Karle, E. and Gehring, F. (1975). *Dtsch. Zahnärztl. Z.*, **30**, 356.
38. Karle, E. J. and Gehring, F. (1978). *Dtsch. Zahnärztl. Z.*, **33**, 189.
39. Karle, E. J. and Gehring, F. (1979). *Dtsch. Zahnärztl. Z.*, **34**, 551.
40. Keyes, P. H. (1958). *J. Dent. Res.*, **37**, 1077.
41. Keyes, P. H. and Jordan, H. V. (1964). *Archs. Oral Biol.*, **9**, 377.
42. Kleiner, F. W. and Kinkel, H.-J. (1960). *Dtsch. Zahnärztl. Z.*, **15**, 664.
43. König, K. G., Marthaler, T. M. and Mühlemann, H. R. (1958). *D. Zahn-Mund-Kieferheilk.*, **29**, 99.
44. König, K. G., Savdir, S., Marthaler, T. M., Schmid, R. and Mühlemann, H. R. (1964). *Helv. Odont. Acta*, **8**, 82.
45. König, K. G., Schmid, P. and Schmid, R. (1968). *Archs. Oral Biol.*, **14**, 13.
46. Larje, O. and Larson, R. H. (1970). *Archs. Oral Biol.*, **15**, 805.

47. LARMAS, M., SCHEININ, A., GEHRING, F. and MÄKINEN, K. K. (1975). *Acta Odont. Scand.*, **33** (Suppl. 70), 321.
48. LEACH, S. A. and GREEN, R. M. (1980). *Caries Res.*, **14**, 16.
49. LEACH, S. A. and GREEN, R. M. (1983). *Caries Res.*, **17**, 157.
50. LEACH, S. A., GREEN, R. M. and APPLETON, J. (1981). *Caries Res.*, **15**, 201.
51. MÜHLEMANN, H. R. (1976). *Helv. Odont. Acta*, **86**, 1339.
52. MÜHLEMANN, H. R., REGOLATI, B. and MARTHALER, T. M. (1970). *Helv. Odont. Acta*, **14**, 48.
53. MÜHLEMANN, H. R., SCHMID, R., NOGUCHI, T., IMFELD, T. and HIRSCH, R. S. (1977). *Caries Res.*, **11**, 263.
54. MUKASA, T. (1977). *Nihon Univ. J. Oral Sci.*, **3**, 266.
55. MUNDORFF, S. A. and BIBBY, B. G. (1977). *J. Dent. Res.*, **56** (Spec. Issue B), Abstract 339.
56. NARKATES, A. J., NAVIA, J. M. and BATES, D. (1976). *J. Dent. Res.*, **55** (Spec. Issue B), B175.
57. NAVIA, J. M. (1977). *Animal Models in Dental Research*, Univ. Alabama Press, Birmingham, USA.
58. OOSHIMA, T., SOBUE, S., HAMADA, S. and KOTANI, S. (1981). *J. Dent. Res.*, **60**, 855.
59. OOSHIMA, T., IZUMITANI, A., SOBUE, S., OKAHASHI, N. and HAMADA, S. (1983). *Infect. Immun.*, **39**, 43.
60. OOSHIMA, T., IZUMITANI, A., SOBUE, S. and HAMADA, S. (1983). *Jap. J. Med. Sci. Biol.* **36**, 219.
61. SCHEININ, A. and MÄKINEN, K. K. (1975). *Acta Odont. Scand.*, **33** (Suppl. 70).
62. SHAW, J. H. (1976). *J. Dent. Res.*, **55**, 376.
63. SHAW, J. H. (1985). *Proc. Scientific Consensus Conf. on Methods for the Assessment of the Cariogenic Potential of Foods, San Antonio, USA.*
64. SHAW, J. H. and GRIFFITHS, D. (1960). *J. Dent. Res.*, **39**, 377.
65. SPENGLER, J. (1960). *Helv. Physiol. Acta*, **18**, 50.
66. STEPHAN, R. M. and HARRIS, M. R. (1955). In: *Advances in Experimental Caries Research*, R. F. Sognnaes (Ed.), American Assoc. for the Advancement of Science, Washington, DC, 47–65.
67. UFAW (1972). *The UFAW Handbook on the Care and Management of Laboratory Animals*, 4th edn, Churchill Livingstone, London.
68. VAN DER HOEVEN, J. S. (1979). *Caries Res.*, **13**, 301.
69. VAN DER HOEVEN, J. S. (1980). *Caries Res.*, **14**, 61.
70. VAN REEN, R. and COTTON, W. R. (1968). In: *Art and Science of Dental Caries Research*, R. S. Harris (Ed.), Academic Press, New York, 277–312.
71. VIRTANEN, K. K. and MÄKINEN, K. K. (1981). *Proc. Finn. Dent. Soc.*, **77**, 228.

Chapter 8

SWEETENERS IN SPECIAL FOODS FOR DIABETIC, SLIMMING AND MEDICAL PURPOSES

K. G. Jackson, J. Howells and J. Armstrong
The Boots Company PLC, Nottingham, UK

SUMMARY

Special diets are used to mitigate many human diseases. Where these diets require changes in carbohydrate content, then sweetness becomes an important characteristic. The first part of this chapter describes the materials which are used to modify carbohydrate content, and the legal and practical restrictions which are imposed on those materials. Illustrations are given of some typical product formulae, and there is a brief market survey of the sweeteners currently in use. The second part describes the sensory evaluation of dietetic foods, and examples are given from the development of table-top sweetener products. The basic principles outlined can be applied to all sensory aspects of food products.

'Things sweet to taste prove in digestion sour'[1]

'Not so, we cry, 'tis but a change in fashionable hour'[2]

1. INTRODUCTION

Not least among the complexities of human nutrition is the psychological aspect of consumer acceptability and patient compliance. To this, the physical aspects of the food add their own importance, with particular regard to taste and texture. Of the four mouth-detected sensations of taste, i.e. saltiness, acidity, sweetness and bitterness (the fifth characteristic of flavour, namely aroma, is another complex subject altogether), sweetness has a special place in that it is a typically 'desirable' characteristic adding to the 'niceness' of food, yet at the same time is affected by degrees of personal preference, often dictated by expectation, in so far as the intensity of sweetness is concerned.

Since the introduction of refined sugar to the Western diet in the early 1700s,[3] sweetness has had a special influence on nutritional intake. Not only was the sweetness itself an attractive characteristic which made sugar a desirable ingredient, but sugar also provided many useful attributes to food, including preservation at high concentration, and colour when slightly burnt (caramelised) during baking, allowing a visual check of cooking effectiveness.

A particular property of sugar is its contribution to calorie intake due to the relatively concentrated nature of refined sugar compared with its much lower natural concentration in fruits and vegetables (see Table 1).

The other specific influence of sugar in the diet is in metabolic disorders. In the majority of the population sucrose and glucose are metabolised easily and utilised for their energy content, but in 2% of the population[5] the absorption of glucose by the cell is interrupted.[6] This condition, diabetes, when attributable to lack of the enzyme insulin, results in the build-up of ketone-bodies following intake of glucose, and these have a toxic effect on various parts of the body.[6]

In addition, for otherwise 'normal' members of the human population, sweetness may be a significant factor in obesity or overweight. To avoid weight gain the energy content of food taken in must balance the energy expended by the body in utilising the food, the work done by and the heat lost from the body, plus the energy content of the unabsorbed food excreted. If the energy input exceeds output the body deposits (undesirable) adipose tissue, i.e. excess fat. Carbohydrate forms a substantial proportion of this energy input, and where this carbohydrate is sugar, the sweetness and appeal of the food may be significant factors.

In seeking to control these undesirable influences on well-being, there is a need to understand, control and modulate the intake of carbohydrate, and to do this moreover in the context of 'normal living' for the patient, rather than under the tightly controlled artificial circumstances of a scientific clinical study. Even then, of course, it is quite likely that the psychosomatic effect of treatment (placebo effect) plays a part in influencing the results. It is important therefore, to 'blind' the trial of a particular regime of calorie control by making the various foodstuffs

TABLE 1
CARBOHYDRATE (SUGAR) CONTENT OF REFINED FOODS[4]

Food	*Carbohydrate (g/100 g) (as monosaccharide)*	*Sugars (g/100 g)*
Sugar, white	105	100
Sugar, demerara	104·5	100
Syrup, golden	79	79
Honey	76·4	76·4
Jam	69	69
Sultanas, dried	64·7	64·7
Apples, eating	11·9	11·8
Apricots, fresh raw	6·7	6·7
Carrots, raw	5·4	5·4
Grapefruit, raw flesh	5·3	5·3
Peas, fresh raw	10·6	4·0
Flour, household plain	80·1	1·7
Potatoes, boiled	19·7	0·4
Cornflour	92	Trace

appear identical to the patient in texture and taste, allowing the investigator to make a more objective judgement of the results himself. The sweet taste of a food is of major significance in assessing its value and is not just a trivial means of achieving its acceptance.

2. DIABETIC DIET

In diabetes the intake of glucose, from whatever source, including complex carbohydrate, results in the build-up of toxins which can rapidly block normal energy provision pathways and lead to diabetic coma. Even when this acute distress is avoided by a limited supply of insulin, the more gradual deterioration of certain organs still represents an additional hazard.

Once insulin had been discovered as the key factor, then treatment of the condition was 'simply' a matter of balancing the carbohydrate intake against the administration of insulin. Unfortunately, the insulin must be administered parenterally, with all the attendant social disadvantages and microbiological infection hazards that injection entails. Pharmaceutical efforts then turned to the development of oral hypoglycaemic drugs which would achieve a similar effect to insulin but by a more convenient route of administration. After 30 years of use these have not had the universal value that had been hoped for.

Recent ideas on treatment therefore have turned to consideration of the management of diabetes by careful control of the diet itself,[7–9] stimulated particularly by the work of Jenkins *et al.*[10] showing that the influence of non-digestible dietary fibre on the absorption of glucose from carbohydrate-containing foods could have a useful bearing on the progression of the disease, and also recognising the effects of food sources low in fat and protein. Both of these, if consumed disproportionately, have their own disadvantages, and it has gradually come to be accepted that a diet at least 'normal' in carbohydrate may after all be suitable for the diabetic patient when used with proper care and understanding.[11]

However, glucose and sucrose are still contra-indicated, palatability is still significant, and sweetness of the food is still of importance.[12] In order to match the care-free quality of life of the non-diabetic, the diabetic patient, who otherwise has to consider every mouthful for its potential glucose content, needs to have some relief in the form of foods which have a nil, or very low, glucose-liberating potential, yet are just as acceptable as normal foods.

A number of materials are available for the preparation of foods with these characteristics. Because of the commercial nature of the food industry and the special technical knowledge needed to use these materials properly, they are in the main not available for use in the home of the patient or the kitchen of the hospital nutritionist, but are probably more widely used by the manufacturer of prepared foods.

2.1. Legal Restrictions on Claims

The manufacturer and retailer of special foods need to disclose by advertising to the public the particular nature of their product. Obviously such advertising should be strictly proper in the information disclosed and the persuasion exerted. The claim that a product is 'suitable for diabetics' is restricted by the Food Labelling Regulations,[13] which specify the following conditions.

(1) A given quantity of the food must not have a higher energy content than the same quantity of a similar food in relation to which no diabetic claim is made.

(2) A given quantity of the food must not have a higher fat content than the same quantity of a similar food in relation to which no diabetic claim is made.

(3) A given quantity of the food must not have a readily absorbable carbohydrate content greater than 50% of the readily absorbable

carbohydrate content of the same quantity of a similar food in relation to which no diabetic claim is made.

(4) The food must not contain a greater quantity of mono- or di-saccharides, other than fructose, than the quantity that is technically necessary to retain the essential characteristics of the food while having regard to its claimed suitability for diabetics.

The way in which such advertising can be used is also influenced by the advice of the Food Standards Committee in the *Second Report on Claims and Misleading Descriptions.*[14] A comprehensive survey of the legal acceptability of various sweeteners in different countries throughout the world is published by the Leatherhead Food Research Association,[15] but as new regulations are continually being issued local restrictions must always be checked.

2.2. Sweeteners

The implication of this legislation and advice is that while the formulation of diabetic foods has always been concerned to eliminate, or at least reduce considerably, the amount of glucose and sucrose, the materials which can be used instead are also restricted, in the UK by *The Sweeteners in Food Regulations*[16] and *The Soft Drinks Regulations.*[17]

Until 1983 saccharin and its salts were the only permitted artificial (high-intensity) sweeteners in the UK, and sorbitol was the only generally-used carbohydrate substitute, providing the bulk and cooking characteristics of sugar without the metabolic insulin demand.

Gradually over the last ten years, however, the characteristics of other high-intensity sweeteners have been investigated and submissions regarding their properties and safety made to the various regulatory authorities, culminating in the acceptance of saccharin, aspartame, acesulfame-K and Talin® for general use. The bulk sweeteners accepted at the same time were xylitol and hydrogenated glucose syrups.

Additionally, fructose has been further considered after use for several years in 'diet' foods in Continental Europe where the labelling suggested that the products were suitable for those requiring special diets, but left it to the consumer to judge the appropriateness of the use of fructose.

In the UK, medical opinion was not convinced of the general acceptability of fructose to diabetics, and there was for a time a proposal that a factor of 70% of the weight of fructose be used to calculate its carbohydrate equivalence. This would have been inconvenient for the patient and was of questionable scientific accuracy, so eventually it was

TABLE 2
SWEETENERS USED IN DIABETIC FOODS

Product	*Sweetener (g/100 g food)*			
	Sorbitol	*Fructose*	*Saccharin*	*Other*
Beverages				
Boots Orangeade, Lemonade, Cola			Yes	Aspartame
Roses Lemon Squash, Orange Squash			Yes	
Sionon Whole Orange	0·1		Yes	
Sionon Blackcurrant Health Drink	2·5		Yes	
Renpro Orange and Pineapple			Yes	
Renpro Blackcurrant Cordial			Yes	
Boots Chocolate Drink	Yes			
Baked goods				
Boots Tea Biscuits			Yes	Polydextrose
Boots Muesli Biscuits		13		
Boots Lincoln Biscuits	9·2		Yes	
Boots Bourbon Creams	19		Yes	
Boots Chocolate Digestives	19		Yes	
Boots Custard Creams	23		Yes	
Boots Ginger Creams	34		Yes	
Boots Sandwich Wafers, orange, lemon, chocolate		20		
Boots Sandwich Cake Mix, plain, chocolate	37		Yes	
Health & Diet Carob Chip Cereal Bar		2·9		
Health & Diet Coconut & Bran Cookie		3·9		
Health & Diet Ginger Nut Cookie		5·4		
Health & Diet Fruit & Nut Cereal Bar		13·3		
Sionon Sandwich Wafer	24		Yes	
Sionon Shortcake	18		Yes	
Rite Diet Fruit Cake	12·5			

Confectionery				
Boots Pastilles, fruit, blackcurrant	25		Yes	
Skels Fruit Pastilles	31·5		Yes	
Jackson's Pastilles, fruit, lemon–menthol	27·5		Yes	
Amurol Fruit Drops	90			
Boots Fruit Flavour Drops	98			
Vivil Mints, Lemon Sweets	99			
Sionon Peppermints	99			
Sionon Fruit Bon-bons	87			Mannitol 7·1
Boots Chocolate, plain, milk, hazelnut		27		
Boots Milk Chocolate Drops (earlier formula)	32			
Boots Milk Chocolate with yogurt, Coffee, Strawberry Creme		30–36		
Boots Praline Truffles		29		
Boots Cherry Truffles		48		
Boots Continental Chocolate Assortment	3·5	28·8		
Boots Luxury Chocolate Assortment	7	34		
Special Recipe Plain Chocolate		29		Polydextrose
Special Recipe Milk Chocolate		21·4		Polydextrose
Special Recipe Plain with Almonds		24·5		Polydextrose
Special Recipe Milk with Nuts		18·2		Polydextrose
Special Recipe Krispi		19·9		Polydextrose
Sionon Plain Chocolate	6	37		
Sionon Crispy Bars	6	28		
Holex Plain Chocolate		43		
Holex Milk Chocolate		30		
Sionon Caramel, Coconut Bar	5	24		
Preserves				
Boots Jams and Marmalade	Yes			
Boots Liqueur Jams and Marmalade	60–62			
Stute Jams and Marmalade	52–54			
Others				
Boots Tinned Fruits			Yes	
Boots Jelly Crystals	66	16	Yes	
Dietade Dessert Mixes			Yes	

agreed[13] that fructose could be regarded as 'not readily absorbable carbohydrate' provided advice was given to the patient that he should not consume more than a certain amount (25 g) per day. This was meant to ensure that with the inevitable patient-to-patient variation in the utilisation of energy sources, it was extremely unlikely that an unusually sensitive individual would be adversely affected. The recommendation of a daily maximum also applies to the other bulk sweeteners, sorbitol, xylitol and hydrogenated glucose syrup (Lycasin®).

Fructose-containing products are now much more widely available in the UK, and indeed for use in chocolate and sweet biscuits, fructose has largely replaced sorbitol as the sweetener. Table 2 gives the sweeteners used in a number of readily available diabetic foods.

The use of sweeteners in conventional food systems was reviewed by Lindley,[18] taking sucrose as the standard material. Special diabetic foods tend to be those which are normally high in sucrose and which therefore would be 'avoided' in this conventional form.

2.3. Diabetic Foods

2.3.1. Carbonated Beverages

These ready-to-drink products are formulated with high-intensity sweeteners such as saccharin and aspartame at low concentrations, and with very little energy-giving carbohydrate, so that they are 'low-calorie' as well.

2.3.2. Dilutable Drink Concentrates

The conventional product has a high (up to 66%) sugar content not only to provide sufficient sweetness after dilution, but also to provide adequate preservation from microbial spoilage during the repeated opening-and-shutting of the bulk container in use in the home, as well as a certain amount of 'mouth-feel'. The sweetness can of course be achieved by the use of high-intensity sweeteners. Saccharin and aspartame are again the two most widely used, saccharin because of its still-acceptable flavour when properly formulated, and its chemical stability, particularly under the relatively strongly acid conditions of the concentrate, so that the product retains its sweetness for an acceptably long shelf-life, and aspartame because of its marginally improved flavour and its connotations of being more 'natural' than the thoroughly artificial saccharin. The disadvantages of high cost and some chemical instability can be allowed for in formulation.

The requirement of preservation could be satisfied by the use of permitted agents such as benzoates and sulphites, but the current demand for natural foods without additives presents a challenge to the formulator that is not easily overcome. Pasteurisation during manufacture, small—even single-shot—cartons, short in-home shelf-life and chill-temperature storage after opening can all help to provide a safe product. Meanwhile the use of a certain amount of sorbitol (up to 20%) not only aids presentation but also assists in flavour and mouth-feel as the diabetic product seeks to avoid the 'thin' character of some low-calorie drinks.

2.3.2.1. Low-calorie beverage mixture.[19] The following properties of fructose can be utilised in the manufacture of beverage powder mixes.

(1) Increased sweetness compared with sucrose.
(2) Rapid solubility in cold water.
(3) Flavour accentuation of fruit, citrus and berry flavours.
(4) Flavour masking of bitter and metallic flavours.

Recipe for lemonade:

Crystalline fructose	88 kg 600 g	Riboflavin	2·5 g
Anhydrous citric acid	9 kg 200 g	Silicon dioxide	500 g
Ascorbic acid	700 g	Clouding agent	320 g
Carrageenan	160 g	Lemon flavour	530 g
		Sodium saccharin	290 g

Usage rate: 140 g per gallon of water

Process:

(1) Pre-mix all ingredients except fructose and silicon dioxide.
(2) Pre-mix fructose and silicon dioxide.
(3) Mix stages (1) and (2).

Manufacture and packaging should be conducted at a relative humidity below 60% and a temperature below 25°C.

2.3.3. Alcoholic Drinks

Dry wines and beers are, by definition, generally low in insulin-demanding carbohydrate since sugars have been fermented to alcohol. However, some malt fractions can be high in unfermentable carbohydrate, although beers claimed to be suitable for diabetics would be expected to be low in these components.

Home brewing of special beers for diabetics is not usually recommended because the variable nature of the fermentable must means that the supplier of the ingredients cannot guarantee the finished carbohydrate content. Sweet wines and cider, however, can often be prepared more satisfactorily in the home since saccharin can be added whereas it may not be permitted in commercially available drinks.

2.3.4. Baked Goods

Sorbitol as the non-carbohydrate 'filler', with saccharin to boost the sweetness, is giving way at the moment to fructose or a combination of polydextrose and saccharin. The acceptance of starch as a 'non-readily absorbable carbohydrate'[13] means that the use of about 20% fructose and 10% extra flour instead of about 30% sucrose, in an otherwise conventional recipe for a biscuit or cake, is now acceptable.

2.3.4.1. Reduced-calorie cake mix. Baking trials indicated that simply replacing all of the sugars in a cake recipe with crystalline fructose would not produce a reduced-calorie cake with acceptable texture, appearance and overall quality.[20]

In order to reduce calorie intake whilst maintaining acceptable flavour, mouth-feel, taste, appearance and volume, and to reduce salt and cholesterol and increase fibre, it was necessary to:

(1) reduce the shortening content from 50% to 6·25% of the flour weight;
(2) add 4% emulsifier;
(3) use fructose as the only sweetener, reducing the amount of sweetener used from 125% of the flour weight to 62·5%;
(4) reduce added salt from 4% to 0·5%;
(5) increase egg-white solids from 8% to 12% to increase batter aeration capacity;
(6) use 14% powdered cellulose, a non-nutritive bulking aid, to provide body and water-holding capacity;
(7) Use 0·2% xanthan gum to provide additional water-holding capacity.

Recipes:

	Reduced calorie white cake	*Reduced-calorie chocolate cake*
High-ratio bleached cake flour	47·3%	44·0%
Soybean oil	3·0%	2·8%
Crystalline fructose	28·5%	27·4%
Salt	0·25%	0·25%
Baking powder	3·8%	3·5%
Non-fat milk powder	1·9%	1·75%
Egg-white solids	5·7%	5·3%

Vanilla powder	0·014%	0·014%
Powdered cellulose	6·6%	6·2%
Xanthan gum	0·1%	0·1%
Emulsifier	1·9%	1·75%
Sodium bicarbonate	—	0·45%
Cocoa powder	—	6·6%

Process:
Mix all powdered ingredients in a ribbon blender for 20 min. Add a mixture of the soybean oil and emulsifier and blend for 10 min.

Usage:
(1) Blend 225 g of cake mix with 90 ml water. Mix for 5 min.
(2) Add 65 ml water and mix for 5 min.
(3) Pour into 8-inch cake tin. Bake for 22 min at 190°C.

2.3.5. Confectionery

The use of certain non-sucrose bulk sweeteners in confectionery technology has been reviewed by Dodson and Pepper.[21] They conclude that using the materials studied it is not possible to reproduce exactly existing confectionery products. Although some stable products can be made, they are different from (and possibly inferior to) existing products. Processing is more complicated and may require different manufacturing plant and increased boiling temperatures requiring more energy. The increased hygroscopicity of certain sweeteners requires improved packaging.

The typical transparent glass of the boiled sweet is a special property of the sugar–glucose mixtures used to make drop-formed or deposited sweets. Only sorbitol so far has been used significantly as an alternative for this purpose. Although Lycasin is also physically suitable it has only been introduced so far in the UK as a lollipop with a claim for reduced cariogenicity to teeth.[22]

The use of mannitol, sorbitol and hydrogenated glucose syrup in confectionery has been reviewed by Sicard and Leroy.[23] The use of fructose and isomalt in hard candies is discussed below.

2.3.5.1. Fructose. The conventional process for manufacturing sucrose-based hard candies involves dissolving sucrose and other carbohydrate syrups and then boiling to remove the water. This cannot be applied to fructose-based hard candies as fructose does not readily form a glass and, due to its hygroscopicity, tends to become sticky.

A recent patent application[24] disclosed a process which overcomes these problems. The process involves the addition of powdered fructose to a fructose melt. The mass produced is formed into hard candies by conventional methods. The candies may contain conventional additives (e.g. edible fats, emulsifiers, acids, colours, flavours) up to a total inclusion level of 3%.

An example formed by drop-rolling is shown below:

(1) Fructose 492 g. Heat to 110°C. Add, with stirring:
(2) Anhyd. citric acid 15 g. Cool to 105°C. Add:
(3) Powdered fructose 492 g
Lemon flavour 1 g
Colour 0·02 g
Stir and cool to 48°C. Cool and form into candies by drop-rolling.

An example formed by moulding is:

(1) Fructose 699 g. Heat to 120°C. Add with stirring:
(2) Peppermint flavour 1 g
Powdered fructose 300 g
Transfer to a pre-heated hopper (105°C) and deposit into moulds.

2.3.5.2. Isomalt. Isomalt[25] is an equimolar mixture of β-D-glucopyranosido-1,6-mannitol and β-D-glucopyranosido-1,6-sorbitol. Its properties are summarised below.

(1) Sweetness level 0·5–0·6 times that of sucrose (10% solution).[26] Synergistic effects with sugar alcohols (e.g. xylitol, sorbitol, hydrogenated glucose syrup) and with intense sweeteners (e.g. saccharin, aspartame).
(2) Heat of solution 39·5 kJ/kg.
(3) Non-cariogenic.
(4) Suitable for diabetic use.
(5) Caloric content claimed to be approximately 50% that of sucrose.

Recipe for moulded candies:

Isomalt	74·4%
Water	24·8%
Citric acid	0·7%
Flavour	0·1%
Colour solution	As required
Saccharin/aspartame	As required

Process:

(1) Boil isomalt and water in a suitable cooker at 142°C.
(2) Apply vacuum (730 Torr) for 3 min. Boil without vacuum at 160–165°C.

(3) Cool to 110–115°C on a cooling table.
(4) Add acid, colour solution and sweetener. Pour into Teflon-lined moulds.

2.3.6. Chocolate Confectionery

Conventional chocolate contains about 40% sucrose (plain) or sucrose–lactose mixture (milk). Until about 1982 this was substituted in 'diabetic' chocolate by about 30% sorbitol and 10% other chocolate ingredients (mostly cocoa butter). Although this reduced the absorbable carbohydrate it simultaneously increased the fat content, which was undesirable for other dietary reasons. The medico-legal acceptance of fructose[13,27] and polydextrose[28] meant the introduction of more attractive diabetic chocolates.

Similarly for filled chocolates a sorbitol-based fondant filling was not satisfactory and products relied on a fat-rich praline-type centre. Again, the use of fructose allows diatetically acceptable and attractive fillings to be developed.

Maltitol and lactitol have both been investigated experimentally for use in special chocolate. They have the triple advantage of being low in readily absorbable carbohydrate (like fructose), low in energy content (like polydextrose), and sweet in taste (unlike polydextrose), but they have not yet been used on a commercial scale, partly perhaps owing to high cost and partly because lactitol is not yet accepted as a permitted food additive in the UK.

The use of fructose in chocolate manufacture requires special treatment:[29]

(a) Fructose crystals pulverise less easily under shock or shear than crystals of sucrose. It is necessary to use more energy to obtain a particle-size range equal to that of icing sugar.
(b) Fructose crystals, and especially ground crystals, have a tendency to agglomerate accentuated by their hygroscopicity.
(c) It is necessary to avoid the presence of moisture in the ingredients and in the manufacturing plant and atmosphere.
(d) The particle size of fructose influences the specific surface area which in turn influences hygroscopicity. A particle size of approx. 200 μm is recommended.
(e) Temperature, especially during conching, is important in determining whether fructose will agglomerate. A conching temperature of approx. 40°C is recommended.
(f) Manufacturing should be carried out as quickly as possible to reduce the chance of moisture pick-up.

2.3.7. Preserves

Jams and marmalades typically contain not less than 66% sucrose in order to achieve satisfactory preservation from microbes at room temperature. Substitution of this by sorbitol, usually the 'non-crystallisable' variety (syrup), provides a very similar product.

The use of preservatives such as benzoates, plus single-shot containers and low-temperature storage after opening, has allowed the use of low-sugar 'jams' containing about 40% sucrose, the bulk being made up with extra fruit and water. These are more properly suitable for a low-energy diet rather than a diabetic diet since they still contain a considerable proportion of readily absorbable carbohydrate.

Honey is another carbohydrate-rich product that the diabetic would be denied, yet the development of a honey-flavoured sorbitol-based gel today gives an acceptable spread.

2.3.8. Canned Fruits

The growing popularity of fruits canned in water as opposed to heavy (66%) syrup has widened the availability of dessert fruits to both the diabetic and slimmer alike, although saccharin is still used in those types specifically for the diabetic in order to maintain a sweet flavour.

3. LOW-CALORIE DIETS

Obesity is a general term with medical implications such as raised blood pressure and heart disease, but there are many Tables published giving the expected weight according to varying height and frame-size, for interpretation both by the professional adviser and by the consumer anxious to retain a slim outline.

Nearly all the theories on weight control are based on the premise that energy input must be balanced against energy output and that where there is an imbalance with excess energy input, then fat is deposited with all its medically undesirable complications. Thus to reduce weight the input must be below the minimum requirement so that stored energy (fat) is utilised. The problem, of course, is in quantification, since energy output is not simply direct physical work by the body, but includes maintenance metabolism, and energy input is not that which is ingested, but that which is absorbed and utilised.[29a]

Nevertheless, energy consumption is a major factor, and carbohydrate, at approximately 3·75 kilocalories (16 kilo Joules) per gram is one of the three principal nutrients; the others being fat (9 kcal, 39 kJ per g) and protein (4 kcal, 17 kJ per g). For the majority of people, carbohydrate forms the largest proportion of the diet—it is recommended[30] that at least 60% of the energy requirement should come from carbohydrate for a healthily-balanced diet—so that it is popularly considered that a weight-reducing diet should limit excess carbohydrate.[31]

Dietary habits vary considerably in different societies, but there is a large element of hedonic pleasure associated with eating. Sweet foods are pleasant to eat and are said to encourage over-eating. Accordingly, there is a demand for sweet foods with a lower-than-usual caloric value.[32]

3.1. Beverages

Beverages generally form part of the diet most associated with 'casual' intake, and because the base is only water, lend themselves to being calorie-reduced by the substitution of sucrose by high-intensity sweeteners.

3.1.1. Ready-to-Drink Carbonates

Saccharin is still widely used in these products, but aspartame, under its brand name NutraSweet, is also enjoying a considerable vogue at the moment. Typical concentrations would be 0·1% w/v for saccharin and 0·2% w/v for aspartame. Sodium cyclamate, which was well-received for its flavour, is not used much due to legal restrictions in many countries prompted by uncertainty as to its safety. Talin also has not gained much in importance in this respect, perhaps because its flavour profile has not yet been adapted satisfactorily to products of this nature. Since these products together with the dilutable fruit drinks (below) are consumed particularly by children, formulators and legislators are much more concerned about safety hazards than with adults, so that there is more concern over the introduction and concentration of sugar substitutes.

3.1.2. Dilutable Fruit Drinks

Again saccharin and aspartame are the principal non-caloric sweeteners employed (see Table 3). Aspartame suffers from loss of sweetness on storage in acid products over periods of about six months. For relatively dilute carbonates at pH 4–5 and with a fairly rapid turnover in sales (and hence a required shelf-life of about six weeks) this is not significant, but for fruit concentrates with a pH below 4, when a much longer shelf-life is required, it is the practice at present to support aspartame with saccharin so that sweetness is maintained. Manufacturing costs, too, encourage the use of saccharin since the present price of aspartame is about 20 times that of saccharin on a sweetness-for-sweetness basis.

TABLE 3

SWEETENERS USED IN LOW-CALORIE FOODS

Product	*Sweetener*		
	Saccharin	*Aspartame*	*Other*
Drinks			
Diet Coke	Yes	Yes	
Diet Tango	Yes	Yes	
One-Cal (Appleade, Cola, Orangeade, Lemonade Blackcurrant, Lime & Lemonade)	Yes	Yes	
One-Cal Plus Crush (Orange, Tropical, Grapefruit, Pineapple)	Yes	Yes	
Schweppes Slimline Shandy	Yes	Yes	
7-up			Sugar
Diet Quosh (Orange, Lemon)	Yes	Yes	
Bitter Sweet (Orange, Lemon & Lime)	Yes		
PLJ (less sharp)	Yes		
Boots Low-Calorie Carbonated Drinks (Bitter Lemon, Lemon & Lime, Shandy, Orangeade, Lemonade, Cola)	Yes		
Boots Low-Calorie Health Drinks (Blackcurrant & Apple, Tropical Fruit, Cloudy Lemonade, Citrus Fruit)	Yes	Yes	
Boots Low-Calorie Drink Concentrates (Orange, Lemon, Lemon & Lime)	Yes		
Foods			
Slender Slimsoup (Chicken, Mushroom, Tomato)	Yes		Glucose syrup
Slender Slimchoc	Yes		Sugar, glucose syrup
Slender Diet Meals (Strawberry, Chocolate, Coffee, Raspberry & Vanilla)			Lactose, sugar, glucose syrup
Slimmer Hot Chocolate Drink			Sugar
Slimway Tablets	Yes		Sugar, dextrose
Ayds (Mint, Vanilla)			Sugar, glucose syrup
Bisks (Chocolate–Orange, Chocolate–Peppermint)			Sugar
Bisks Chocolate Nut Cookies			Sugar, invert syrup
Bisks (Muesli & Almond)			Sugar, honey
Limits Lunchpacks			Honey
Limits (Orange Creams, Vanilla Creams, Chocolate Mint Creams, Coffee Creams, Wheatmeal Biscuits)			Sugar, invert syrup

TABLE 4
SWEETENER INGREDIENTS USED IN TABLE-TOP SWEETENERS

Brand	*Sweetener*	*Amount (mg)*
Tablets		
Sweetex	Sodium saccharin	15
Hermesetas	Sodium saccharin	15
Natrena	Sodium saccharin	10
Shapers	Sodium saccharin	15
Boots	Saccharin	12·5
Saxin	Saccharin	
Canderel	Aspartame	18
Sweetex Plus	Acesulfame K	25
Hermesetas Gold	Acesulfame K	18
Granules		
Sweet 'n' Low	Saccharin, lactose	
Sweetex Granules	Saccharin, maltodextrin	
Sugar Lite	Saccharin, sucrose	
Sweetex Powder	Saccharin, sorbitol	
Sucron	Saccharin, sucrose	
Canderel Spoonful	Aspartame, maltodextrin	
Sprinkle Sweet	Saccharin, maltodextrin	
Liquid		
Sweetex	Saccharin	

3.2. Table-Top Sweeteners

Tea and coffee form a major group of adult beverages in Western society and it is common for them to be sweetened at the time of serving by the so-called table-top sweeteners. Sugar, still the principal sweetener, is generally used in granular or lump form and therefore is easily substituted by tablet or granular forms of high-intensity sweeteners,[33] since it is only the dispensing of the sweetener which needs to be controlled.

The sweetener ingredients of commercially available table-top sweeteners are listed in Table 4.

3.2.1. Tablets

3.2.1.1. Saccharin/sodium saccharin. Saccharin has been available in tablet form for many years.[34] Each tablet contains 12·5 mg of saccharin together with sodium bicarbonate and tartaric acid as disintegrants.

Due to its increased solubility, it is possible to present sodium saccharin in tablet form without the need to use disintegrants. The resulting tablets are small (typical weight 15 mg, diameter 3 mm) and therefore allow complex packaging (one-by-one dispensers holding a large number of tablets).

3.2.1.2. Acesulfame-K. Acesulfame-K became a permitted sweetener in the UK in 1983,[16] and is available in tablet form.

Due to its lower solubility and lower sweetening power (about half) compared with sodium saccharin, acesulfame-K tablets are larger and require disintegrants. Sodium carboxymethylcellulose has been used for this purpose. This has a 'wicking action' pulling water into the tablet, swelling and disrupting the tablet structure.

3.2.1.3. Aspartame. The use of aspartame has, like acesulfame-K, been permitted in the UK since 1983.[16]

As with acesulfame-K, aspartame's solubility and sweetening power are such that a disintegrant is required in tablet formulations.[35] Also, because of its poor flow and dissolution properties, lactose is used as an excipient. Such formulae cannot be labelled 'calorie-free'.

Leucine has been used as a water-soluble lubricant. It has been suggested[36] that aspartame is unstable in the presence of the much more commonly-used tablet lubricant magnesium stearate.

3.2.2. Powders

Sweetener powders may be presented either as relatively dense products (e.g. admixtures of an intense sweetener and sucrose such as Shapers Sugar Lite) or as low-density co-spray-dried mixtures of an intense sweetener and a carrier (as in Candarel Spoonful and Sweetex Granulated).

The most common forms of packaging for this type of product in the UK are cartons or glass jars. Individual sachet packs, providing the same sweetness as one teaspoonful of sugar, are commonly used in the catering trade but less often in the retail trade.

3.2.2.1. High-bulk-density sweetener powders. The forms of such preparations that are easiest to produce and most convenient to use are simple admixtures of an intense sweetener and a caloric carrier (e.g. sucrose, sorbitol). It has been suggested[37] that certain carriers, e.g. fructose, can overcome the saccharin aftertaste that is detected by some people. Due to the increased sweetness of such mixtures compared with sucrose, less needs to be used. A calorie saving is therefore achieved. By selecting appropriate ratios of intense sweetener to carrier, significant calorie reductions can be achieved while maintaining convenience (e.g. a sorbitol/sodium saccharin product can be formulated so that quarter of a teaspoonful provides the same sweetness as one teaspoonful of sugar, achieving a 75% calorie reduction).

Such mixtures typically contain low levels of intense sweetener so that dispersing the sweetener through the bulk of the carrier may be a problem. To ensure even distribution, the intense sweetener can be micronised prior to blending with a coarse carrier material. An ordered mix can be formed, the fine sweetener particles adhering to the coarse carrier particles. An alternative means has been described,[38] in which sodium saccharin solution is sprayed onto the crystalline sucrose, followed by drying. This results in a layer of sodium saccharin on the surface of each sucrose particle.

Cellulose has been used as a carrier in admixture with acesulfame-K. Although the resulting product is calorie-free, the insoluble nature of cellulose means that it is not suitable for use in hot drinks, etc.

3.2.2.2. Low-bulk-density sweetener powders. Low-bulk-density sweetener powders may be prepared by co-spray-drying solutions of an intense sweetener and maltodextrin. Hollow spheres are formed in the process. It is therefore possible to make a product with the same volume : sweetness ratio as table sugar, but which weighs considerably less. Typically, such products are one-tenth as dense as table sugar and therefore achieve a 90% reduction in energy content per unit volume.

Such products may be used by many diabetics provided that the carbohydrate content is taken into account.

3.2.3. Liquids

Concentrated liquid products (typically containing about 10% sweetener) that can be measured dropwise are more useful than tablets for many home-cooking recipes.

3.2.3.1. Sodium saccharin. Sodium saccharin can be formulated as a stable liquid preparation by dissolving in propylene glycol and glycerol. Such a preparation does not require preservatives. Sodium saccharin has also been formulated as an aqueous solution, but requires the use of a preservative. The following may be used:[39]

Preservative	*Maximum usage level (mg/kg)*
Benzoic acid	750
and either methyl 4-hydroxybenzoate	250
or ethyl 4-hydroxybenzoate	250
or propyl 4-hydroxybenzoate	250

3.2.3.2. Acesulfame-K and aspartame. Both of these sweeteners can be presented in solution in non-aqueous solvents.[40,41] The instability of aspartame in water suggests that an aqueous formulation would be unsatisfactory.

4. MEDICINAL PRODUCTS

Medicinal products have traditionally been formulated to include sugars, e.g. sucrose (syrups, lozenges) and lactose (tablets) to increase palatability and ensure patient compliance. For patients prescribed medical syrups to be taken during the day (and night), the presence of sucrose in the product poses a potential risk of dental caries.[42] The growth in popularity of chewable tablets (e.g. multi-vitamin preparations) means that this problem can also occur when such products contain sugars. These disadvantages can be overcome by formulating products using non-cariogenic sweeteners.

4.1. Tablets

Tablet manufacture involves the flow of powder into a die followed by compaction by punches at speeds of up to 1·5 million tablets per hour. It is obvious, therefore, that the flow properties of the powder are critical, and it must flow quickly into the die. The die must be filled consistently and evenly, otherwise the medicament dosage will vary. Many powder mixes have inadequate flow properties. To overcome this, a proportion of the formulation is often subjected to a granulation process, which improves flow properties. The granules are then blended with the remaining excipients prior to compaction. The granulation process can be carried out either wet or dry, shown schematically in Fig. 1.

The tableting process is simpler if the need to granulate can be avoided,

TABLE 5
HEAT OF SOLUTION (ENDOTHERMIC) OF BULK SWEETENERS[43,44]

Sweetener	*Heat of solution*	
	(kJ/mol)	*(kJ/kg)*
Sucrose	6·21	18·16
Sorbitol	20·2	110·99
Mannitol	22·0	120·88
Xylitol	23·27	153·07
Isomalt	14·6	39·4

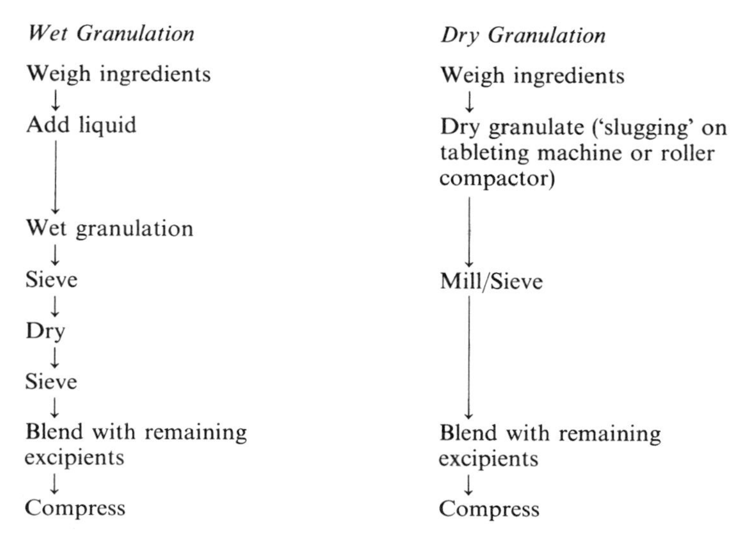

FIG. 1. Granulation processes.

but only certain excipients are suitable for direct compression (illustrated in Fig. 2).

Although directly-compressible excipients are available, the compaction properties of a mixture will depend on the properties of the individual ingredients.

The palatability of chewable tablets or lozenges can often be improved by the inclusion of ingredients which have negative heats of solution, which impart a cooling effect in the mouth. Some of these are listed in Table 5.

4.1.1. Sorbitol

Sorbitol is a commonly used tableting excipient suitable for direct compression. Its hygroscopic nature sometimes causes problems of poor

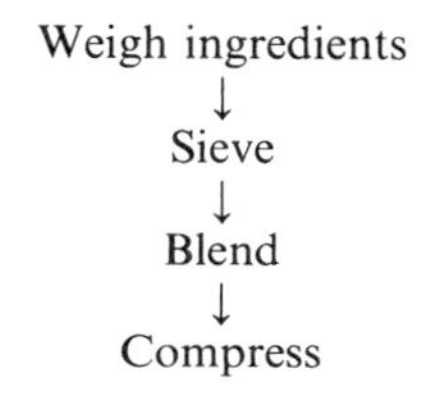

FIG. 2. Direct compression process.

TABLE 6
PROPERTIES OF SPECIAL GRADES OF SORBITOL

Property	*Sorbit Instant*	*Tableting grade*
Particle-size range (90% of distribution)	180–620 μm	160–1 200 μm
BET surface area	0·96 m^2/g	0·77 m^2/g
Microscopic examination	Agglomerated Rough surface	Regular particles Smooth surface
Bulk density	0·42 g/cm^3	0·69 g/cm^3
Tapped bulk density	0·50 g/cm^3	0·77 g/cm^3

flow. There have been attempts to produce grades of sorbitol with improved tableting characteristics.[45,46]

Schmidt[47] compared two types of directly compressible sorbitol (Table 6). One grade was produced by spray-drying (Sorbit Instant, E. Merck); the other, a commercial tableting grade, was prepared from a melt. Tablets were produced from a mixture of sorbitol (99·5%) and magnesium stearate (0·5%). They were tested for hardness, bending strength and friability and were also examined by scanning electron microscopy. The results indicated that Sorbit Instant produced harder, less friable tablets. At higher compression forces both grades of sorbitol showed a sintering effect.

The results indicated that Sorbit Instant produces tablets with high specific volumes (cm^3/g). The practical implications of this are that less Sorbit Instant (approx. 8–12% less) is required to produce tablets of a given volume.

The particle structure (agglomerated, large surface area) of Sorbit Instant may offer advantages in direct compression of formulations containing a small quantity of a highly active drug. Using a micronised grade of the drug, it might be possible to produce an ordered mixture with the fine drug particles lodging in the surface irregularities of the larger sorbitol particles.

4.1.1.1. Hard coating with sorbitol. It is often necessary to coat tablets for aesthetic reasons (to disguise surface mottling), for stability (to protect active substances from air, light, moisture) or simply for ease of identification. It is possible to coat sugar-free tablet cores with sorbitol. In addition to pharmaceutical applications, the technique has also been applied to confectionery products (Diabetic Mint Imperials).

The hard coating process involves covering the cores with a film of syrup which is then evaporated to leave a crystalline layer on the surface. Layers

TABLE 7
PROCESSING CONDITIONS FOR TABLET COATING

	Sucrose	*Sorbitol*
Concentration of syrup (%)	75–80	65–70
Temp. of core bed (°C)	55	30
Temp. of drying air (°C)	60–70	35–45

of the coat are built up by numerous additions of syrup, interspersed with hot-air drying, while the cores are tumbled in a circular coating pan.

Coating with sucrose is a well-established process in the pharmaceutical industry. Due to differences in the solubility of sucrose and sorbitol and the viscosity of their saturated solutions, it is necessary to use different processing conditions when coating with sorbitol.[48] These are summarised in Table 7.

A 300% solution of sorbitol is obtained at 35°C whereas a temperature of 65°C is required to produce a 300% solution of sucrose.[48] To promote quick and even coating of the cores the viscosity of the coating syrup should be low (e.g. 200 cps).

4.1.2. Fructose

Although fructose is not a non-cariogenic sweetener its flavour properties make it a good excipient for chewable tablets. Moreover, it is suitable for diabetics.

Fructose is the sweetest natural sugar, particularly in its crystalline form as fructopyranose (Table 8), much sweeter than other commonly-used tableting excipients.[49]

This comparatively high sweetening power is of benefit in obtaining the

TABLE 8
COMPARISON OF THE SWEETENING POWER OF COMMONLY-USED SWEETENERS

Fructose	173·7
Sucrose	100
Glucose	74·3
Sorbitol	60
Mannitol	40

maximum effect from a limited quantity of excipient when presenting bitter-tasting ingredients in tablet and lozenge form. Crystalline fructose has a cooling effect on the tongue and enhances fruit flavours. Osberger[50] reviewed the use of fructose as an excipient in vitamin C, vitamin E, multi-vitamin, dextromethorphan hydrobromide and plain fructose tablets.

Direct compression of mixtures of active ingredients, crystalline fructose, lubricants and binders was unsuccessful. Tablets with acceptable hardness and friability characteristics could be produced only at slow machine speeds. Fructose was therefore granulated with polyvinylpyrrolidone (PVP), a commonly-used binding agent, and the granules used to prepare various formulations. An example of a chewable multi-vitamin tablet is shown below.

	Ingredient	*(mg/tablet)*
1.	Folic acid	0·52
2.	Thiamine mononitrate Rocoat, $33\frac{1}{3}\%$	5·46
3.	Riboflavin Rocoat, $33\frac{1}{3}\%$	5·61
4.	Nicotinamide Rocoat, $33\frac{1}{3}\%$	65·50
5.	Pyridoxine hydrochloride	8·82
6.	Calcium pantothenate	13·6
7.	1% Biotin triturate	33
8.	Vitamin B_{12}, 0·1% SD	6·9
9.	Vitamin C, 90% granulation	43·3
10.	Sodium ascorbate Type AG (Roche)	30
11.	Vitamin E acetate, 50% SD (Roche)	66
12.	Syloid 74 (silicon dioxide)	4
13.	Microcel C	4
14.	Magnesium stearate	8
15.	Avicel PH102	40
16.	Flavours	1·5
17.	Flavours	6·5
18.	Flavours	6·5
19.	Fructose/PVP (above)	660·9
20.	Dry vitamin A acetate beadlets Type 500 (Roche)	17
21.	Dry vitamin D_2 beadlets Type 850 (Roche)	0·54
22.	FD & C Red No. 3 Lake (jet milled)	5

Whilst the use of these granules produces tablets of acceptable quality, the method is not so straightforward or economical as direct compression. Osberger therefore investigated the effect of blending fructose with tablet-grade sorbitol. Trials indicated that a 3:1 ratio of crystalline fructose to tablet-grade sorbitol provided an optimum combination of good flavour and direct compression capability.

An example of a direct-compression formula for a chewable vitamin C tablet (250 mg) is shown below.

	Ingredient	*(mg/tablet)*
1.	Sodium ascorbate AG (Roche)	177
2.	Syloid 74 (silicon dioxide)	4
3.	Vitamin C, 90% granulation (Roche)	117
4.	Crystalline sorbitol	153·5
5.	FD & C Yellow No. 6 (jet milled)	6
6.	Orange flavour	10
7.	Avicel PH102	20
8.	Crystalline fructose	460·5
9.	Magnesium stearate	5
10.	Stearic acid	47

Process:

(A) Mix 5, 6 and 7. Sieve 40-mesh.
(B) Mill 3, 4 and 8 (Fitzmill, No. 1 plate, medium speed, knives forward).
(C) Mill 1 (Fitzmill, No. 0 plate, medium speed, knives forward).
(D) Mix A and C. Add B. Mix.
(E) Mix 2, 9 and 10 with a portion of D sieved 40-mesh. Resieve 40-mesh. Add to the remainder of D. Mix.
(F) Compress with $\frac{1}{2}$-inch sugar coating punches.

Due to the hygroscopic nature of fructose, Osberger recommended that manufacture should be conducted in air-conditioned rooms.

In a recent study, Staniforth[51] examined ordered mixtures of a coarse fructose-based tablet excipient (Tabfine type F94M, Finnish Sugar Company) and fine-particle pyridoxine hydrochloride. Pyridoxine hydrochloride was incorporated at a level of 1%. Prior to mixing, the Tabfine type F94M was stored for 48 h at 0% RH/20°C or 55% RH/20°C. Final moisture contents were 0·24% (w/w) and 0·74% (w/w) respectively.

The Tabfine and pyridoxine hydrochloride were blended and the uniformity of the mixture assessed. Samples were then subjected to vibration, a treatment that might cause segregation. It was found that the Tabfine F94M containing 0·74% (w/w) moisture produced stable mixtures, whereas the material containing 0·24% (w/w) moisture did not. It was felt that the increased mix stability was due to the presence of water as a surface film on the fructose particles.

This study indicates the importance of environmental factors in direct-compression formulation.

4.2. Syrups

4.2.1. Hydrogenated Glucose Syrup/Sorbitol

The use of hydrogenated glucose syrup and sorbitol in place of sucrose in pharmaceutical syrups is increasing. This is particularly desirable for children's medicines. A number of points need to be considered.

(a) The preservative systems used in pharmaceutical syrups generally require a low pH. It is claimed that these conditions will not give rise to caramelised by-products from hydrogenated glucose syrup.

(b) The absence of reducing groups in these materials suggests that stability may be superior to that of more chemically active sugars, e.g. fructose and glucose.

(c) Commercial hydrogenated glucose syrup (Lycasin® 80/55) does not crystallise. It is claimed to have an anti-crystallising effect on other ingredients. This can lessen the 'cap-locking' that takes place with sucrose-based syrups.

(d) The viscosity of a syrup containing hydrogenated glucose syrup and sorbitol can be adjusted by varying the proportion of the two ingredients. For example, Table 9 shows the viscosity of syrups containing hydrogenated glucose syrup, sorbitol syrup and glycerol.

(e) High levels of dry substance can be achieved in syrups composed of hydrogenated glucose syrup and sorbitol. This has important implications in stability towards micro-organisms. It has been observed that a syrup of sorbitol solution and hydrogenated

TABLE 9

VISCOSITY OF SYRUPS CONTAINING HYDROGENATED GLUCOSE SYRUP, SORBITOL SYRUP AND GLYCEROL

Formulation			*Viscosity at 20°C (cps)*
Hydrogenated glucose syrup 70% (g)	*Sorbitol syrup 70% (g)*	*Glycerol (g)*	
200	300	25	280
190	310	25	256
185	315	25	244
175	325	25	231
165	335	25	220
160	340	25	214
140	360	25	196

glucose syrup in the ratio of 1·7:1 [dry substance 72% (w/v)] will meet the requirements of the BP 1980 Challenge Test[52] in the absence of preservatives. This self-preserving effect is thought to be mainly due to the osmotic pressure, but may also be a function of the relatively poor properties of sorbitol and hydrogenated glucose syrup as microbial substrates.

4.2.2. *Xylitol*

Xylitol occurs naturally in various plants, but at too low a concentration to allow economic industrial extraction. It is prepared commercially, however, from D-xylose, which is widely distributed as a polymer, xylan, a major component of hemicelluloses. The most attractive sources[26] are maize kernels, corn cobs, seed husks of Plantago species, seaweed (*Phodymenia palmata*), nut shells, cotton seed hulls, birch wood chips and bagasse.

Xylitol is relatively expensive compared with polyols such as sorbitol, but it is approximately as sweet as sucrose, whereas other sugar alcohols are less sweet.

An example of a xylitol-based vitamin syrup formulation is shown below (N.B. The heating, mixing and filling procedures should be protected from light and oxygen).

Ingredient	*Label claim*	*Overage (%)*	*Quantity (mg/5 ml)*
Thiamine hydrochloride	10 mg	75	17·5
Riboflavin 5-phosphate sodium	10 mg	10	13·98
Pyridoxine hydrochloride	10 mg	10	11
Cyanocobalamin	6 μg	50	0·009
Biotin	0·3 mg	25	0·375
Nicotinamide	40 mg	10	44
Ascorbic acid	100 mg	50	150
Panthenol	15 mg	30	19·5
Vitamin A palmitate (1·7 m-i.u./g) (Roche)	10 000 i.u.	100	11·76
Vitamin D_2 Type 500 CWS (Roche)	400 i.u.	25	1
dl-α-Tocopheryl acetate	30 i.u.	10	33
Xylitol			1 500
Emulsifier (Cremophor RH40)			200
Glycerin			1 500
Ethanol			405
Benzoic acid			10
Flavour, Colour			As required
Distilled water			To 5 ml

It will be noted that even though the vitamins are presented in a non-reducing base which might be more stable than a conventional sucrose syrup, some of them require the inclusion of an 'overage', above the level declared on the label (i.e. an excess needed to allow for the chemical degradation of the vitamins during manufacture and storage).

Stable preparations with smaller vitamin overages can be formulated as dry powders in different packaging, for example:

(1) A powder packed in individual sachets to be dissolved in water prior to use, in packaging materials offering good protection against moisture ingress (e.g. foil:polythene laminate). Such a presentation, whilst offering good stability characteristics, may be more expensive in packaging than a bottle of syrup.
(2) A powder packed into a glass bottle which the consumer makes up to a predetermined volume. The vitamin overage can be selected as appropriate for the required shelf-life.

5. SENSORY EVALUATION

With the introduction of alternative sweeteners to sugar, but with a slightly different 'sweet' taste, there was inevitably considerable commercial competition to claim 'improved sweetness', or 'just like sugar'. Accordingly the subtleties of different tastes become significant and the techniques of evaluating flavour to describe these nuances in a scientific way become ever more sophisticated. It is therefore necessary to devote considerable attention to evaluation procedures in order to achieve reliable conclusions.

In general, sensory evaluation is used in four main areas.

(1) Quality Assurance, to maintain product quality and identity over long periods.
(2) Quality Control, to maintain product quality and identity with changing production methods. This could include changes in manufacturing or packing methods, and substitution of, reduction in, or removal of, ingredients for many reasons, often for the purpose of cost reduction.
(3) Evaluation of competitors' products, to define differences, generate meaningful terminology and provide guidelines for product improvement.
(4) Product Development, to confirm changes are in the direction indicated by consumer testing, to support advertising claims, to

reduce risks in decision making, and to investigate ageing effects in storage and use, and establish shelf-lives.

Sensory evaluation is frequently used to help provide answers to the following questions:

—Is there a difference present? If so,
—How large is the difference?
—What is the nature of the difference?
—What effect does the difference have on preference?
—Is it detected by all or only some consumers?

5.1. The Development of Table-Top Sweetener Products

Sensory evaluation is required to help resolve a number of issues in table-top sweetener development. Each requires its own type of panel evaluation. Methods must be adapted, or specifically designed, to provide the information required.

5.1.1. Sweetness Level to be Chosen?

Tablet sweeteners are usually described as equivalent in sweetness to a 'teaspoonful of sugar'. This is not a rigid measure—justification can be made for any weight in the range 4–8 g. Until *The Sweeteners in Food Regulation* (1983)[16] allowed the use of alternative intense sweeteners in the UK, the content of sweetener tablets was required to be at a saccharin level equivalent to approximately 6 g of sugar. Tablet sweetness levels are now subject to commercial pressures and consumer preference. Relatively high raw-material costs and lower potencies have forced manufacturers to move to lower sweetness levels in tablets using the new sweeteners, though equivalence to a 'teaspoonful of sugar' is still claimed. The manufacturers' dilemma is that too low a level may result in consumer dissatisfaction, whereas too high a level may give a product that fails because it is over-priced or does not generate enough profit to justify development, production and marketing costs.

Granular sweeteners combining an intense sweetener and a caloric base offer a more natural, food-like presentation than tablets. The versatility and ease of use, however, are offset by a higher-energy/more readily-absorbable carbohydrate content than tablets. The sweetness level chosen, a compromise between these factors, is usually twice or four times that of sugar.

Consumer research can be used to identify the sweetness level/format preferred by the target market. Sweetness matching and ranking evaluations can be used to determine the sweetness level chosen by competitors.

5.1.2. Which Sweetener or Combination of Sweeteners is to be Used?
Sweet compounds often do not provide an identical sweetness to that of sugar. The sweetness can have additional qualities such as bitterness or a licorice note, and the rate at which the sweetness is perceived increases and then decreases (the intensity–time profile) may be considerably different from that of sugar. In developing a sweet-tasting chemical into an accepted marketable sweetener, describing and quantifying the sweetness–quality–time profile is a vital task.[53–60]

Sensory profiling[61–63] is the panel evaluation method appropriate to this. It is also pertinent to a manufacturer incorporating a sweetener or sweetener combination into a food product. It is, however, a very expensive method[61–63] in terms of both staff and time, and may only be affordable for a major brand or by a manufacturer with a major interest in a restricted product area.

For many purposes difference testing will be sufficient to decide whether or not two sweetener products (or foods containing different sweetener systems) are perceived as different from each other. Descriptive or ranking methods can then be used to quantify and describe a difference which has been revealed. Preference testing can then determine whether or not the difference is desirable, though to be valid information on preference *must* be sought from a panel representative of the target market.

Many other factors also influence the choice of sweetener apart from that of the most acceptable taste:

—The nutritive or dietetic purpose of the product.
—The relative cost of the sweetener(s).
—The public or target market image of the sweetener(s) in terms of safety and taste quality.
—The effect of packaging, branding, advertising and retail price upon the consumer.

5.1.3. How can the Desired Sweetness Level be Achieved?
The sweetness level or potency of a sweetener compound relative to that of sugar varies with application, concentration and temperature. When two or more sweeteners are mixed the resulting sweetness level is not always what would be expected from the known sweetness levels of the constituents. Synergism is not always observed in mixtures of sweeteners but a significant positive effect can allow a reduction in total sweetener content.

Variation in sweetness level with concentration[53] of most sweeteners has been extensively studied by both the companies manufacturing the sweeteners and in food-industry/academic Research Institutes. Most information is available for simple model food systems. Paired comparison, or other difference testing against a range of sucrose standards, and threshold determinations are the usual methods of evaluation.

Magnitude estimation can be used to establish psychophysical connections between sweetener content and perceived sweetness level in food systems.[64] The method can be very time-consuming, however, and is more generally used in research than in the food industry. A more direct approach would be to ask panellists to match a known concentration of the sweetener(s) to a series of sugar-containing standard samples.

Synergism between sweeteners[53] has been less well studied than the variation of sweetness level with concentration, but the above methods can also be applied to its investigation.

5.1.4. Do Excipients Alter the Taste of the Product?

To check whether or not tableting or other processing affects the flavour of sweetener systems, difference testing comparing the product and the sweetener system should be undertaken. If a difference is shown using the standard methods, preference testing will then show whether or not the excipients have an adverse effect on taste.

It is often claimed that flavour modifiers can be used to improve the taste quality of sweeteners, especially that of saccharin. Difference testing of the modified and plain sweetener, followed by preference testing if a difference is found, will give information on the effectiveness of the taste modifier. Sensory profiling methods can be used to define the nature of the difference. Flavour enhancers can be investigated as a way to reduce the content of an expensive sweetener in a food system. Difference testing would be used to compare the enhanced lower level and standard level.

5.1.5. Is the Product Stable in Storage and Use?

Accelerated ageing trials are used to establish product shelf-lives and give advanced warning of problems that might arise in storage and use. Difference testing can be used to compare both samples stored under different conditions and stored samples with freshly prepared controls. Changes in packaging may also require storage/usage trials.

5.1.6. How does the Product Perform Relative to Competitors' Products?

Difference testing followed by preference testing should be used to compare two products. A preference ranking method can be used to compare several products. Descriptive or profiling methods can be used to gain more information about the nature of the difference and to help identify the reasons for the preference.

Laboratory panel information is normally obtained under 'blind' conditions. Consumer research, however, often includes aspects of packaging, ease of use, a comparison with product concepts, branding and likelihood of purchase, in addition to product taste. It is vital that the panellists or consumers accurately represent the target market if the results are to be valid.

5.2. General Arrangements for Panel Evaluations

The optimum conditions for sensory evaluation are widely discussed in the literature.[53,64–70] Particular care should be taken with sample preparation, presentation and coding to avoid bias. Order effects are known to operate in many standard evaluation methods, so sample order rotation should be practised.

For 'analytical' panel evaluation methods, panellists should undergo both a screening and a training programme.[53,64–66,71,72] The quality and extent of the training is mainly dependent upon the type, quality and quantity of information required, its commercial importance, and the resources available. Jellinek gives extensive details in all aspects of panellist selection and training.[64]

5.2.1. A Selection and Training Programme for a Panel to Evaluate Table-Top Sweetener Products

What follows is an example of such a programme.

All members of staff were sent a questionnaire requesting cooperation in a screening/training programme and the formation of a panel to evaluate table-top sweetener products. Staff were also asked about their usage of sugar and table-top products. A 72% return rate was achieved, but unfortunately a large percentage of staff proved to be non-users of any sweetener in beverages (45% of coffee drinkers and 66% of tea drinkers did not sweeten their beverages). It was immediately apparent that preference testing of sweeteners was going to be very difficult indeed.

Two screening tests, both based on identification of the basic taste sensations, were undertaken, the details of which are listed below. Each screening test was run on two separate occasions in order to maximise the number of staff able to take the tests. In a department where approximately 180 staff were 'nominally' available, 91 panellists took screen 1, 61 of whom passed and 60 people took screen 2, 51 of whom passed. This illustrates the law of 'diminishing returns' when trying to set up a sensory evaluation panel.

Screen 1:
Panellists were presented with seven samples, numbered 1–7, and asked to identify the basic taste in each by circling options on a questionnaire. The samples consisted of three blank tap water samples and one each of the following solutions:

SWEET	Sucrose, 2·0% w/v
BITTER	Quinine sulphate, 0·0015% w/v
SALT	Sodium chloride, 0·2% w/v
SOUR	Citric acid, 0·1% w/v

The order of presentation was:

SWEET, BLANK, SOUR, BLANK, BITTER, BLANK, SALT

Screen 2:
Panellists were presented with seven samples, numbered 1–7, and asked to identify the basic taste in each (no prompts were given on the questionnaire).

The sample concentrations were as above except:

SWEET	Sucrose, 2·5% w/v
SOUR	Citric acid, 0·08% w/v

The orders on the two occasions this panel was used were:

SALT, BLANK, SWEET, BITTER, BLANK, SOUR, BLANK
SALT, BLANK, SWEET, SOUR, BLANK, BITTER, BLANK

Training exercises were required to give panellists experience in ranking, matching, detecting and describing differences in sweetness, all routine tasks in sweetener product development. Practical considerations limited the choice of exercises to a programme of three. It would have been desirable to have extended the training period presenting panels with relatively simple tasks at first, then increasing the difficulty slowly as skill and confidence increased. Preliminary experiments amongst small numbers of staff were used to refine testing methods and assess the practicality and level of difficulty of the tasks.

The final choice of exercises was:

(1) Ranking coffee samples sweetened to a level of 4 g, 5 g and 6 g of sucrose per 200 ml 'standard cup', followed by a matching experiment of a 6 g per cup standard to the samples already ranked.
(2) A triangle test to distinguish between coffee sweetened with sucrose or sodium saccharin at matched sweetness levels.
(3) A triangle test to distinguish between tap-water solutions of 3% sucrose with either 0·03% or 0·06% caffeine (i.e. two supra-threshold levels of bitterness against a sweet background).

Panellists performed well in exercise (1), 86·4% correctly ranking, 63·6% correctly matching and 59·1% correctly ranking and matching the samples. That fewer people matched correctly than ranked correctly may be an indication of failure of the panellists' concentration during the second task.

A very significant result was achieved in triangle-test exercise (2); 61% of the 36 panellists taking the test correctly identified the odd sample, significant at the 99·9% confidence level ($P = 0{\cdot}001$). Although the panel as a whole could discriminate between the two samples they were not consistent in describing the nature of the difference.

A very significant result was achieved in triangle-test exercise (3); 56·8% of the 37 panellists taking the test correctly identified the odd sample, significant at the 99% confidence level ($P = 0{\cdot}01$). Panellists were also accurate and consistent in describing the nature of the difference present.

Of the 51 panellists taking all three exercises, only seven passed all of them, while 33 panellists passed at least two of the four aspects of the three training exercises. If the training exercises were used as screening exercises the panel would cease to exist! The usage of sweeteners by the 51 accepted panellists was such that it would also be impossible to set up evaluations at the preferred usage rate of the panellists, so a compromise level of one teaspoonful of sugar per cup of coffee/tea was chosen. Specialist sub-group panels, such as 'lapsed artificial sweetener users' were not feasible because of lack of numbers.[72]

5.2.2. Maintaining Panel Performance

Panellists are believed to perform better if given encouragement and feedback both on their personal performance and the value of their efforts in the product's development. Whether or not their performance improves, encouragement and feedback certainly help to keep up their morale, and more important, to keep up the number of panellists prepared to make themselves available. They should not be given information that might bias their future performance. Panellists like to find differences, however, and a series of very difficult triangle tests in which they continually fail to identify the odd sample is very bad for morale. Interspersing different testing methods and the occasional easier or standard panels in a long series of difficult evaluations may ease the situation (and confirm that the difference investigated is very small if panellists still succeed in discriminating the standard differences). Panellists' performance during a series of tasks ought to be monitored so that poor performers can be encouraged, given special help or eventually excluded from panels.[53,64–67] Sequential difference testing can be used as a screening method for panellists, checking on the consistency of their judgements.[53,64,66]

5.3. Evaluation Methods

Evaluations are of two main types:

(1) *Analytical*, where panellists are acting as objective analytical instruments, and where selection and training is required to produce consistent results;
(2) *Preference*, where panellists' subjective opinions are sought, where training and selection are a hindrance and large numbers of consumers representative of the target market are required.

Analytical methods include:

—Difference testing,
—Ranking and matching,
—Rating and scoring,
—Grading,
—Descriptive methods,
—Sensory profiling.

5.3.1. Difference Testing Methods

The triangle test is the most widely used difference test method in the food industry.[53,64–67,73,74] In the simple triangle test, panellists are required to determine which of three coded samples (two identical and one different) is the odd one. In the extended form panellists may be required to comment on the nature and magnitude of the difference or to express a preference between samples. Positional bias and lack of panellist attention may affect the secondary tasks, however.

The triangle-test method can be adapted into a sequential method, where panel sizes are not fixed, but testing continues until a decision can be reached.[66,74] This can be further adapted for use as a method of panellist screening.[53,64]

Other difference methods include the Duo/Trio test,[53,64–66] the Two-out-of-Five test[53,65,66] and the A/Not-A test.[53,66] Matching methods can be used to measure the degree of distinctiveness of several samples, providing information on the magnitude of difference between them.[56,65,66]

Paired comparison tests are used to detect a difference between two samples in terms of a directional pre-defined characteristic. The test is widely used in market research.[53,64–67,73,74]

5.3.1.1. Illustration of a difference test. Figure 3(a) shows an extended triangle-test questionnaire for a coffee evaluation comparing sodium-saccharin-sweetened coffee with sugar-sweetened coffee. Figure 3(b) shows the code permutations for the evaluation and Fig. 3(c) shows the completed master sheet, used as a training exercise for a sweetener panel.

Triangle Test *Questionnaire No. 7*

You are asked to taste three samples of coffee, labelled:

64 81 92

Please taste them in the order above and answer the following questions.

A. Please indicate which two samples are the same.
B. Please indicate which sample is the odd one.
C. Try to describe the way in which the odd sample is different.

____

____

____

D. Is the difference between the paired sample and the different one:

Slight		Moderate		Large	

If you can detect no flavour difference between the samples please guess which sample is the different one (by placing its code in the box below).

[ ]

Name .... Date ....

FIG. 3(a). Questionnaire for triangle test.

Combinations	*Permutations*	*Q. No.*	*Combinations*	*Permutations*	*Q. No.*
1. C T T	56 81 92	1. 25	1. C T T	56 92 81	13. 37, etc.
2. T C T	92 64 81	2. 26	2. T C T	81 64 92	14.
3. T T C	81 92 56	3. 27	3. T T C	92 81 56	15.
4. C C T	56 58 81	4. 28	4. C C T	64 56 81	16.
5. C T C	64 92 56	5. 29	5. C T C	56 92 64	17.
6. T C C	81 56 64	6. 30	6. T C C	81 64 56	18.
1. C T T	64 81 92	7. 31	1. C T T	64 92 81	19.
2. T C T	92 56 81	8. 32	2. T C T	81 56 92	20.
3. T T C	81 92 64	9. 33	3. T T C	92 81 64	21.
4. C C T	56 64 92	10. 34	4. C C T	64 56 92	22.
5. C T C	64 81 56	11. 35	5. C T C	56 81 64	23.
6. T C C	92 56 64	12. 36	6. T C C	92 64 56	24.

C1 = 56, T1 = 81
C2 = 64, T2 = 92

FIG. 3(b). Triangle-test sample permutations.

Date 5.2.84 Evaluation Reference ____________
Control Codes 56, 64
Test Codes 81, 92
Control Identity Sodium saccharin, 15 mg per 200 ml cup coffee
Test Identity Sugar, 6 g per 200 ml cup coffee

No.	*Combination*	*Name*	*Correct (C) incorrect (I) no difference (ND)*	*Comments—odd sample is (difference is)*
1.	C T T		I	—
2.	T C T		I	—
3.	T T C		C	Slightly sweeter
4.	C C T		I	—
5.	C T C		C	Sweeter
6.	T C C		C	Much sweeter
7.	C T T		I	—
8.	T C T		I	—
9.	T T C		C	Slightly bitter
10.	C C T		C	Moderately sweeter
11.	C T C		C	Moderately harsher and sweeter
12.	T C C		I	—

(continued)

FIG. 3(c). Triangle-test master sheet.

13.	C T T	C	Moderately more bitter
14.	T C T	C	Slightly less sweet
15.	T T C	C	Moderately less sweet
16.	C C T	C	Different profile
17.	C T C	C	More acceptable
18.	T C C	C	(Slight)
19.	C T T	ND	—
20.	T C T	C	Slightly less sweet
21.	T T C	ND	—
22.	C C T	C	Moderately more bitter
23.	C T C	C	(Moderate)
24.	T C C	C	Moderately less sweet
25.	C T T	C	Moderately more bitter
26.	T C T	I	—
27.	T T C	I	—
28.	C C T	I	—
29.	C T C	I	—
30.	T C C	C	More bitter
31.	C T T	C	(Slight)
32.	T C T	C	Slightly sweeter
33.	T T C	C	Moderately sweeter
34.	C C T	ND	—
35.	C T C	C	Tastes sweeter, less bitter
36.	T C C	I	—

Correct selections	22	Correct $+\frac{1}{3}$ (No difference)	23
Incorrect selections	11		
No difference	3	Thresholds for significance:	
Total panellists	36	95% 18	
		99% 20	
		99·9% 22	
		Significance level obtained	99·9%

FIG. 3(c)—*contd.*

5.3.2. Ranking and Matching Methods

In ranking tests panellists are asked to order samples according to the intensity of some specified attribute.[53,64–67] Ranking is less informative than rating or scoring methods. If the basis of a scale is undefined or poorly understood, however, it is wiser to apply a ranking analysis to the data than simply to work out the arithmetic means.

5.3.2.1. Illustration of a ranking test. To demonstrate the ranking method a small panel evaluation comparing the sweetness levels of commercial sweetener tablets in water (1 tablet per 200 ml cup) is detailed. The sweetener tablets used were:

Code	*Tablet*
J	Sweetex Tablets, Crookes
R	Natrena Tablets, Bayer
P	Canderel, G. D. Searle
G	Experimental formulation containing 30 mg of acesulfame-K

Panellists were given samples in a random order and asked to rank them in order of decreasing sweetness. The results are shown in Fig. 4.

This indicates that sample J is sweeter and sample R less sweet than the average sweetness level of the four samples at a significance level of 95% ($P = 0{\cdot}05$), i.e.

Decreasing sweetness order

J > P = G > R

Not significant

95% significance 95% significance

5.3.2.2. Matching evaluations. In a direct approach to evaluating the intensity of a stimulus, unknown sample(s) are matched within a range of known controls. From the consensus of the matched samples the unknown sample's intensity is deduced. This method can be used for determining the sweetness equivalent in terms of grams of sugar of a table-top sweetener product. A sample questionnaire is shown in Fig. 5. Matching experiments can be used to investigate the variation of sweetness potency with concentration of various sweeteners.

5.3.3. Rating and Scoring Methods

Rating is similar to ranking but a scale is used, the points on which are clearly defined and understood by the panellist. A scale can be unipolar or bipolar, structured or unstructured, and anchored in one or several places using standards or qualifying adjectives.[53,64–67]

It is often assumed that category scales are linear so that categories can

Panellist	*Order*	*J*	*R*	*P*	*G*
1	J > G > P > R	4	1	2	3
2	J > P > R > G	4	2	3	1
3	J > G > P > R	4	1	2	3
4	J > G > P > R	4	1	2	3
5	G > J > P > R	3	1	2	4
6	J = G > P > R	$3\frac{1}{2}$	1	2	$3\frac{1}{2}$
7	P > J > R > G	3	2	4	1
8	P > R = J > G	$2\frac{1}{2}$	$2\frac{1}{2}$	4	1
9	R > J > P > G	3	4	2	1
10	P > G > J > R	2	1	4	3
11	J > G > R > P	4	2	1	3
	Sum	37	$18\frac{1}{2}$	28	$26\frac{1}{2}$

Rank significance sums (any sample, 11 tastings, four samples)
5% 19 36
1% 17 38

FIG. 4(a). Samples ranked in order of decreasing sweetness.

Panellist	*Order*	*G*	*P*
1	J > G > P > R	2	1
2	J > P > R > G	1	2
3	J > G > P > R	2	1
4	J > G > P > R	2	1
5	G > J > P > R	2	1
6	J = G > P > R	2	1
7	P > J > R > G	1	2
8	P > R = J > G	1	2
9	R > J > P > G	1	2
10	P > G > J > R	1	2
11	J > G > R > P	2	1
	Rank sums	17	16

Rank significance sums (any sample, 11 tastings, two samples)
5% 13 20
1% 12 21

FIG. 4(b). Re-ranking the samples P and G in order of decreasing sweetness.

SWEETNESS MATCHING

You are presented with four standard samples of known sweetness level labelled F, G, H and I, sweetness level increasing in alphabetical order. You are presented with a test sample, labelled ____________

Please match the sweetness level of the test sample as closely as possible within the range of the standard samples.

Test sample is closest in sweetness to:

Less than F	F	Closer to F	F–G MIDWAY	Closer to G	G	Closer to G	G–H MIDWAY	Closer to H	H	Closer to H	H–I MIDWAY	Closer to I	I	More than I

Comments: ....

....

Name .... Thank you

FIG. 5. Sweetness matching questionnaire.

be assigned scores, and arithmetic means can be worked out. These assumptions can only be checked by magnitude estimation techniques.[75]

Scoring differs from rating in that numbers are assigned to the scale positions and the intervals form a linear or ratio scale.[53,66] Scoring methods can be used for evaluating intensity, or for preference.

5.3.3.1. Illustration of rating test. To demonstrate the use of rating scales a granulated sweetener evaluation is given below.

The granular sweetener products were:

Code	*Product*	*Sweetness level (relative to sucrose)*
J	Granulated sugar	1 ×
Q	Caster sugar	1 ×
G	Sugar Lite	4 ×
W	Sugaree	10 ×

Panellists were presented with four bowls of cornflakes, milk and separate bowls of granular sweeteners with sweetness levels marked on them. They were asked to sweeten the cornflakes to the same level, taking account of the products' sweetness level (measuring spoons were provided). Panellists commented on appearance, taste, and overall preference, using the following scales.

Appearance (bipolar, structured, five-point):

Very poor–Poor–Adequate–Good–Very good

Taste and overall preference (bipolar, structured, 11-point, three anchors):

0 = Extremely unpleasant
5 = Acceptable
10 = Extremely pleasant

Figure 6 shows the results.

This evaluation was rather naive; the panellists were laboratory personnel and probably deduced that samples J and Q were sugar, since spoon-for-spoon sweeteners had not at that time been widely marketed. They also probably reasoned that the 10 × product contained more saccharin than the 4 × product, and would therefore taste worse! What is interesting in this evaluation is the information on appearance. Granulated

Product	*Appearance*					*Total*
	V. Poor	*Poor*	*Adequate*	*Good*	*V. Good*	
J		1	1	3	14	19
Q		1	5	6	7	19
G		2	10	6	1	19
W		4	5	8	1	19

Product	*Taste*											*Total*
	0	*1*	*2*	*3*	*4*	*5*	*6*	*7*	*8*	*9*	*10*	
J	1					2		1	4	6	5	19
Q						1		2	8	5	3	19
G						2	4	2	8	1	2	19
W			1	2	6	1	2	5	2			19

Product	*Overall performance*											*Total*
	0	*1*	*2*	*3*	*4*	*5*	*6*	*7*	*8*	*9*	*10*	
J					1	1	1		1	12	3	19
Q		1				1	2	1	6	5	2	19
G			1		1	3	1	7	2	2	2	19
W		1	3	4	2	2	4	1	1			19

FIG. 6. Rating test results.

sugar was rated higher than caster sugar, and the comments of panellists indicated that sample Q was thought of as 'saltlike'/'like washing powder'. Sample G has a base particle size similar to caster sugar, whilst sample W has a base particle size similar to granulated sugar, and this also affected the appearance ratings of G and W. In addition to illustrating the use of rating scales, this evaluation indicates some of the pitfalls in running sensory evaluation panels. In further granulated sweetener evaluations panellists were asked to rate ease of use on a five-point scale like that used for appearance. This was introduced into the evaluation following panellists' comments during the early tests.

5.3.4. Grading

Grading methods also imply a fitness for purpose of the sample and involve the evaluation and judgement on the relative importance of multiple attributes.[64,66]

5.3.5. Descriptive Methods

These ask panellists to describe the order and nature of the taste sensations associated with a product. Results are collated by the panel leader or reached in a controlled discussion chaired by the panel leader. For a sweetener solution a panellist may be asked to describe the taste of the product in terms of:

(1) Immediate onset of sweetness/delayed onset of sweetness.
(2) Sweetness decreases smoothly/sweetness increases after initial intensity, peaks, then 'decreases'.
(3) Sweetness over quickly/sweetness lingers for considerable time.
(4) Sweetness clean/additional flavours present.
(5) Aftertastes present (their nature and duration).
(6) 'Mouth-feel' effects such as cooling or dryness.

This type of evaluation is very useful for establishing the quality of new sweeteners where limited quantities of materials may be available. For a sweetener compound under development and not covered by legislative approval great care must be taken in ensuring that adequate toxicological information exists and that specialist medical advice and full panellist consent is available before asking panellists to taste samples.

5.3.6. Sensory Profiling Methods

A sensory profile[53,64,65] is a reproducible description of a product involving analysis of the intensity and nature of each component in the stimulus in the order in which they are perceived.

The method uses a small panel who, through controlled group discussions, generate terminology and valid intensity scales for the major sensory components in a product. Profiles are generally presented as 'star diagrams' and comparison between samples is possible by overlaying diagrams and assessing the agreement in fit. Magnitude estimation scales may be used, producing powerful models with many uses in research and product development.[75] Profile methods require a very high commitment of time in training and maintaining a panel.

5.4. Preference Testing and Consumer Research

Difference testing is relatively easy to interpret, requires relatively small selected and trained panels, and is relatively inexpensive. Preference testing in food development laboratories is used as a cheaper alternative and precursor to market research but the interpretation of the results may be difficult and the results may not be valid or predictive of the market place.

Preference testing[53,66,68,69,76] can be riddled with bias. The validity of conclusions is entirely dependent upon the design and administration of the trial. In terms of design all sample pairs about which relative preference information is required must be directly compared; it is not safe to make conclusions about the relative merits of samples A and C, if each sample was compared directly only with sample B. The design and wording of the questionnaire is vital: it should not lead the panellist to deduce the desired outcome. Preference testing without large numbers of representative consumers is a very risky business. There is a danger that laboratory panellists may focus on attributes unimportant to consumers, or overlook or underestimate factors important to consumers.

Directional questioning may limit the consumers' choice or responses and cause important information to be missed. Consumers generally display the 'error of leniency', so that very high percentages of satisfaction, or strong interest in purchase, may be required to indicate that a product will succeed in the market place. Contrast errors may occur and the overall acceptability of a product may be enhanced or reduced depending on whether or not it is being compared with a very poor or a very good product. Scales can suffer from bias depending upon how many points they have and how the points are anchored. The serving order of the samples, and order and length of home placement, can all result in bias. The relative merits of hall testing and home placement have also to be considered. Above all special note should be made of all circumstances which might affect the results of the trial, such as local advertising campaigns, local promotions, regional preferences, the weather or simply the malfunctioning of the air conditioning!

As an interesting example a preference test designed to determine the most acceptable level of aspartame in a sweetener tablet is reproduced from some market research reported by Beck.[77] Panellists were provided with standard sweetened samples of coffee and asked to indicate whether or not the coffee was over-sweetened, just right, or undersweetened. According to their responses further samples were given and the same information sought. The preferred sweetness levels of 200 panellists were collected and then analysed according to their relationship with various dose levels of aspartame proposed for a tablet, with the reservation that panellists must be satisfied with either one or two *whole* tablets, and at least half the test population must be satisfied by one whole tablet. Panellists' responses were

ANALYSIS OF ADVERSE CONSEQUENCES OF A GIVEN LEVEL OF ASPARTAME IN TABLET FORMULATION

Two tablets	*One tablet too sweet*	*One tablet just right*	*One tablet not sweet enough*
Too sweet	Dissatisfaction: Must break tablet or quit	Acceptance	Dissatisfaction: must break tablet or quit
Just right	(Inconsistent)	Acceptance	Acceptance
Not sweet enough	(Inconsistent)	(Inconsistent)	Dissatisfaction: must use 3 or quit

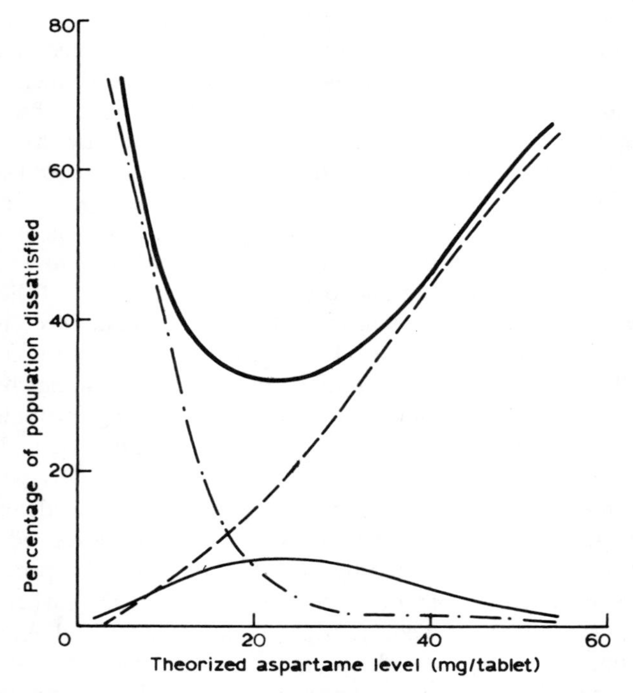

FIG. 7. Investigation to determine the optimum aspartame level for tablets. The graph shows the effect of aspartame content on predicted tablet acceptance (based upon a coffee model). (– – – –), One tablet too sweet; (– · – · –), two tablets not sweet enough; (———), one tablet not sweet enough; two too sweet. Source: adapted from Beck.[77]

converted to a satisfaction/dissatisfaction matrix and percentage dissatisfaction was plotted against tablet aspartame content (Fig. 7).

6. CONCLUSION

We have tried to show the significance of sweetness as a flavour characteristic in special diet foods, and in particular, its influence on the acceptability of those foods, and hence its value for both patient compliance in adverse circumstances, and in bringing a degree of normality where the strain of keeping to an otherwise restricted diet becomes an effort.

The sweeteners in use are changing as a result of increasing understanding of the dietetic requirements of the patients, of increasing awareness of the need for, and safety of, the sweet materials available, and as the legislative requirements of various health authorities respond to that awareness.

Because of the small differences in actual 'sweetness', and the differences in flavour profiles between the various competing artificial sweeteners, it is necessary to use relatively sophisticated evaluation techniques in order to develop satisfactory products and to substantiate claims for the acceptability of one product versus another.

REFERENCES

1. SHAKESPEARE, W. (1594). *Richard II*, Act. I Scene III.
2. JACKSON, K. G. (1986).
3. YUDKIN, J. (1972). *Nature* (*London*), **239**, 197.
4. PAUL, A. A. and SOUTHGATE, D. A. T. (1978). *McCance and Widdowson's, The Composition of Foods*, HMSO, London.
5. TALBOT, J. M. and FISHER, K. D. (1978). *Diabetes Care*, **1**, 231.
6. LEHNINGER, A. L. (1975). *Biochemistry*, 2nd edn, Worth, New York, Part 3, Chapter 30, p. 847.
7. BAUER, J. R. and HORWITZ, D. L. (1979). *Compr. Ther.*, **5**, 12.
8. WOODS, H. F. and BAX, N. D. S. (1982). *Diabetologia*, **23**, 213.
9. AMERICAN DIABETES ASSOCIATION. (1979). *Diabetes*, **28**, 1027.
10. JENKINS, D. J. A., GOFF, D. V., LEEDS, A. R., ALBERT, K. J. M. M., WOLEVER, T. M. S., GASSUL, M. A. and HOCKADAY, T. D. R. (1976). *The Lancet*, (24 July), 172.
11. ANON. (1983). *Dietary Recommendations for Diabetics for the 1980s*, British Diabetic Association, London.
12. METCALFE, J. (1986). *Balance*, (February), 35.
13. ANON. (1984). *Food Labelling Regulations 1984*. SI 1984, No. 1305, HMSO, London.

14. FOOD STANDARDS COMMITTEE. (1980). *Second Report on Claims and Misleading Descriptions*, HMSO, London, pp. 112–41.
15. WYATT, E. J. (1984). *Food Legislation Survey No. 1, Artificial Sweeteners*, 3rd edn, Food RA, Leatherhead.
16. ANON. (1983). *The Sweeteners in Food Regulations 1983*, SI 1983, No. 1211, HMSO, London.
17. ANON. (1964). *The Soft Drinks Regulations 1964*, as amended SI 1964, No. 760.
18. LINDLEY, M. G. (1983). *Developments in Sweeteners—2*, Applied Science Publishers, London, p. 225.
19. US Patent 4 277 504 (1981). Ledges Associates, Rosco, Illinois, USA.
20. Adapted from *Roche Technical Publication* RCD 3254/0880.
21. DODSON, A. G. and PEPPER, T. (1985). *Food Chem.*, **16**, 271.
22. GRENBY, T. H. (1984). *J. Royal Soc. Health*, **104**, 27.
23. SICARD, P. J. and LEROY, P. (1983). *Developments in Sweeteners—2*, Applied Science Publishers, London, pp. 1–25.
24. UK Patent Application GB 2 161 360 (1986). Hoffmann la Roche.
25. ANON. *Palatinit® Infopac*, Palatinit Subungsmittel GmbH.
26. PAULUS, K. and FRUCJER, A. (1980). *Ztschr. Lebensmittel-Technologie und Verfahrungstechnik*, **31**, 128.
27. ANON. (1980). *Drug and Therapeutic Bulletin*, 67.
28. ANON. (1982). *The Miscellaneous Additives in Food (Amendment) Regulations 1982*, SI 1982, No. 14, HMSO, London.
29. ZIMMERMAN, M. (1974). *Manuf. Confectioner*, (August), 39.
29a. BRITISH NUTRITION FOUNDATION. (1986). *Energy Balance in Man, Facts and Fallacies.*
30. ANON. (1983). Report, National Advisory Committee on Nutritional Education.
31. LABUZA, T. P. (1985). *Cereal Foods World*, **30**(12), 829.
32. FIAL, A. Z. (1978). *J. Amer. Diet. Ass.*, **73**(December), 658.
33. ANON. (1986). Artificial sweeteners. In: *Which?* (April), 155.
34. ANON. (1949). *British Pharmaceutical Codex.*
35. MOLARD, F., MEDARP, A., LHERITIER, J. and BLANC, F. (1981). *Lab. Pharm. Problemes Tech.*, **310**, 475.
36. EL SHATTAWAY, H. H. (1981). *Drug Dev. Ind. Pharm.*, **7**(5), 605.
37. ANON. (1978). *Properties, Manufacture and Use of Fructose as an Industrial Raw Material*, Technical publication, Xyrofin Ltd.
38. British Patent 977 482 (1964). Ashe Laboratories Ltd.
39. ANON. (1979). *The Preservatives in Food Regulations 1979*, SI 752, HMSO, London.
40. US Patent 4 153 737.
41. Canadian Patent 1 026 987.
42. MUHLEMANN, H. R. and FIRESTONE, A. (1982). *Swiss. Dent.*, **3**(6), 21.
43. COUNSEL, J. N. (1978). *Xylitol*, Applied Science Publishers, London.
44. *Kirk-Othmer Encyclopedia of Chemical Technology*, Vol. 1, Part 3, p. 756.
45. German Patent Application 3 009 875 (1980). Huckette, M. and Devos, F. (Roquette Freres S.A.).
46. US Patent 4 252 794 (1981). Duvoss, J. W. (ICI Americas Inc.).
47. SCHMIDT, P. C. (1983). *Pharmaceutical Technol*, **7**(11), 65.

48. Whitmore, D. A. (1985). *Food Chem.*, **16**, 209.
49. Schallenberger, R. D. J. (1963). *Food Sci.*, **28**, 584.
50. Osberger, T. F. (1979). *Pharm. Technol.*
51. Staniforth, J. N. (1985). *Powder Technol.*, **45**, 73.
52. Anon. (1980). *British Pharmacopoeia*, Vol. 2, p. A192.
53. Amerine, M. A., Pangborn, R. M. and Roessler, E. B. (1965). *Principles of Sensory Evaluation of Food*, Academic Press, New York.
54. Moskowitz, H. R. and Klarman, L. (1975). The tastes of artificial sweeteners and their mixtures. In: *Chemical Senses and Flavour*, **1**, 411.
55. Moskowitz, H. R. and Klarman, L. (1975). The hedonic tones of artificial sweeteners and their mixtures. In: *Chemical Senses and Flavour*, **1**, 423.
56. Schiffman, S. S., Reill, D. A. and Clark, T. B. (1979). Qualitative differences among sweeteners. In: *Physiol. Behav.*, **23**, 1.
57. Schiffman, S. S., Crofton, V. A. and Beeker, T. G. (1985). Sensory evaluation of soft drinks with various sweeteners. In: *Physiol. Behav.*, **34**, 369.
58. Swartz, M. (1980). Sensory screening of synthetic sweeteners using time–intensity evaluations. In: *J. Food Sci.*, **45**, 577.
59. Powers, N. L. and Pangborn, R. M. (1978). Paired comparison and time–intensity measurements of the sensory properties of beverages and gelatins containing sucrose or synthetic sweeteners. In: *J. Food Sci.*, **43**, 41.
60. Du Bois, G. E., Crosby, G. A. and Stephenson, R. A. (1981). Dihydrochalcone sweeteners, a study of the atypical temporal phenomena. In: *J. Med. Chem.*, **24**, 409.
61. Harper, R. (1982). The Littlejohn Lecture, Part 2. Art and science in the understanding of flavour. In: *Food*, (February), 17.
62. Lyon, D. (1985). Product profiling in R and D. In: *Food Man.*, (January), 43.
63. Williams, A. A. and Arnold, G. M. (1985). A new approach to the sensory analysis of foods and beverages. In: *Progress in Flavour Research*, Proceedings of the 4th Weurman Flavour Research Symposium, Dourdan, France, 9–11 May 1984) J. Adda (Ed.), Elsevier Science Publishers, Amsterdam.
64. Jellinek, G. (1985). *Sensory Evaluation of Food Theory and Practice*, Ellis Horwood, Chichester.
65. Anon. (1983). *Course T021—Operation of Taste Panels*, British Food Manufacturing Industries RA, Course Notes, Unpublished.
66. Anon. (1980). Methods for Sensory Analysis of Food Part 1. Introduction and General Guide to Methodology, BS5929: Part 1, 1980, British Standards Institution, London.
67. Anon. *An Introduction to Taste Testing of Foods*, Merck Technical Bulletin, unpublished.
68. Pearson, A. J. (1982). How to manipulate a taste test without really trying. In: *Beverage World*, (November), 148.
69. Kramer, A. (1979). Conducting consumer taste panels. In: *Quick Frozen Foods International*, (April), 135.
70. Health, H. (1985). Sensory evaluation: an update. In: *Food*, (September), 28.
71. Spencer, H. R. (1971). Taste panels and the measurement of sweetness. In: *Sweetness and Sweeteners*, G. G. Birch, L. F. Green and C. B. Coulson (Eds.) Applied Science Publishers, London.
72. Armstrong, J. (1985). *Sweetener Panel—Screening and Training 1984–85*,

Development, Department Report No. 116, The Boots Co., Nottingham, unpublished.
73. Roessler, E. B., Pangborn, R. M., Sidel, J. L. and Stone, H. (1978). Expanded Statistical Tables for Estimating Significance in Paired-Preference, Paired-Difference, Duo–Trio and Triangle Tests. In: *J. Food Sci.*, **43**, 940.
74. Steiner, E. H. (1964). *Schemes for Triangular and Paired Comparison Tasting Tests with Controlled Risks of Making Wrong Decisions*, Technical Circular No. 274, British Food Manufacturing Industries RA.
75. Moskowitz, H. R. and Chandler, J. W. (1977). New uses of magnitude estimation. In: *Sensory Properties of Foods*, G. G. Birch, J. G. Brennan and K. J. Parker (Eds), Applied Science Publishers, London.
76. Blair, J. R. (1978). Interface of marketing and sensory evaluation in product development. In: *Food Technol.*, (November), 61.
77. Beck, C. I. (1974). Sweetness, character, and applications of aspartic acid based sweeteners. In: *Symposium Sweeteners*, G. E. Inglett (Ed.), AVI Publishing Co., Westport, Connecticut.

Chapter 9

THE FOOD MANUFACTURER'S VIEW OF SUGAR SUBSTITUTES

DONALD A. M. MACKAY

President, Applied Microbiology Inc., Brooklyn, New York, USA

SUMMARY

Collaboration between the producer and the user of sugar substitutes is discussed from the point of view of the food industry. Disadvantages for the producer exist, particularly in the patent area, but are probably inevitable.

The ideal sugar substitute is described but almost certainly does not exist. It is proposed that sweetness is not critically important, but a water-soluble bulking property combined with biological stability permitting non-cariogenic, non-glucogenic and non-caloric claims is. Achieving this combination of features without laxation effects and intestinal distress will be difficult.

In view of the current high regard for sugar as a safe substance (other than in caries involvement) it is necessary to reconsider the purchase rationale for food products containing sugar substitutes. Sorbitol, the current major sugar substitute, may itself become a target of substitution.

Product purchase depends upon advertising a benefit or rationale. Consequently, greater emphasis is needed on the process whereby claims and labels are brought to a legally permissible status, especially in countries where health claims are subject to rigorous regulation.

Finally, the failure to date of sugar substitutes to affect sugar consumption is discussed. The need for producer–user collaboration in strategic planning is emphasized in the light of this failure.

1. INTRODUCTION

The term 'push–pull' has been used over the years in various senses. At one time, it referred to the electron forces exerted in chemical reactions required by two different molecules to approach each other in order to get over or through a seemingly impossible energy barrier and thereby react with each other. In similar vein, it serves as a useful concept to describe the coordinated, simultaneous, cooperative effort needed by both the producer and the user of a sugar substitute in order to make a successful product out of it in the face of great difficulties. Yet all too often, it seems that the intended user is pulling for one product, while the sugar substitute producer may be pushing for another.

Usually the reasons are fairly obvious, but sometimes explanations have to be sought on a sociological and psychological level where differences in operating philosophy of the companies involved, or commitments made at the personal level by the negotiators involved, tend either to obscure or to promote the goals of each organization.

Basically, a producer of a sugar substitute wants a long-term monopoly of a highly profitable item. The user (i.e. the food manufacturer) of the sugar substitute wants a long-term monopoly of a highly profitable item *containing* the sugar substitute. At once, we have two problems: the definition of 'item', and the effect of the user's monopoly, if present, upon the potential market size of the sugar substitute. A producer's monopoly of an item controlled by one user's monopoly of a particular product form is not likely to raise by much the prospects and share prices of the producing company, unless it is clear that monopolies are temporary and serve the process of orderly market-making in the introductory phases of product development.

2. THE ROLE OF PATENTS AND PROPRIETARY INFORMATION

The usual form of monopoly, as industrialized states have long intended, is the patent, with another type of monopoly (not so obviously intended) being afforded in the form of regulatory clearances for safety and designated product application classes. The patent monopoly can be for a new composition of matter (e.g. a newly synthesized artificial sweetener); or for a new use (e.g. discovery of novel sweetening properties of an existing compound—as with aspartame); or for some novel process to extend the benefits or by-pass the disadvantages of that artificial sweetener in a particular product form.

The original patent on synthesis or use spawns a crop of patents mushrooming in the fields of product application or development. Thus, to take aspartame (APM) as our example, we see how the original report of its synthesis in 1966 was followed by the report of its sweetening property in 1969, then by two basic US patents awarded in 1971 describing its general application and utility in a number of foods. In turn these were followed by a rapidly expanding number of patents in detailed areas such as extended sweetness; specific foods; or specific processes for incorporating APM into foods for timed release of sweetness; or for protection of APM in stored foods against the effects of decomposition caused by heat, water or other components that may be present in the food.

There comes the time, sooner or later, when market expansion of a successful sweetener, even if not limited by the types of food classes permitted by regulation, becomes seriously hindered by the constraints placed on the food manufacturers of the later products by the patent structures raised by food manufacturers of the early entries. The economic value of a monopoly on synthesis of a sugar substitute eventually becomes capped by the economic and legal difficulties faced by later entries into the food market. These can be due to normal market factors such as saturation, or to highly technical ones where one food manufacturer has found patent coverage for economical use of a high-priced ingredient, or novel ways for sweetness and flavor duration or timed release or other benefit. In the USA at least, there is a new regard for the economic clout of patents and the penalties that might be incurred in ignoring them.

This description of possible events is not meant to bewail the sad fate of the sugar-substitute producer. Indeed, patent saturation at the application end may prove to provide the entry point for another manufacturer with another sugar substitute, no better than an earlier one, but unencumbered with the cloud of patents arising from successful solution of problems that only presented themselves at the time of application.

Reported impending entries of other dipeptide sweeteners besides APM suggest that their manufacturers are willing to pay the exceedingly stiff entry price for toxicological testing and regulatory clearance not only because of possible price and stability advantages over APM, but because a new sweetener, even one slightly inferior to APM in taste and potency, could prove highly interesting to food companies now barred from effective competition in specific foods using APM itself. The five-year extension recently awarded in the USA for the life of the manufacturer's basic APM patents—also covering about another 30 related patents obtained by the food industry (mainly the General Foods Corporation), whose utility had been affected by the Food and Drug Administration's initial decision to rescind approval of APM pending re-evaluation of the toxicity data—has probably also been a goad for competitive action by other chemical manufacturers with potential dipeptide sweeteners waiting in the wings.

It is of course impossible for the sweetener manufacturer to be aware of problems with a given ingredient which surface only in the product-development phase, and to develop a broad umbrella of patent coverage for their solution such that the widest possible market for usage of the sweetener ingredient is preserved. And while not impossible, it is not likely that the new ingredient manufacturer will have such intimate connections with a wide array of potential users that deficiencies of the currently available sweeteners become known to him and are eventually overcome by resorting to the new sweetener. There is sometimes a touching naïveté shown by chemical producers in believing that the food industry will tell them all, when in reality one part of this industry rarely talks to another. Even within one company, the Research and Development and Manufacturing Departments convey the minimum necessary information to each other and to Marketing. Product success, at least in the confectionery business, hinges so often on minor, arcane points that provide the key to economical production, or improved shelf-life, or decreased wrapping costs, that it is most unlikely that any food company will or can give a completely honest and candid appraisal of any sweetener, especially the one currently in use, compared with the new sweetener being offered by the manufacturer. Whilst taste and other sensory properties may be fairly and adequately known by the sweetener manufacturer, it is improbable that he will appreciate the role played by functionality of sweeteners, particularly bulk sweeteners, in products like boiled sweets (hard candy), toffees, caramels, chocolate and chewing gum.

The sweetener manufacturer is looking for a one-on-one comparison of sweeteners. The food manufacturer, however, has to consider the comparison of both products and processes since he may be giving up the fine tuning of products, process machinery and packaging equipment that has kept him in business since the day he started. As a result, he may know that only a certain kind of sugar from certain ports will give the required degree of cold flow in glassy candies; or that his fine-tuning of production equipment has tied him so tightly to one source of sorbitol that he cannot take advantage of competitive pricing by other suppliers, nor readily take advantage of a 'better' sweetener.

Development and refinement of a new sweetener, and development of its market position by the manufacturer, need the closest possible feedback from the potential user so that its utility is properly appreciated, both as an ingredient and in the final food product, in both factory manufacture and performance in wholesale and retail distribution. For this purpose, collaboration between the manufacturer and the user is highly desirable and beneficial to both parties. However, two problems lurk beneath the surface. The user is reluctant to tell all his findings, for what they may reveal about his current operation, and because he knows that later the new sweetener will have to be offered to his competition. The manufacturer knows he may have lost some degree of control over his product and its uses since the new findings made by the user's technical staff are necessarily highly susceptible to the patenting process, which surely has the potential of limiting the new sweetener's sales to a broader market.

The chemical manufacturer with a new sweetener in hand (or in mind) has the problem of timing his first contacts with the food industry, or a particular member of it, and of determining the extent and nature of the contact. At one extreme there is a company that will not initiate any contact until it has determined, refined and costed the manufacturing process for the new sweetener; until it has completed the required safety testing for the country (or countries) of interest; and until it has determined the food product categories of interest, developed the necessary patent coverage in those categories, and developed the analytical methodology needed for reliable analysis, monitoring and recovery in the food classes of interest. At that point the manufacturer can open his door and sell his approved product to the world.

At the other extreme, there is a different kind of company. Perhaps because of limited resources, perhaps by philosophical choice, a very early approach to the food industry is made, often to the second- or third-largest company in each category, with intentions of leaving visions of eventual category leadership firmly implanted there. Fields of cooperation are laid out from the beginning. It might be decided that costs should be shared for toxicology testing, or the user's portion of these placed in escrow against eventual license payments. Responsibility for analytical development, especially for the methodology required in each food class, is assigned. Agreement is reached as to which categories of foods should be put forward for regulatory clearance, and what product claims are eventually to be made and substantiated, even—or especially— if not intended to be made explicit at the time. Decisions on food categories are important for controlling and measuring the potential intake of the new sweetener into the diet, and for providing the regulatory authorities with means to control exposure of various population groups. Abuse potential, especially by children in attractive foods like confections, must be carefully considered.

On one hand, these operations are entirely mechanical, and a company could perhaps find the resources to do all the work by itself, or pay for the safety testing and product development work to be carried out under contract. On the other hand, it may lack the marketing acumen, the sales and distribution forces, and the ability to carry out in-depth motivational analysis of the various consumer populations within product franchises, especially brand-name product franchises.

3. PRODUCT DIMENSIONS OF NEW SWEETENERS

Is the new sweetener to compete simply on taste, or on fear (justified or not) of the safety of sugar and other sugar substitutes, or is it intended to develop new markets based on dental, diabetic or other dietetic grounds? Early decisions whether to provide explicitly or implicitly for these product benefits will have major importance on steering both the ingredient and the product categories in which it will be utilized through the regulatory maze created by needs for safety data, exposure potential, abuse potential (possibly pertinent in each of these product categories) and (most importantly) the intended claim structure that will have to be utilized in labels and advertising in order to motivate purchase by the consumer.

In the USA at least, the design of the Food Additive Petition, or the GRASness* Affirmation Petition (whose requirements are now similar to those of the Food Additive Petition) should be carefully considered, bearing in mind that the health benefit claims which underlie most sugar substitution rationales should be very much in the forefront of the petitioner's thinking if not of his petition. The amount of the new sweetener to reach the consumer will be governed not only by the food categories and ingredient limits set out in the Food Additive (or GRASness Affirmation) Petition, but by labeling requirements set out in Special Dietary Foods regulations, e.g. the caloric standards for low- and reduced-calorie foods, and by its eventual designation as either a nutritive or a non-nutritive sweetener, a distinction which can be very important for legal even more than for psychological reasons.

* Generally Recognized As Safe.

4. THE JOINT VENTURE APPROACH TO ASPARTAME DEVELOPMENT

Presumably considerations such as these led the Searle Company, a drug company with no prior experience in the food industry, to form a joint venture group with General Foods, whose analytical and product development skills, safety testing expertise, market research, and knowledge of food regulatory procedures were beyond question. It takes little imagination to guess that the Food Additive Petition for aspartame (APM) was designed and prepared almost in its entirety by General Foods. Only with the recent amendments to this Petition allowing APM usage in carbonated beverages (and very recently in breath mints), has APM been permitted in the USA in a food category not known to be important to General Foods. Early access to the APM needed for the development of both new food products and the associated analytical methodology obviously was of some advantage to General Foods, for about 35 US patents on APM usage have been issued and assigned to that company. Although the joint venture was later disbanded in some disarray (presumably due to the severe financial losses incurred by Searle in abandoning their as-yet-uncompleted plant facility for APM manufacture when FDA withdrew its initial approval), there is little doubt that General Foods, as a result of its early access to APM, enjoyed a huge advantage in maximizing the availability of APM throughout its line of dry packaged foods. Interestingly different, however, was the tack taken by the US National Association of Soft Drink Manufacturers who (before an adequate supply of APM was available to all bottlers) requested the FDA to *delay* implementing the regulation permitting APM usage in carbonated soft drinks until all stability questions had been evaluated. To give more perspective to the impact of the General Foods patent coverage, it is worth noting that a database search (using keywords based on the various names for APM and dipeptide sweeteners) showed that 49 patents had been issued worldwide by the end of 1981. Since then another 199 patents have been issued, though this is doubtless due in large part to the increased emphasis being placed on finding new types of dipeptide sweeteners. Without physically obtaining and reading all 248 patents, it cannot be said for sure that these are all different and all involve APM directly, rather than as just one name in the incidental, obligatory recitation of all sweeteners that will be found in nearly every patent on *any* sweetener. But it is clear that patent coverage has gone a long way since APM's synthesis and the discovery of its sweet taste was first reported, and that the free license to use APM as a sweetener, implicit in the purchase of APM from Searle, may not be very useful to a new user in the face of such tremendous patent activity by entrenched competition.

5. THE REAL NEED: ARTIFICIAL SWEETENER OR SUGAR SUBSTITUTE?

Up to this point little has been said of the nature of the sugar substitute, yet whether it is an intense sweetener or a bulk sugar replacement is of crucial importance to the product form, or to the health benefits that may be inferred (correctly or otherwise) from the absence or replacement of sugar.

Although the choice offered by this question is in some ways artificial, the terms employed have a certain utility, in spite of their semantic insufficiencies, in communicating food processors' needs to the ingredient manufacturer or supplier. Sweetness is a property inherent in the molecule and, as judged by a set of taste buds, either exists or does not. What is usually sought is an intensity of sweet taste that, when diluted, is comparable with that of sucrose. Thus a given weight of sucrose in a food product can be replaced without taste penalty by a smaller weight of intense sweetener. In theory, then, any other attribute of sucrose (other than taste) can be minimized or avoided in the product, with the potential of creating noteworthy advertising if the missing attribute is disadvantageous, provided of course that the taste of the sucrose is replaced by the intense (artificial) sweetener, and also provided that the size (weight or bulk) of the food product is restored to its previous value in some other way.

In practice the restoration of the product bulk is nearly always made with water; this is fine for certain products like carbonated beverages, which contain water anyway, and whose chemical and microbiological stability requirements can be met easily by pasteurization or ultrafiltration and resorting to chemical stabilizers such as low pH, carbonation, and sodium benzoate. And since water is non-caloric and non-cariogenic, products made with water have far better opportunities to make health and reduced-calorie claims than the sugar-containing products they seek to displace.

But once beyond the beverage category the problem of replacing the non-sweet attributes of sugar becomes acute, since food products are usually sold by weight or in some cases, like ice-cream, by volume. Legal and practical reasons limit the volume that can be obtained by pumping air into ice-cream (or whatever name has to be invented where 'ice-cream' is legally protected); and there are few products outside table-top sweeteners and beverage powders in which agglomerated powders based on hydrolyzed starch, lactose or dextrose can be used to maintain the illusion of bulk in the short time between package opening and adding the contents to hot or cold water, i.e. until the bulking agent is provided by the consumer.

Sugar in foods, of course, has many functions beyond providing sweetness and bulk,[1] and no intense sweetener can possibly fulfill them if water is a necessary prerequisite for its use in meeting weight requirements or providing for reduced-calorie claims. There are indeed borderline cases where the viscosity conferred by sugar in jams, preserves, gelatin desserts, puddings, etc., can be approximated by use of small amounts of pectins, vegetable gums and modified starches in the presence of large amounts of water, but the legal definition of jams, jellies, preserves, etc., usually means new names and product classes have to be invented, and recourse to chemical preservatives is needed to compensate for the lost preserving action of concentrated sugar solutions.

From the point of view of new product development, whether with the intent to hammer away at sugar's supposed deficiencies or to create products never accessible through usage of sucrose and related sugars, it is—or ought to be—evident that there is a far greater need for new agents to substitute for the bulk and function of sucrose than for the taste of sucrose. Excluding water and air, the only major alternative bulking agent yet discovered is sorbitol, and even here the use is severely limited to two categories. The first is in 'sugarless' confections;[2] the second as a less sweet version of the sugar needed in surimi (frozen, well-washed fish muscle extruded and flavored to simulate crab or lobster) to stabilize the protein of the comminuted fish muscle against the syneresis induced by freeze–thaw cycling. Sorbitol usage in surimi, even in Japan, is a very recent development depending upon its osmotic and complexing properties. Sorbitol is now provided in comparatively cheap form as the syrup obtained by hydrogenation of glucose syrup. As noted, the decreased sweetness of sorbitol is considered an advantage over sucrose, the previous sole stabilizer in this product.

By contrast, the use of sorbitol in confections is a vastly more complex matter, beginning with the absolute requirement for sorbitol in expensive crystalline form for the manufacture of 'sugarless' chewing gum and pressed mints. In order to illustrate this complexity, and the opportunities it offers for collaboration between food manufacturers and ingredient manufacturers, it seems advisable to devote a further section to the problems peculiar to sugar substitution in the confectionery area.

6. CONFECTIONERY NEEDS FOR SUGAR SUBSTITUTES

Confectionery means different things in different countries, and even in the same country at different times. It can range from a broad meaning of high-sugar-content foods, such as chewing gum, sweets, chocolate, ice-cream, jams, jellies, candy bars, biscuits, cream-filled cakes, etc., to a narrow one restricted to sweets. Here sweets mainly means toffees (caramels), pressed mints, hard candies (boiled sweets), chewing gum, and slab and filled chocolate. A country's definition may exclude candy bars with a baked fabricated center (and call them chocolate-covered biscuits) or include purported nutritional snacks like granola bars if the sugar content is high enough. The definition usually applies to both sugar and 'sugarless' types of confectionery.

By 'sugarless' confections is meant those calling attention to an absence of easily fermentable carbohydrate, which permits making or implying health claims relative to teeth or some disease state like diabetes. However, inspection of these claims reveals that the dental claim refers to not promoting tooth decay or not producing acid by bacterial action; and the diabetic claim may be only implied and linked to a dietetic claim based on the absence of salt or presence of vitamins since better defensive ground may be found there. Generally the key ingredient is sorbitol (and related polyols) which provides resistance to easy fermentation by oral bacteria and an alternative to sugar for diabetics advised to avoid it.

Like sucrose, sorbitol is colorless, crystalline, soluble in water, and available in the liquid, particulate and solid forms necessary for factory production of boiled sweets, pressed mints, and chewing gum. In all cases, however, very substantial changes in factory processing conditions are required in order to change from the standard sugar product to the sorbitol analogue. This is particularly true for the chewing gum product area, and it may be instructive to dwell on some of the intricacies of sugarless gum manufacture in order to appreciate the concatenation of events that make substitution of sugar bulk in these products a much more technologically demanding feat than substitution of sugar taste. Additionally, it should be easier to demonstrate both the opportunity for manufacturers to supply other bulking agents besides sorbitol, and the necessity for early and intense collaborative efforts between confectionery maker and ingredient supplier if the deficiencies of sorbitol are to be overcome, or if new product claims not available to either sugar or sorbitol are to be exploited.

It is ironic that the very success of sorbitol 'sugarless' gums has provided the opportunity for substitution of sorbitol as well as for substitution of sugar. With sugarless gum either close to or even exceeding sugar gum in consumption in a number of countries, the deficiencies of sorbitol are well worth some study.

Crystalline sorbitol is expensive. It could be sweeter, and in countries where artificial sweeteners are not available or are restricted to certain product forms xylitol can exploit a major taste advantage over sorbitol. Xylitol is even more expensive, so that xylitol/sorbitol mixtures are the norm in countries where xylitol is associated with positive dental benefit, or legally available. Xylitol is legal in the USA as a food additive for sweetening of Special Dietary Foods, but unlike sorbitol it is not GRAS, and major confectioners in the USA are reluctant to use it until its regulatory situation has been clarified.

Lycasin®, the trade name of hydrogenated high-maltose corn syrup, has recently been Affirmed as GRAS in the USA and is accordingly a possible substitute for the liquid forms of sorbitol (70% syrup) used in place of the traditional corn syrups needed in chewing gum to provide the plastic phase for its manufacture.

A plastic phase is necessary to soften, plasticize and emulsify the gum base (these are probably overlapping functions) in order to convert a rock-hard gum base (the mastic component) to a flexible stick of chewing gum, or a coated pellet or cube form that can be bitten through without breaking one's teeth. Unlike sugar gum (consisting of powdered sugar, molten gum base and corn syrup), sugarless gum rather rapidly becomes brittle and inedible, but the brittleness is not simply due to the rapid loss of water from the colloid-poor sorbitol syrup. The nature of the solid sorbitol is critical, since it exists in polymorphic forms which vary according to the process used by the various sorbitol manufacturers to obtain solid sorbitol crystals from highly concentrated sorbitol solutions. It seems that some forms are highly susceptible to crystal bridging, probably facilitated by but not necessarily directly related to the water content introduced by the use of sorbitol syrup, so that product failure as measured by adverse consumer reaction to brittle sticks of chewing gum occurs long before the sugar product fails, probably by a different mechanism.

For this and other reasons a strange situation exists where three major US gum manufacturers are almost completely locked into sole-source situations of sorbitol supply, probably because their own drive to optimize product quality and process economics has maximized their ability to handle one familiar form of sorbitol and removed the flexibility of operations associated with working with varying types of sorbitol from multiple suppliers. As far as the patent literature shows, some efforts to avoid sugarless gum staling involve use of a liquid polyol such as glycerol or propylene glycol, with or without the concomitant avoidance of aqueous sources such as those provided by sorbitol solutions. Avoidance of water and reliance on glycerol to soften the gum base is not without its own associated problems of expensive ingredients, novel processing and need for protective packaging so that the crystalline sorbitol difficulty has been circumvented at a cost rather than solved.

Added to this, we may have the further complication due to the artificial or intense sweetener when APM is used by choice or necessity to replace saccharin or cyclamate as sweeteners for chewing gum. Aspartame is an excellent sweetener, but is surprisingly labile in chewing gum due to the alkaline mineral filler (chalk), or to the active moisture content, or to chemical reactions with flavor and gum base constituents. As a result there is a mass of patent literature bearing on anhydrous conditions which, as it turns out, bear as much on short shelf-life due to APM instability as on short shelf-life due to rapid embrittlement.

This describes a perhaps extreme but not atypical situation where the advantages and disadvantages of close cooperation between ingredient supplier and food product manufacturer are readily illustrated. It can be confidently asserted that no matter how well the supplier thinks he knows his (prospective) customer's requirements, he in fact knows very little of the prejudices, particularities and peculiarities of the major customer with established brands and product franchises to consider. The very obviousness of the factors that the customer chooses to discuss (price, supply, yield, performance, etc.) is more than likely to disguise the apparently trivial factors (and sometimes they are) which are truly determining in new-product or new-process decisions.

While by no means an always entirely satisfactory solution, there is much to be said for the patenting process which forces, or at least permits, the emergence of the real issues governing decisions to move from one sweetener to another; or to replace sugar with sorbitol; or one form of sorbitol with another form; or of sorbitol with another polyol or non-fermentable carbohydrate free from sorbitol's problems.

The area of sugar bulking replacements seems much more worthy of joint exploration than simply the development of a new sweetener with improved price, taste, stability and sweetening power, since this usually means further travel down the well-trodden path of diet beverages and aqueous systems, and low-calorie claims for foodstuffs of high water content.

7. POTENTIAL VALUE OF REDUCED-CALORIE BULKING AGENTS

The most recent assessment of the health risks of sugar[3] makes it clear that sugar replacement strategies can be based justly only upon its involvement in tooth decay as a possible frequent contributor to the metabolic activity of oral bacteria. Less justly, the possible association of the sweet taste of sugar with an excessive calorie intake might be taken as a strategy to produce foods for those who wish to restrict their intake of sugar for whatever reason is personally cogent to them. But in this case, in order to move to new food classes, it will be necessary to develop bulking agents (for fats more logically than sugars) that are themselves reduced in calories.

Since polyols, the current major sugar substitutes, are usually judged to be the caloric equivalents of carbohydrates, new bulking agents that are not isocaloric with carbohydrates will be needed to permit the exploitation of intense sweeteners in non-aqueous systems on reduced-calorie grounds; or, to be more exact, the intense sweeteners will be needed to permit the exploitation of low-calorie bulking agents if these agents are unable in themselves to meet the appropriate sweetness level.

Two examples that come to mind are Polydextrose®, a randomly cross-linked tasteless dextrose polymer at 1 kcal/g, and Palatinit® (isomalt), a fairly sweet isomeric mixture of glucosyl-sorbitol and glucosyl-mannitol which under some conditions may yield only 2 kcal/g. A third example may exist in Japan where low-calorie claims seem possible for maltitol, but these claims have been disputed in the USA as inapplicable to man. Polydextrose® is not really a polyol, but is very water-soluble. Lack of this property puts other low-calorie bulking agents like cellulose and various gums outside the scope of this discussion.

Some may consider Polydextrose® to be only a bulking agent rather than a sugar substitute, since it does not taste sweet. But it is water-soluble and, like sugar, its solutions can be boiled down to give candy glasses. Also, the FDA recognizes that a value of 1 kcal/g may be assigned to it in making up a number of confectionery-type foods (as specified in its Food Additive Petition) designed for calorie-reduction purposes.

Palatinit®, which is a polyol sugar substitute made by a German sugar company, is in various stages of regulatory clearance in different countries and, while definitely sweeter than sorbitol, is believed to provide only 2 kcal/g (though this value may not yet have official sanction), i.e. only half that given by sugar or sorbitol. Non-cariogenic and non-glucogenic claims are also possible. Whilst lacking the high solubility of sucrose needed for certain types of confections, it can form a candy glass on boiling off water from its solution, giving it the potential advantage in some confectionery applications of being non-hygroscopic, non-sticky, and non-graining in products with long shelf-life.

Apart from reduced calories, low-calorie bulk replacements for sugar might have application in 'high-fiber' foods, since one type of dietary 'fiber' providing nutrients to bacteria of the large intestine may be describable in terms of both reduced caloric utility to the host and increased bacterial masses, causing reduced intestinal transit times and corresponding benefits to the surrounding tissues.

Naturally, there is a wide gulf between a tentative medical theory and a food product making such claims as an inducement to purchase. Until recently, the FDA in the USA reacted strongly to foods making claims not contemplated by the Special Dietary Foods category; and while they doubtless view with extreme disfavor the implied medical claims made by some cereal manufacturers in calling attention both to their content of bran and to the National Cancer Institute's statement on the value of high-fiber foods, it has spurred them to attempt to formalize their philosophical and regulatory approach to so-called Medical Foods, by extending this category beyond individually doctor-prescribed foods to foods making health claims other than those now made for nutrients, minerals, vitamins, reduced calories, and absence of dental harm.

In such an event it would be advantageous for suppliers and potential users of reduced-calorie bulking agents to be prepared with products, facts and arguments. Undoubtedly, patent considerations will be a factor in product-development strategies, and one that a new ingredient supplier will have to consider carefully before committing to joint development efforts. Short-term advantages will have to be weighed against potential long-term problems when market broadening is attempted. But bird-in-hand considerations may well dominate in both supplier and user company boardrooms in view of the enormous expense involved to the supplier in obtaining safety clearance and regulatory approval of the new class of ingredient; and the possibly equally large commitment required from the user food company in working up formulations, protecting them with patents, devising new processes, buying machinery, hiring staff, and building factories, even before the need for the massive advertising expenditure required to mobilize consumer interest and motivate their purchases on a scale sufficiently large to provide economic justification for the venture.

Under such circumstances user patenting will have to be considered an unavoidable nuisance to the ingredient supplier. He will have to be sure his own basic patent coverage is broad and deep enough to permit the eventual expansion of a market for far less dramatic caloric reduction than the consumer may have come to expect based on common experience with diet beverages.

8. TRENDS FOR THE FUTURE

The situation today is becoming further confused as the newer sugar substitutes are in some cases being put forward as substitutes for older ones such as sorbitol and mannitol. The field is therefore now made up of such disparate items as: (a) intense sweeteners requiring solution in water or saliva for bulking, or requiring bulking with corn syrup solids, dextrose or lactose for use as table-top sweeteners; (b) sweet-tasting, caloric bulk substitutes such as sorbitol; (c) sweet-tasting, reduced-calorie bulk substitutes for either sugar or sorbitol such as Palatinit®; and (d) non-sweet, reduced-calorie bulking agents such as Polydextrose®. All these classes of compounds can claim indirect dental benefits, based either on the absence of fermentable carbohydrates or on the presence of non- or poorly-fermentable polyols.

The time is coming, however, when sugar substitutes will be required to justify themselves in their own right. Misconceptions based upon sugar are not the proper basis for marketing these types of products; and emphasis should be put on what they are rather than what they are not.

Except for caries, the health case against sugar has been progressively weakened over recent years almost to the vanishing point as shown by the very recent FDA report on 'Health aspects of sugars contained in carbohydrate sweeteners'.[3] The decline in caries in developed countries, said to be due partly to fluoride intake, can be used to stress the sensible use of sugary foods in combination with proper dental hygiene, so even this rationale for sugar avoidance is weakened. On a behavioral basis, it has yet to be shown that any intense sweetener has by itself caused weight loss in the average consumer in spite of calorie reduction in products formulated with it. Reduced calories in a product mean little if increased *per caput* consumption results. On a US national basis very little, if any, sign of reduction in total sugars consumption has been seen, whether cyclamate was used (equal in taste to 0·25 million tons of sugar in 1969) or saccharin (equal in taste to 1·2 million tons of sugar in 1984) or now APM (equal in taste to 1·3 million tons of sugar in 1985). The USDA data[4] are sensitive enough to show that in the first half of this decade the growth of saccharin and APM consumption was equal to 1·2 million tons of sucrose, while total caloric sweeteners usage also increased by 1·3 million tons over the same five years. Sucrose consumption has certainly decreased in recent years in the USA because of the heavy inroads of high-fructose corn syrup (as a replacement for sucrose or invert syrup in soft drinks), but total average *per caput* caloric carbohydrate sweetener consumption has remained static or increased. Even consumption of caloric soft drinks has been almost unaffected in spite of the huge increases in directly competitive new-product categories like diet soft drinks, 'light' beers, and 'wine coolers'.

What we seem to be approaching (at least in the USA) is the saturation of 'negative sugar' products. Eventually, every sugary product may have a mirror 'sugarless' one to accompany it, but not necessarily to compete with it. If new sugar substitutes are to be successful, positive attributes of these new substitutes will have to be addressed. Avoidance of calories when tens of thousands of runners are on starch-loading regimens, or avoidance of tooth decay in a fluoride environment, or avoidance of insulin production when sweet taste alone is sufficient to provoke it, seem to be the basis of weaker and weaker product arguments.

Positive health features should be stressed in their own right. The opportunity may be here since the FDA is expected soon to draft a philosophical regulatory framework (similar to that proposed for Medical Foods) that will permit the development of guidelines for labels and advertisements and statements that are factual in setting forth health benefits.[5] This must be done without precipitating the definition of 'new drug' that the FDA has used so effectively as a weapon in policing nutritional claims outside the mineral/vitamin arena.

In such an event, non-digested or poorly absorbed carbohydrate bulking agents might be as useful as the various types of 'fiber' now thought useful, directly or indirectly, in avoidance of the development of diverticulitis and colonic cancers. Perhaps candy bars, biscuits, etc., where only the sucrose (among many nutritive ingredients) has been exchanged for a polyol, will be permitted to draw attention (if proved) to a reduced potential for cariogenicity, so that there could be a range of 'dentally improved' products other than those consisting of 100% of sorbitol and related polyols. Unlike Switzerland, where sugarless gum, candy and cough syrups can make a 'kind-to-teeth' claim based upon model *in vivo* experiments certifying minimal plaque pH responses to the actual product under test, claims in the USA for the conjoint statement 'sugarless—not non-caloric—does not promote tooth decay' are based upon the inherent non-cariogenic property of polyols like sorbitol, mannitol and xylitol. They are not based on the results of clinical trials (or animal studies or plaque pH studies) of the actual products containing the polyols. Thus no nutritive ingredient besides polyol can be present in the product without losing the right to this claim, unless the product itself has been proven non-cariogenic in a clinical trial. The American Dental Association has been active in coordinating efforts with certain parts of industry and universities to obtain a generally recognized measure of cariogenic potential but there is as yet no regulatory recognition of such a measure.

It might also be possible to follow up the various reported leads of the dental utility of saccharin[6] and other sweeteners in retarding growth or modifying the acidogenesis of dental plaque; or changing plaque's adhesivity to teeth; or in suppression of acid-forming activity of other carbohydrates. There also are research reports that polyols—of which xylitol has been most studied—can induce remineralization of incipient decay lesions, and suggestions that these effects may occur best in mixed diets where fermentable carbohydrates are also consumed. The positive effects of salivation also deserve more study, whether due to chewing action or to the taste of the sweetener employed.

Development of such positive attributes will require more push–pull in producer–user relationships, rather than less. These attributes should be sought in all fields of food technology, however, not just that of health benefits. Sorbitol has long had major use as a humectant in foods, and as a nutritive sweetener has had no problem in being used with foods also containing sugar. (In the USA, sweetening sorbitol with saccharin is another matter, however.) A new major use of sorbitol–sugar mixtures is currently being developed as a texture preserver for surimi, as mentioned in Section 5.

The problem of 'user' patenting in the developmental phase has to be addressed, perhaps by pooling of royalty-free patents, and alternative incentives developed for those innovators in the food industry willing to work with new materials. This they will do gladly in expectation of protecting their own vital interests in their own areas of technology, and especially so if early market development can be facilitated through obtaining some legal means of guaranteed access to the new sweetener when it does finally become available through completion of the regulatory process.

Many, if not most, of the developments in food technology that are based on new-ingredient exploitation come about through formal or informal relationships being developed which provide incentives for the user while guaranteeing the rights to privacy regarding the specifics of uses that may be developed. Fear of losing these rights is one good reason for seeking patent protection; but, on the other hand, some way should be sought to limit or restrain this activity so that patents protecting the user will not encroach unduly on the producer's ability to develop his market in an orderly fashion.

9. SPECIAL NEEDS IN INTERNATIONAL RELATIONSHIPS

These problems are especially pronounced when producer and user are in different countries or even continents. It is difficult to perceive the logic of the FDA in their various decisions (and there is a logic) without a firm grasp of the historical influences that went into creating the FDA and giving it regulatory powers for only certain foods and drugs. Prohibitions against metal trinkets in children's candy, joined with fears (by the public *and* the sugar industry) of economic adulteration of sugar with saccharin, led to the ban in foods of nutritive/non-nutritive mixtures, which includes nutritive/non-nutritive sweeteners. The flavor advantage of such mixtures has never been possible in the USA despite common usage in many other countries.

Whilst there are sound reasons for any food ingredient manufacturer to join forces with a food manufacturer in each of the likely areas of ingredient usage, the rationale is even stronger for an international manufacturer to try to develop his markets abroad via use of a similar strategy. He would be prudent, however, to limit the enlistment of potential overseas users to one or two per country.

The 'native son' factor is not to be overlooked. In times when the consumer activist is pressuring the regulatory authorities to restrict the entry of new additives or ingredients to the food supply, the regulatory agency may not be moved to immediate action even by masses of impeccable safety data brought to them by an overseas ingredient manufacturer.

Even in a hoped-for universal society, they seem more likely to be responsive to a need expressed by a native food maker for a new additive or ingredient, especially when backed by data supporting the economic need for the new and documenting the deficiencies of the old. The overseas manufacturer, in order to minimize lengthy procedural delays in processing of Food Additive Petitions, might be well advised to form alliances with native sons who are better placed to negotiate with their own public servants. A poor application will not be more likely to succeed, but an adequate one might get prompter attention if a national concern can be identified.

The domestic food manufacturer is also in a position to inform the regulatory agency what ingredient or additive he will cease to use if given the opportunity to employ a better one. Given the present unease over food color additives, antioxidants and some other ingredients in prepared foods, any change in the course of sugar substitution that can be brought to bear on the decreased need for such additives (and other items of inadequately characterized health potential) can now be made the focus of a possible trade-off for regulatory action to decrease concerns over health issues. A sugar substitute that at the same time permits decreased use in the food supply of sulfites, nitrites, modified starches, colors or antioxidants that might otherwise be required, is likely to be viewed with increased sympathy if such replacement properties are included among the technological attributes of the proposed substitute.

10. DESIRABLE NON-SWEET PROPERTIES OF SUGAR SUBSTITUTES

For the reasons discussed it seems that sugar substitutes with advantages beyond sweet taste will need to be developed in order to provide the benefits related to the bulk, preservative, physical and chemical functions that have until now been conveyed by the use of sugar. These substitutes can be nutritive and digestible, or non-nutritive and non-digestible, or somewhere in-between, in order to minimize osmotic laxation or the generalized discomfort associated with enhanced bacterial action in the large intestine subsequent to incomplete digestion and absorption of nutrients in the small intestine.

Ideally, the substitute will be as sweet, safe and soluble as sucrose, and as non-cariogenic, non-caloric and non-glucogenic as cellulose. In addition, it will be non-laxative and resistant to bacterial action in the lower gut, making up in total a disparate group of properties most unlikely to be found in any one molecule.

Of these properties the least important is sweet taste, since this is already adequately supplied by available intense sweeteners. Taking safety and stability for granted, the most important property is good water solubility, followed closely by a reduced or non-caloric property, since non-caloric to the body almost certainly means non-cariogenic to the teeth and non-glucogenic to the blood.

The next most valuable property will be for a laxative potential lower than that of sorbitol, since the total market size and variety of product forms will eventually be determined by the ADI (Acceptable Daily Intake) set for this new substance. In the USA the ADI for sorbitol is 50 g/day for adults, and 20 g/day for children, barely sufficient for one sugarless candy bar; and since it is already being approached by some people through the current consumption of 'sugarless' confectionery, the real need in this area will be for sorbitol rather than sugar substitution.

Since laxation and other forms of intestinal distress can also be the result of bacterial action in the lower gut, as well as that of the osmotic effect due to non-absorbed small molecules in the upper gastrointestinal tract, the new substitute will have to be reasonably inert both to bacteria already resident in the lower intestine and to those that might eventually exist through adaptation. Availability to gut bacteria compromises the non-caloric claim, even though the overall process is less efficient in providing energy to the host.

In fact, this is the nub of the problem. How to get a substance through the mouth without affecting or being affected by oral bacteria; how to get it through the upper gastrointestinal tract without being taken up by passive diffusion or hydrolyzed by brush-border enzymes; and finally how to get it through the lower intestine without the kind of bacterial fermentation that produces calories, flatulence, excessive stool size and other discomforts to the host. It is the water-insolubility of cellulose that enables it to meet these tests, since, of course, a 'water-soluble cellulose' would at once become vulnerable to enzymes from adaptive organisms.

11. NEED FOR DEVELOPMENT OF PURCHASE RATIONALE

On another, non-biological level, the product advantages of the food using the new sweetener must be considered, and the legal implications of drawing attention to sugar avoidance (or to the advantage of new sweetener usage) through advertising and label statements very carefully thought through by the parties involved. The target consumer has to be identified, his purchasing power estimated, and plans laid to coax a total market of the required size into existence if it is not already there.

The sweetener should be tailored to the advertising claims or labels needed to induce product purchase by the consumer, rather (as seems often the case) than assembling product claims as best possible once the sweetener has been discovered. It is most important that such claims be known to be in accord with the health and nutrition policies of the country involved. If a radical change or new departure is seen to be needed, it may be unrealistic to expect such a policy change without the mediation of food-industry groupings and the support of pertinent public bodies like the dental and medical professions. Though the direct one-to-one relationship between producer and user may be less appropriate in these situations, the producer still has the task of choosing a user likely to show the needed skill and initiative in mobilizing the interest and support of food-industry groupings that in turn eventually ensure the legal presence of a new sugar substitute in the marketplace.

With the sugar industry itself in the van of efforts to alter the sucrose molecule to make it much sweeter (chlorosucrose) and less caloric (isomalt), it is logical to expect somewhat less than the usual monolithic resistance to sugar substitution shown in the past by this industry. If only the sugar industry could come up with a tasteless form of sucrose, then at least part of the bulking problem would be solved.

12. IDENTIFICATION OF POTENTIAL MARKET SIZE

Finally, it must be emphasized that joint efforts will be necessary to focus much more carefully on the targets of sugar substitution strategies. Up to now it can be said with some truth that the real target has been water, i.e. to convert water to sweetened water, and not to substitute for sugar. Even the use of sorbitol, by far the largest weight of sugar substitute, is minuscule compared with sucrose, for which it can be a direct replacement.

The size of the sugar market must also be rethought and new substitution strategies developed. As long known or suspected in the food industry, US sugar disappearance or 'consumption', based on industrial production and shipment figures of all caloric carbohydrates,[4] is twice the estimated average consumption of all sugars in foods as measured by dietary surveys.[3] It may be close to three times that amount of sugar which can be accounted for as *added* sugar (by individuals or by food companies), and close to five times the average dietary consumption of added sucrose, since this value has been greatly diminished recently by the use of high-fructose corn syrup in beverages.

Since trends are ascertainable only from the annually reported USDA 'disappearance' data, and individual consumption of added sugar (as obtained by dietary surveys) varies extremely widely from the calculated averages, it cannot be said with certainty what effect past sugar substitution strategy has had on that part of the sugar supply potentially available as a target, i.e. that part responsive to conscious actions and decisions by the food processor or individual in deciding to substitute for sugar.

However, on average it appears to be nothing. Sugar substitution so far has created new products which supplement but do not directly compete with sugar-sweetened products, though they may have prevented or impeded growth of sugar products that might otherwise have occurred in the absence of artificial sweeteners.

But the basic facts remain. It is very difficult to document sugar substitution and very easy to document sugar substitute growth. How long can this go on before new sugar substitution strategies are required, especially those not involving health scares about sugar? Since sugar at current usage and consumption patterns could be implicated in only one disease (dental caries) out of 14 health issues,[3] it is obvious that careful thought is needed to find new market niches, i.e. to target what opportunities may yet exist for either real or apparent sugar substitution benefits. Corn-derived sweeteners, in taking half of the US market for sucrose, clearly showed that direct sugar substitution is possible, given suitably benign regulatory and economic climates, through utilizing physical factors which offered cost and functional advantages of liquid syrups over crystalline sucrose.

It would be an oxymoron to suggest that sucrose or dextrose are clearly the best 'sugar substitutes' in the sense of providing bulking agents suitable for use with intense sweeteners in non-aqueous systems. United States food law, however, would have to be changed to permit this combination of sweeteners—an unlikely prospect, at least for non-nutritive sweeteners—and in any case, 'less cariogenic', as opposed to 'non-cariogenic', would have to be established as a new provable and significant product claim.

13. CONCLUSION

Factors permitting further major changes and market opportunities will become operative only after careful planning by both the prospective producer and user of new sugar substitutes. This will necessarily involve a detailed study of the regulatory, political and economic issues in expanding the use of intense sweeteners with new bulking agents. The need for close producer–user collaboration is stronger than ever, particularly in order to escape the present preoccupation with beverages and other foods depending on water as the real substitute for sugar.

A bulking agent other than water is needed by the food industry if sugarless (and particularly reduced-calorie) claims are to become possible for more food classes. A bulking agent other than sorbitol will probably be needed by the confectionery industry if the benefits of dentally harmless products are to be found beyond their present very highly concentrated positions in sugarless chewing gum and candy.

The designation 'sugarless' now rests uncomfortably on sorbitol-containing confectionery products, but the 'sugarless' class of products, even in confections, is not likely to spread until the reduced-calorie sense of 'sugarless' becomes as sustainable as that of 'not promoting tooth decay'. This will undoubtedly require new bulking agents more biologically inert than those now available. The exculpation and disassociation of sugar from all disease causation except that of tooth decay is likely to leave only tooth benefits and calorie reduction as rationales for products based on sugar substitutes.

REFERENCES

1. MACKAY, D. A. M. (1985). Factors associated with the acceptance of sugar and sugar substitutes by the public. In: *Int. Dent. J.*, **35**, 201.
2. MACKAY, D. A. M. (1979). Sorbitol as a sugar substitute. In: *Health and Sugar Substitutes*, *Proc. ERGOB Conf. Geneva 1978*, B. Guggenheim (Ed.), Karger, Basel, pp. 124–9.
3. Sugars Task Force. (1986). *Evaluation of Health Aspects of Sugars Contained in Carbohydrate Sweeteners*, US Food and Drug Administration, Washington, DC.
4. ANON. (1986). *Sugar and Sweetener Situation and Outlook Yearbook*, SSRV11N2, Economic Research Service, US Department of Agriculture, Washington, DC.
5. DENSFORD, L. (1986). FDA sets guidelines on health-claim labeling. In: *Food and Drug Packaging*, (Oct.), 94.
6. LINKE, H. A. B. (1980). Inhibition of dental caries in the inbred hamster by saccharin. In: *Ann. Dent.*, **39**, 71.
7. SCHEININ, A., MAKINEN, K. K. and YLITALO, K. (1975). Turku Sugar Studies V. Final Report on the effect of sucrose, fructose and xylitol diets on the caries incidence in man. In: *Acta Odont. Scand.*, **33**, 67.

Chapter 10

EVALUATION OF THE USEFULNESS OF LOW-CALORIE SWEETENERS IN WEIGHT CONTROL

D. A. BOOTH

Department of Psychology, University of Birmingham, UK

SUMMARY

There is no scientifically adequate evidence nor theoretically solid reason to assert that sugar is any more of a stimulus to energy intake than any other constituent of attractive energy-containing foods or drinks, or that indiscriminate cutting back on sugar intake helps to reduce body weight or to prevent obesity. There is evidence, however, that energy consumed in a drink or a food item by itself an hour or more before a normal eating occasion would not contribute to control of total energy intake as effectively as such energy consumed in the early part of a meal or mixed snack. Bulk sweeteners such as sucrose contribute to the energy that many people habitually consume between meals or at the end of meals. Thus their replacement by intense (low-calorie) sweeteners in drinks or food items frequently used at those times might in principle contribute to weight control. It is suggested that, although such designs have yet to be used in consumer research or in epidemiology, investigations can be carried out that could provide evidence for or against a contribution to weight control from this specific pattern of use of intense sweeteners.

1. RAISON D'ÊTRE FOR 'ARTIFICIAL SWEETENERS'

Low-calorie sweeteners exist chiefly because they are used in efforts to control body weight. Yet there is no scientifically adequate evidence that the substitution of intense sweeteners such as saccharin, cyclamate, aspartame and acesulfame-K for bulk sweeteners such as sucrose helps the reduction of weight or the prevention of obesity.

This extraordinary situation has arisen because it seems so obvious that satisfying a liking for sweetness without consuming sugar should reduce energy intake. Even some nutritional authorities commend the notion. As a result, there is a demand for these substitutes and it is regarded as legitimate to meet it, so long as customers are not misled.

The issue remains as to whether there is any benefit from 'diet drinks', except presumably to the teeth. Research into dietary aspects of obesity is rather unfashionable, especially if it concerns ordinary food. Productive investigation of mechanisms operative in everyday life may not be easy either. Yet those barriers to the resolution of scientific issues such as the role of sweeteners are arguably a major reason why unhealthy overweight is so prevalent. The public interest makes it incumbent on both health research and commercial research to address the issue with urgency. This chapter considers the existing assumptions and evidence with a view to the design of such investigations.

2. CURRENT ASSUMPTIONS

2.1. The Myth that Sugar Causes Obesity

The usual rationale for the use of intense sweeteners in weight reduction is that sugar causes obesity or that, even if it does not, sugar is a good candidate for exclusion from the diet when energy intake has to be reduced.

This rationale is largely mythical. 'Myth' in this context is not a word chosen for effect, but the correct technical term from anthropology. The story that a sweet tooth or cravings for sugary things cause obesity is used for moral purposes in our culture, regardless of its empirical truth or falsity. These ideas are promulgated by authorities to encourage healthy eating. They are internalised in the experiences and anecdotal observations of most ordinary citizens. Yet the official doctrine in Britain, for example, that halving the nation's sugar intake would reduce the prevalence of obesity, is admitted to be without scientific foundation by expert groups that recommend the policy.

Myths about behaviour are not readily recognised as such by professionals or by the general public because what people think, feel and do are matters of everyday experience. However, this does not legitimise intuitive pronouncements on psychological mechanisms, neglecting behavioural science, by those not expert in the field.

2.2. Craving for Sweetness

The main supposed fact on which this myth rests is the idea that fat people suffer from abnormally strong compulsions to consume sweet foods and drinks. This psychological hypothesis has been tested repeatedly over the last decade or so, with no support for it emerging.

Nevertheless, official advice is still based on little else. Even if the premiss were true, the advice would be unrealistic. If it were so difficult to do without sweetness, avoidance of sugar would have to rely on intense sweeteners to such an extent that toxicologically acceptable limits would often be exceeded, even with wide use of sweetener mixtures.

2.2.1. Measurement of Motivation

There is no doubting the common experience of wanting something sweet for its sweetness. Furthermore, most if not all of us know what it is to crave for a sweet item when not wanting anything else to eat. Nevertheless, the claim that such experiences and the actions resulting from them contribute to obesity (or make difficulties for weight control when the sweetener is caloric) is a causal hypothesis with a crucial psychological component: that sweetness causes eating that would not otherwise occur.

Claims about psychological causation are no different from other mechanistic statements in their susceptibility to objective test. The relevant evidence to evaluate this hypothesis must derive from systematic observations of what people do in everyday life when faced with foods and drinks containing sweeteners. Experiments or surveys have to be designed and analysed to yield measurements of the vigour with which each individual reacts to sweetness. The testimony of a dieter, even a survey of the feelings of large numbers of typical dieters—or the informed opinions of leading clinicians, marketers, food legislation experts or academic psychologists—do not distinguish public mythology from psychological processes in the way that controlled, theory-testing behavioural investigation of representative individuals does.

No adequate direct or indirect tests of the idea that a sweet tooth causes problems in weight control have yet been carried out. Nevertheless, experimentally-based theory of the psychological mechanisms constituting appetite for foods is beginning to provide some highly relevant background. There is now good reason to suppose that sweetness plays only a modest role in the motivation to eat many sugar-containing foods: social conventions of eating and drinking served by the marketers of food and drink have at least as much to do with the craving for sweetness as does sugar itself.

To see this, let us consider first the nature of likings for sweetness in food and drink and then the roles of sweet foods and beverages in the diet.

2.3. Innate and Acquired Likings

The innate liking for sweetness must be distinguished from acquired likings for sweetness in foods. Conceptual confusion between the two mechanisms and a lack of adequate psychological measurements of their relationships have been major factors fostering the myth that sugar causes obesity.

2.3.1. Congenital Sweet Preference

Sweetness is the only sensory quality (apart from 'fluidity') that elicits, invigorates and rewards ingestion in newborn human infants. It is not feasible to test infants' preferences by choice behaviour, but from tests of differential acceptance it is evident that the stronger the sugar solution, the more vigorous is the infant's ingestive response.[1,2] The higher concentrations of sweetener elicit ingestive motor patterns more powerfully via a brainstem reflex mechanism.[3] These patterns produce a beatific expression on the infant's face,[4] which suggests pleasure to the observer.

In a more developed human being, anticipation of such responses is likely to be felt as a desire to ingest, and their elicitation would be experienced as satisfaction of that desire. The vigour with which the tendency to respond was being stimulated would no doubt also be expressible verbally, e.g. as the degree of pleasantness of the sampling, or as a choice between different samples.

2.3.2. Acquired Likings for Sweet Foods and Beverages

Experimental evidence has been accumulating over the last two decades that the appetising action or palatability of a food in man, and in animals such as monkeys, rats and birds, is normally a learned reaction to a specific complex of sensory characteristics that has become familiar[5] and has been associated with preference-inducing social[6] and physiological[7] consequences. Sometimes that learned liking for a type of food may be relatively restricted to a specific eating context, such as breakfast foods,[8] but sometimes not, like bread (which appears, in some English cuisine for example, as breakfast toast, luncheon or teatime sandwiches, and dinner rolls).

It is of course not new to recognise the many sources of influence on the collection of acts and reactions that we label 'eating motivation', hunger, appetite, craving or palatability (all different words for one and the same objective tendency to ingest). Sensory characteristics of foods are usually attractive. Bodily sensations of food deprivation are agitating and may put the sufferer in mind of food. External cues, like mealtime on the clock or somebody else starting off to an eating place, provide incentive to go for food. However, these sensory, somatic and social stimuli to eat and drink have usually been considered to contribute independently to the strength of the motivation.

What is relatively novel is the idea that not only interaction among the sensory factors need not be additive, but also the contributions of palatability, deprivation and situation do not simply add up to acceptance of a food. In fact, contextual interdependencies in appetite have seldom been considered, let alone measured. When tested, much of the motivation to eat is found to be acquired by attachment to distinctive combinations of dietary, bodily and environmental stimuli.[7,9–11] That is, a food identity in context—an ideal configuration, i.e. what might be called an objective *Gestalt*—controls the disposition to eat (being hungry, having an appetite), and indeed also controls its suppression as a result of eating (satiety). A practical example would be the effect that a brand name sometimes has on preference for a food formulation.

To the extent that eating and drinking are thus a response to stimulus complexes that have acquired their motivating power by habituation and learning, any deviation from a familiar formulation (or indeed from the familiar eating context) will reduce the strength of appetite. That is to say, a food or a drink which is familiar to someone always has a peak preference level of each of its salient characteristics—including even its sweetness.[12] Formulations not distinguishable from the acquired ideal will be most acceptable. On the other hand, when a largely unfamiliar formulation is presented, little or none of the usual acquired motivation to eat or drink will be elicited. Between these extremes, any sensory characteristic (or indeed any somatic or social factor) will show declining acceptance on either side of its ideal point for the individual. If psychologically equal decreases in an influence on acceptance obtain on each side of ideal, then the hedonic or tolerance function will be triangular—an inverted V.[12]

Our observations confirm that sweeteners are no different in this respect from any other salient influence on acceptability. Every individual whom we have tested with familiar sugar-containing foodstuffs and beverages has given an isosceles tolerance triangle for sugar level. This shows total suppression of the innate preference, which would increase continuously with sweetener level.

2.3.3. Reversion to Innate Sweet Preference

If an unappetising but sweetened food is presented, however, then the acquired *Gestalt* of particular ideal levels of sweetness and of other sensory characteristics may not be elicited strongly enough to suppress the innate motivation that increases monotonically with sweetness. In other words, the crudest sweetening strategy—'the more the better'—is characteristic of a breakthrough of the primitive reaction.

Such a breakdown in the learned decline of preference with excess sweetener could result from low-quality flavouring,[13] low-quality sweetener, or a combination of both (as perhaps seen by Beck[14] with artificially sweetened cola drinks).[12,15] We have also found it with high-quality fruit cordial presented at a dilution too great for a particular assessor.[16]

Saccharin water is the ultimate non-drink: familiar flavour complexes are infinitely diluted and what flavouring there is suffers from bitter sidetaste and undue lingering. Even a cleanly sweet sugar solution (or crystal) is not a normal drink (or food). Hence many people do not show a breakpoint or ideal value in hedonic ratings of sugar solutions.[17,18] As yet, no one has adequately studied the cognitive mechanisms involved, but the task is nonsensical as an assessment of materials the assessor would choose to drink (let alone as a measure of pleasures of the palate, as claimed by Cabanac[19]). All that assessors can do is revert to infantile reflex responding, perhaps moderated by strained extrapolation from a drink of a low-tartness, low-aroma fruit. Such attempts to get subjects to make sense of rating pure sensory characteristics for pleasantness out of the context of real drink or food are an extreme example of failure to appreciate the principle that model systems do not simplify the analysis of sensory influences on acceptability; they destroy it.

2.3.4. Consequences of Reversion to Unsocialised Appetite

This uncontextualised innate liking for sweetness makes it the only sensory characteristic that can sometimes be pleasant to adults when presented by itself or in a novel food. If children's peak preferences have yet to be fixed as strongly as adults', then perhaps the young are even more likely to persist in eating a novel food if it tastes sweet. Furthermore, this ingestive response to sweetness is liable to result in the induction of a preference for whatever level of sweetness caused the persistent consumption.

If a variant of a familiar type of sweet food was sufficiently unfamiliar to have reduced overall acceptability, an increase in sweetness above the familiar level might make it more palatable by the same innate breakthrough effect. As a result, an optimisation test of a new product or reformulation could be distorted to unnecessarily and even uncompetitively high sweetener levels if the consumer panellists were not first fully accustomed to the rest of the new formulation. Such contamination of consumer research data could cause an upward drift of sweetener levels in product ranges subjected to major reformulations. This effect may have contributed to the recent disaster on launch of a slightly sweeter variant of a leading cola drink.

Such susceptibility to sweetening in unfamiliar foods no doubt contributes to the impression that sweetness must be a strong stimulus to intake. However, the inference is invalid because nearly all intake is of thoroughly familiar foods and drinks, in which sweetness determines palatability in the same way as any other salient characteristic.[12,16,20]

Sweetness also balances certain other sensory characteristics, such as non-nutritive bitterness and sourness, or the coarseness of dietary fibre, to yield a mixture that becomes an acquired 'taste'. These interactions are probably perceptual and also involve competition between innate preference and innate aversions. Some sweeteners also augment certain flavourings, although this is mainly of economic significance.

In short, the effects of sweetness on intake do not depend primarily on sugar or other sweeteners themselves. The normal contribution of sweetness to intake depends on which foods or beverages have come into habitual use with a character which is partly determined by some level of sweetener.

However, in order to elucidate the role of sugar or intense sweeteners in energy intake, we must go beyond even the contextualisation of sweetness with other inherent characteristics of foods and next consider the contextualisation of sweetness in varied situations of eating and drinking.

2.4. Contextualisation of Sweetness to Eating Occasions

2.4.1. Sweet Desserts

Many readers of this chapter will be acculturated to the habit of a second and final course in ordinary family meals, 'the sweet'. Many desserts are strongly sweetened, but not all, and every sweet dessert is 'balanced' with tartness or some other strong flavouring, texture and/or colour.

Furthermore, substantial levels of sweetness also appear in some aperitifs, appetisers, soups and meat preparations, and in some of the dressings or sauces used in main courses. In addition, when an elaborate meal is served in a culture that uses 'sweet' desserts, a sugary dessert is not usually the final course, but is followed by cheese with plain or salty crackers, and by drinks that are strongly flavoured but not necessarily sweet, like coffee and some liqueurs.

Nevertheless, many people conceive themselves as liking nothing but the sequence of a 'salty/savoury' course followed by a 'sweet' course. Sweet desserts are regarded as natural, even though they are not universal within cultures where the second course often is sweeter than the first—let alone across human cultures or indeed in the behaviour of other omnivore species. Indeed, this socialisation into the concept that desserts have to be sweet is so strong that Western scientists and gastronomists sometimes speculate about biological mechanisms that make the sweetness or the physiological effects of sugars likely to be desired at the finish of a meal.

Within such a culture, therefore, the lay-person and the expert are liable to confuse the desire for more food after eating the first serving of the final course of a meal with a desire specifically for the sensation of sweetness in such extra food. If we want more food at dessert-time and we conceive of dessert as 'the sweet', then we will be aware of the desire for a bigger meal as a craving for sweetness. It need be no such thing. Someone who habitually has a cheeseboard available after the dessert may feel a craving for a tasty piece of cheese and a positive aversion to a second portion of a sugary dessert.

Indeed, given the evidence for the above theory of acquired appetite and given the total lack of directly relevant evidence, there is no scientific foundation for the belief that the sweetness of many desserts is more of a stimulus to eating than any other salient appetising factor in the foods typically available in that context.

2.4.2. Sweetness between Meals

It is also entirely possible that the use of sweet items of food or drink between meals is purely conventional. Sweetness is a salient component of the habit of taking a drink in the middle of the morning, at mid-afternoon or late in the evening. However, there is no evidence that the sweeteners have a stronger influence on such ingestion than does a dry mouth, other salient sensory characteristics of the identity of the drink (try mixing infusates of tea and coffee) or the time of day (try serving some sweet coffee or alcoholic beverage at 4 p.m. to a conventional Briton who takes sugar in tea).

Soft drinks, flavoured milks and hot chocolate are always sweet and tea and coffee are often sweetened. Yet many people strongly prefer bitter drinks between meals, like beer or unsweetened tea or coffee. Many users moderate the bitterness of tea or coffee by adding creaminess and strongly dislike added sweetness. Some even augment the flavour of tea with a little lemon but no sugar. Thus it is by no means evident that sweetness has any special role in encouraging an adult to take a familiar drink.

The same considerations apply to any food item that is conventionally made available on a between-meals drink occasion. A well-organised office or works canteen in the UK has biscuits to eat in the teabreak, even though many adults can comfortably divide their daily caloric intake between three meals at intervals of about 5 h. A cookie or a candy is also served with the postprandial drink. Even some of the French have been induced to eat chocolate mints after dinner. The innate sweetness preference may make it easier to acquire an initially strange habit but the habit that contributes to daily energy intake in the long term could as well, and does often, involve any other food constituent such as alcohol, fat or the crunchy and juicy structures of fruit and vegetables.

2.5. Palatability, Need and Convention

From such considerations we may conclude that it is misleading to think of sugar as encouraging people to eat (or to drink caloric beverages) when they are not hungry. In normal circumstances, sweetness has no special power to stimulate eating more than any other factor in palatability, and palatability represents but one set of factors in the overall motivation to eat.

Indeed, the notion of 'eating when not hungry' is at best empty and at worst nonsense. Wanting to eat *is* being hungry, even when the hunger is more selective to certain categories of food (e.g. 'snackfoods' if it is not a mealtime) than in some other circumstances. The sight of a snackfood or a second dessert may even elicit an epigastric pang. Conversely, a meal may be needed in the absence of hunger pangs. The ascription is not an empirical psychological one but a moralistic complaint—they want to eat when, on some views, it may be ill-advised or repellent to observe.

The same scientific vacuity attends the notion that snackfoods or sweetness overwhelm satiety. Some may get fat if they acquire habits of eating desserts during sensations of abdominal fullness or of eating snackfoods without any hunger pang. Nonetheless, satiated eating motivation is nothing other than not wanting to eat. So it is nonsense to say that someone is satiated if they want to eat something—even if what they want is more food on top of what others and even they may regard as a 'proper' meal.

Gluttony is a moral concept, not a scientific one. The scientific concepts of overeating are either thermodynamic or psychosocial. Eating more energy than is expended is a precondition for obesity. Consuming large amounts of food at a session and/or a distressing sense of being out of control of the eating can be part of the diagnosis of a disorder of eating behaviour (bingeing). Many instances of overeating in the thermodynamic sense are likely to be entirely normal in the psychosocial sense, because individuals differ so much in resting metabolism, efficiency of food utilisation and physical activity. Indeed, so far from being detectably open to charges of gluttony, weight gain in certain circumstances can involve eating less than average.

The phrase 'eating when not hungry' obscures the scientific issue of whether any particular causal influence stimulates eating of more than is needed for energy balance. Nothing considered so far implicates sweetness as a contributor to obesity.

2.5.1. Relative Strength of Influence of Sweetness on Hunger

In the light of the above considerations, estimating the extent to which sweetness motivates eating is a complex matter of measuring individuals' behaviour in a particular culture. Intuition and anecdote cannot provide such an estimate, nor even survey data or cultural anthropology.

Two separate questions are involved. Does sweetness motivate eating and drinking more strongly than any other influence on appetite? Does sweetness motivate eating more in obese people, or those who have difficulty in losing weight, than in others? Answers can only come from valid measurement of the strength of influence of sweetness and other factors on food choices and amounts eaten, in contexts most relevant to weight control.

We shall turn to potentially obesifying contexts shortly. This is the point at which to register the need for sound psychological measurement of the motivating effect of sweetness and other influences in appetite. Although some psychologists and physiologists have been tackling these issues for some years, measurement techniques adequate for quantitative estimates of sensory motivation have not been used. The field has been held back by the assumption, widespread inside and outside psychology, that references to sensations and other subjective states are descriptions of those private experiences, and expressions of their magnitude are measurements in that world of private objects. Such views were shown to be incoherent by Wittgenstein.[21] The language which refers to subjective magnitudes ('strongly sweet', 'this is twice as sweet as that') are not reports of observations of an inner world but are ascriptions to an objective situation. Objective psychophysics[22] uses the relationship between variations in the stimulus and the assessor's graded judgements about the stimulus to quantify the individual's objective performance, i.e. to measure an effect of the situation on the person's response, such as the sensitivity of ratings of sweetness intensity or food acceptance to changes in level of sweetener in the food.[12] These are measures of perception or motivation, not of sensation or affect. An elaborate cognitive theory built on many complicated experimental analyses of performance will be needed before we could attempt to measure subjective magnitudes, if it is possible even then.[23] Rating responses by themselves measure nothing. This fact has been obscured by polemics in American psychology for the last 25 years, to the effect that asking people to rate characteristics of objects by matching subjective magnitudes with numbers used in ratio has been taken as a direct measurement of the intensity of sensations like sweetness, or indeed of affective events like the pleasures that sometimes accompany eating.

The exponents (slopes) of stimulus–response relationships using ratings of sweetness or pleasantness fail to distinguish measurement of performance from the assessor's use of numbers and from effects of the experimenter's choice of test stimuli and units. All the studies of responses to sweeteners in obesity and dieting to date use these techniques and so fail to assess either the role of sweetness in eating or drinking or the existence of individual differences in sweet tooth.

Most of the studies have failed to confirm some initial claims that obese people liked sweet water more than lean people and that the liking for sweet water was less easily satiated by sugar loads in obese people or formerly obese people.

However, these negative results might be attributable to the introspectionist inadequacies in these uses of ratings. Also, as we have seen, tests of sugar water are contaminated by innate sweet preference, which is irrelevant to normal eating (as also are the satiating or nauseating effects of loads made up of concentrated sugar).

2.5.2. Is Sweet Tooth a Difficulty in Weight Control?

The objective measurement of effect of sweetener level on acceptability of a diverse selection of foods, plus rated preferences between foods that differ in sweetness without the fact necessarily being recognised, has confirmed that people do vary from one another in how much sweetness they like in foods and how much they like sweet foods.[12] Furthermore, the new methods are precise enough for a small sample to distinguish slight differences on average between the sexes and between young and middle-aged adults.

Therefore obesity, weight history and, most relevantly, difficulties in weight control should be re-examined with the new method. There seems to be no strong reason to expect 'sweet tooth' to be stronger in those who gain weight easily. The clinical impression that craving for sweetness creates difficulties in restraining intake is obscured by conflation of the desire for food between meals and at the ends of meals with a desire for the sweetness that is commonly available on such occasions. However, if some eating habits do give sugar a role in obesity, as might be the case with energy intake between meals (see below), then some possessing those habits and having difficulty in losing weight might have a sweet tooth specifically for the foods they use habitually in that fashion. This would of course be a 'culturally-induced' sweet tooth.

In recent Norwegian culture at least, obesity and underweight have a substantial genetic origin.[24] The hypothesis that a strong sweet tooth causes obesity has often carried genetic overtones. However, there is evidence against inheritance of variations of the congenital sweet preference, from comparisons of infant acceptances with parental hedonic ratings of sugar water.[1] Furthermore, as we have seen, any innate variations in sweet preference are likely to be overwhelmed by acquired preferences for particular levels of sweetness in particular foods. However, unsuccessful slimmers have been shown to be poor at acquiring control of eating by complexes of stimuli from both the food and the body.[25] Maybe a low capacity for such learning is one genetic factor in obesity.[10] People who, as a result, are more susceptible to breakthrough of the innate sweet preference may have difficulties with weight control in a sugary food-using culture. However, even if inadequate learning allows infantile reactions to sweetness which contribute to obesity, it remains to be seen whether this poor learning capacity is inherited or has been induced by struggles to restrain intake that were made difficult by other factors.[7]

2.6. Inefficient Satiety between Meals

An individual's energy balance depends on personal variants of culturally influenced uses of energy-containing foods and drinks, and on how the physiological, social and cognitive controls specific to energy intake and expenditure affect that person's habits. The consequences for weight control of the liking for sweetness in foods and drinks will therefore be much more complicated than a simple correlation. It follows from the previous argument that the role of sugar or its low-calorie substitutes in weight control could involve also much more than mere liking for sweetness.

A different possible connection between sugar and obesity is nevertheless quite readily conceptualised in practical terms. Rather than the taste of sugar stimulating eating or overwhelming the feeling of fullness, the energy absorbed from sugars in mistimed food and drink fails to suppress appetite on subsequent eating occasions. Some between-meal drinks and accompanying 'nibbles' contain sugars, starch, fat or alcohol in small amounts that are absorbed before the next eating occasion and so their energy does not stimulate the major physiological mechanism for compensating food intake.

Thus for people who are timing some of their energy intake 'inefficiently' in this sense, replacement of bulk sweeteners by intense sweeteners in the drinks and foods usually involved in such habits might in principle contribute to weight control.

This sort of possibility is sufficiently obvious to have been widely considered in some form or other, e.g. 'snacking makes you fat'. However, there is no evidence for that particular proposition. On the contrary, eating small meals frequently is less associated with overweight than one or two big meals per day.

2.6.1. Suppression of Appetite by Absorbed Energy

It has been shown that modest amounts of readily assimilated energy suppress appetite for only 30–60 min at the most.

People who have drunk 50 g of glucose in a flavoured 50% solution eat less after 20 min, but not after 3 h, than they eat after a similarly flavoured saccharin load.[26] However, substantial amounts of free sugars are not normally ingested at such high concentrations. Indeed, their osmotic action in the human intestine is sufficiently powerful to suppress even the innate preference for sugar water.[27] Nevertheless, similar or smaller doses of concentrated maltodextrin also transiently suppress both food intake and rated interest in eating foods.[28–30]

There is first several minutes' delay for digestion and absorption to get under way. Then the suppression of appetite for food lasts only as long as the concentrated carbohydrate is being absorbed following rapid emptying from the stomach.[28] In animals, intravenous infusion of glucose at the rate it is absorbed from carbohydrate loads or maintenance diet meals causes the same suppression of intake as a load or meal.[31] The appetite-suppression even lasts for a short while after the end of rapid absorption or infusion. This, with comparisons of other more or less readily utilised sources of energy in the rat[31,32] and in human beings,[33] shows that it is the absorbed energy that is inducing the transient satiety, not gastrointestinal action of the food nor the hormonal secretions triggered by absorption.

This appears to be the main mechanism by which a piece of confectionery (whether of the common sort or sold supposedly to aid in weight control) can suppress appetite to some extent when eaten 15–30 min before a meal. However, as much as 200 kcal or more of readily assimilable energy is fully absorbed within an hour of ingestion.[34] Hence, the energy in a drink between meals (or last thing at night), and the energy in a food item taken with the drink, are unlikely to contribute to moderation of food intake on the subsequent eating occasion.[35] As a result, the satiating effect of the energy is ineffective at controlling total daily intake. In contrast, a substantial snack or meal that is still being absorbed or processed by the liver at the time of the next meal is likely to reduce the voracity of appetite for a meal.

Energy taken between meals or even in the latter half of a meal is also likely to be an ineffective suppressor of subsequent intake via a learning mechanism based on physiological after-effects of energy eaten in the earlier stages of a meal.[30,36] When a substantial amount of disguised maltodextrin is included in the first half of a distinctive menu, the total amount eaten on that menu in a meal on a subsequent day decreases relative to another menu to which the maltodextrin had not been added.[25,29,30] Yet when the extra carbohydrate is included in the dessert, the learned menu-specific compensation does not take place.[30] This is apparently because the absorption of glucose from the maltodextrin does not start until 10 min or so after its ingestion. Therefore, with carbohydrate in the dessert, the strong satiating effect of absorption starts several minutes after the end of the meal. In contrast, when the carbohydrate is ingested earlier in the meal, the jolt of satiety begins immediately after or even during the last course. The extra 'filling' effect is therefore more readily attributed to the dessert, even though the carbohydrate was in fact eaten earlier.

The extra satiety conditions a relative aversion to the dessert which is contextualised to the late stage of the meal: in other words, a relative loss of appetite or a satiation is learned.[25,30] This learned satiation is a tendency to be less interested in eating that dessert in that context next time. While the extra carbohydrate continues to be included in the early part of the menu the acquired hesitancy about the dessert will be maintained.[29]

Hence neither energy consumed between meals nor energy eaten near the end of meals will be compensated for in subsequent intake by means of either of the only known energy-related satiety mechanisms.[28,37,38] On the available evidence, the ineffectiveness of these timings contrasts with the compensation for bulk sweetener or any other energy eaten in the earlier stages of major meals or in substantial snacks that are taken at sufficiently short intervals.

2.6.2. Implications for Weight Control and Drink Formulation

Western culture has conventions of using sugar at the time of drinking, both in the beverage and in any accompanying food item. Sugar is also a common constituent of second courses of meals and snacks. On the above evidence, foods and beverages which are commonly used sufficiently frequently at such times are in principle likely to be 'slipping in' enough energy past the physiological controls of food intake to be an insidious or even obvious cause of weight gain, or of difficulty in weight loss.

It follows that there may be a role in weight reduction and even perhaps in the prevention of obesity for properly informed and managed use of drinks and foods that have had their contents of bulk sweetener replaced by intense sweetener. Sugar would have to be the main energy source in the item and the substitution would have to be virtually complete, because otherwise the avoidance of inefficiently satiating energy could readily be vitated by taking larger amounts of the item. It is hard to see any objective merit in reduced-sugar products. There might be some benefit from low-sugar items for eating between meals by those who cannot refrain from nibbling something, but these must be items low in energy overall. The clearest merit in principle lies with zero-energy drinks, replacing sugared drinks and even milk-containing beverages that individuals use frequently on occasions when only fluid is needed.

However, this strategy cannot be effective unless individuals with weight-control problems regularly choose the intense-sweetened products over the bulk-sweetened products. The intense-sweetened products must be readily available and their use must be socially acceptable in the situations where energy-containing drinks are frequently used by the individuals (e.g. young and middle-aged sedentary men). The most relevant health education and product marketing has hardly begun.

On the other hand, as an individual strategy over the months and a social strategy over the decades, sweetener substitution might best be regarded as merely an interim measure. Many sugar-containing foods cannot be reduced in energy, so substitution by intense sweeteners cannot eliminate altogether the potential fattening effect of habitually eating sweet foods between meals or at the ends of meals. Hence, it may be argued that individuals and societies with weight-control problems should encourage the habitual use of energy-free between-meal or end-of-meal drinks and foods that are not regarded as sweet, on the grounds that this should in the long run reduce the abundance and hence the acquired relative attractiveness of sweet items that cannot be substantially reduced in energy.

2.7. Nutrient Density: Bulk Sweeteners as Nutrient Diluents

It is commonly argued that whether or not sugar intake is a cause of obesity, sugar is a prime candidate for exclusion from the diet in the treatment of obesity.

The basis for this is that sugars provide 'empty calories'. More specifically, refined sucrose furnishes no other macronutrient or micronutrient besides readily-assimilated carbohydrate, and it contains no dietary fibre. Sometimes the argument is vaguer and indeed unrelated to reliable scientific principles—namely that anything 'artificial', 'refined' or 'manufactured' is liable to be bad and so the purity of table sugar must be deadly.

It is the principle that is artificial. Nobody, not even a compulsive binger, makes a habit of eating table sugar. One of the least nutrient-dense common uses of table sugar is in hot drinks, but nobody lives off coffee or tea, and most people drink tea, coffee or chocolate with milk. The nutrient density of the diet could be prejudiced if someone frequently took a substantial proportion of his daily energy entirely as confectionery. There is no evidence that this happens, even in children. The risk of nutrient dilution is hardly less in the more likely eventuality of frequent snacking on fried potato or corn starch products, or indeed on any other single type of food. Clearly it is advisable to replace any such high-frequency monophagic meals or snacks with mixtures of different types of food. Yet that does not preclude the use of sucrose as a substantial source of energy.

Average intakes of sugar found in dietary surveys in the UK and USA are in the range of 8–14% of energy, with the upper decile boundary in the USA being about twice the average energy percentage. That is, rather few people seem to be taking more than one-fifth of their energy as sugar. This is notwithstanding the fact that the recorded national sucrose production rates give an average per capita of 20% of energy or more in both the USA and the UK (on the basis of an average energy intake of 2000 kcal/day), a disparity which is currently inexplicable.

What this adds up to is that sugar is a food commodity, not a food. Sweeteners are mostly used as constituents of foods and drinks containing many nutrients, and these complex items are eaten in combination with other foods. That is, sucrose as generally used does not function as 'empty calories'. It is often unlikely to be having a nutrient diluting effect, owing to the 'cultivated palate' or the good sense of users and manufacturers.

For example, sugar contributes readily assimilable energy when added to or incorporated in breakfast cereals, eaten with milk. Indeed, the density of many micronutrients in presweetened fortified cereals used without added table sugar is greater than in uncoated fortified cereals with similar amounts of sugar added by the eater, because the fortification levels are relative to the energy in the packaged product. The sugar put in or with bran cereals, muesli and granola may well increase mass intake of soluble and insoluble cereal fibre more than any other single factor—directly contrary to the 'refined carbohydrate' presumption.

Therefore an undiscriminating presumption that sugar is one of the parts of the diet most suitable to be cut for weight control is nutritionally rather unrealistic. It becomes ridiculous if a fear of nutrient dilution motivates consumers to replace sugar with intense sweeteners in items that contribute little other nutriment in any case.

It would make more scientific and practical sense to rethink the argument in terms of avoidance of energy intake on occasions when what is desirable is the intake of palatable liquid only. Non-caloric sweeteners (and alternatives to alcohol) in beverages drunk by themselves then have, for the interim at least, the important role already described.

Calculation of daily micronutrient intakes after deleting sugar from a notional record is therefore quite bogus. The only reasonable method of addressing this issue is to relate individuals' food choices involving sugar to symptoms of nutrient deficiency or excess, or to failure to attain adequately founded recommendations of daily nutrient allowances.

3. THE EVIDENCE

Clearly we cannot merely assume that cutting down on sugar helps to reduce weight. We must seek real evidence whether or not any such a strategy helps the individual to reduce total energy intake. Even evidence that sugar intake contributes to obesity could be relevant at least to prevention, whether or not weight reduction prospects were implied.

3.1. Sugar Intake and Obesity

3.1.1. Epidemiology

Data on individuals' body weights and heights are available for large populations. However, there is at present no validated method of estimating free-living dietary intakes, especially of a food constituent having some emotional attribute like sugar. Nevertheless, data from food frequency questionnaires (especially when backed by portion-size assessment) may in principle be subject to least bias, although imprecise. The only evidence for the theory that sugar causes obesity that such data could provide would be a positive association between weight-for-height and sugar intake.

Extensive studies of the relation of individuals' sugar intake to obesity in adolescents and adults in the USA, the UK and South Africa have consistently shown negative correlations: the lean people are eating more sugar than the fat ones.[39–42] There must be a causal connection, and it cannot be that sugar intake is a substantial cause of obesity. Assuming (for lack of a scientific basis) that sugar is not a slimming agent, the only possibilities are that obesity causes lower sugar intake or that a third factor both increases sugar intake and reduces weight. There is every reason to adduce the first explanation for cultures where sugar is widely considered to be fattening: in such societies a fat person is likely to try to avoid sugar, and to succeed to some extent. Furthermore, the existence still of a negative correlation amounts to evidence that cutting back on sugar is not an important part of successful weight reduction.

A positive association across populations might be expected. Average intake is likely to be excessive after eating habits have been established during highly active periods, be they during an individual's youth or in previous decades of an increasingly mechanised urban society. The prediction of a positive correlation between sugar intake and obesity within a population, however, is quite unsound. Obesity could be the result of a period of positive energy balance that has long since ended. Therefore current intake may have nothing to do with aetiology. Also, degree of overweight is no measure of relative success or failure in efforts to lose weight or to resist weight gain: some instances of mild obesity could be triumphs over past or potential gross obesity. Total sugar intake could merely correlate with total energy intake, and in any case the large variations in physical activity, resting metabolism and thermogenesis among individuals make the relation between energy intake and body weight or adiposity notoriously blurred. The relevance of the proportion of energy intake as sucrose has not been established physiologically or psychologically, as shown above.

In short, the epidemiological data measure the belief that sugar causes obesity but provide evidence against that belief.

3.1.2. Clinical Correlations

Some case-control studies of sugar intake and weight-for-height have been carried out and also some clinical intervention studies where extra sugar intake was prescribed for investigative purposes. The accuracy and validity of weight records, let alone the recall method used, do not approach those needed for energy balance estimates. Nevertheless, weight increases were observed in groups instructed to increase their sugar intake and who gave dietary recalls that indicated compliance.[43] However, because of the inefficient satiation mechanism, pressure to eat extra sugar is most easily responded to by consuming energy between meals, and this can be extra energy, causing weight gain unless there is also a sufficiently large increase in physical activity. Seen in this light, the weight gain indicates nothing about an effect specifically of sugar: energy consumed between meals could be derived from alcohol, fat, fibre-rich starch and even protein. Also, the consequences of induced extra sugar intake imply nothing about the effect of sugar taken in the absence of the pressure from the investigator.

3.2. Sugar, Artificial Sweeteners and Weight Control

3.2.1. Cutting Down on Sugar to Reduce Weight

Clearly many health professionals are of the opinion that advice to reduce sugar intake should be part of the treatment of obesity. Many anecdotes from patients and indeed from self-help dieters express the feeling that cutting back on sugar has been of help. However, there has been no adequately designed clinical trial. Indeed, that may be a practical impossibility, not least because of the difficulties of undistorted estimation of dietary intakes and the impossibility of disguising sufficient reduction in sugar intake.

Rather than asking directly for an opinion or attempting to measure the effect of an intervention, Lewis and Booth[44] asked 100 female dieters what their weight was before and after their latest diet and what procedures they used in dieting. The respondents were on average 18% above ideal weight. Their weight reports implied an average loss of 15% during the last diet and maintenance of 9% loss to the time of interview. There was no statistically reliable difference in weight losses between those who did and those who did not try to cut down on sugar intake, although the weight lost by the 14 not sugar-cutting during their diet was greater than that by those who were. Relevant also are the facts that those who used low-calorie products and those who did not both showed average weight losses. The four respondents who were still using meal replacements at interview had not lost on average any weight at all since the start of the last diet. That is, persistent use of 'diets' at main meals may be a relative hindrance in dieting. (Alternatively, respondents who knew that they were failing to slim were resorting to these products as a desperate but irrational gesture. As already illustrated, an experimentally uncontrolled relationship by itself does not identify the causal direction(s) involved in the association.)

A majority of the dieters also professed to cut back on intake of starch and of fats. However, unlike the sugar and starch reductions, fat reduction showed a strong statistical association both with slimming (16·5% loss in fat-cutters) and with maintenance of lower weight (10·5% loss). Regular exercise also helped steady-weight maintenance significantly, even though persistence in taking exercise was avowed by only one-third of the respondents.

Thus the evidence for effectiveness of exercising and of cutting out fats, and for the counter-productiveness of meal replacements, shows that the method is sensitive to both positive and negative associations between the use of strategies and success at weight loss. Therefore, we are justified in concluding that the lack of such an association in the cases of indiscriminate cut-down on sugar and the use of low-calorie products is evidence that these strategies are ineffective for weight loss and reduced-weight maintenance after the period of deliberate dieting.

3.2.2. Using Intense Sweeteners to Lose Weight

Dieters and health professionals commonly 'feel' that intense sweeteners are useful in weight control but the heaviest weight of opinion does not amount to evidence. Experimental evidence that has been adduced to the issue is considered in detail below, but no field experiment or adequately-controlled clinical trial has been carried out. Objective use of multiple associations among survey items at the individual level has so far been very limited. Several surveys have at least assessed separately both weight loss and the use of intense sweeteners. Often no relationship is found.[45,46] The Parhams[47] found that saccharin users reduced sugar and starch intake but still ate sugar-containing foods and were more overweight than non-users. In a recent comparison of long-term users and non-users of intense sweeteners (largely saccharin) who had not changed major habits likely to affect weight (such as smoking), the users were no more likely to report a loss of weight over the previous year, and were actually somewhat more likely to gain weight.[48] Other aspects of these data again support attribution of this association to an effect of weight on behaviour rather than the reverse direction of causation. The higher the relative weight of these older American women, the more likely they were to be using artificial sweeteners, and frequently eating vegetables, chicken and fish. This presumably reflects the conventional effort to avoid 'fattening foods'. Such reliance on food selection rather than habit change is unlikely to be productive, particularly if only the consumption of energy in and with drinks between meals is most likely to be fattening. The weight lost by the heaviest women was greater in the users of artificial sweeteners, but these are presumably the people who are most motivated to use all means to lose weight, some of which might be effective even if reliance on saccharin was counterproductive.

So there is no encouragement in the existing data to think that an indiscriminating advocacy of the substitution of intense sweeteners for sugar and other caloric bulk sweeteners has the justification it is generally assumed to have, both by health professionals and by the food industry.

3.3. Do Diet Drinks Make You Hungry?

3.3.1. Compensation for Loss of Sugar Energy

It would be realistic of those who suppose that a general substitution of intense for bulk sweeteners would reduce daily energy intake, to recognise the possibility that the intense sweeteners would leave a person hungrier than the bulk sweeteners did. Indeed, this possibility has been seen, paradoxically, to be a merit of cutting down on sugar intake to treat or to prevent obesity, because the increase of consumption of 'other energy-containing foods' (*sic*) would reduce the supposed risk of nutrient dilution by sugar.[49]

However, if the substitution were confined to occasions when the appetite-suppressant effect of sugar did not affect subsequent food intake, then the absence of that satiating effect would not be so likely to encourage compensatory eating.

Both these positions assume that intense sweeteners neither decrease nor increase the motivation to eat. However, it has been suggested recently that intense sweeteners, or aspartame at least, can positively stimulate human hunger.[50] This report is considered in detail because it raises many of the issues about experiments addressing food and health.

3.3.2. Hunger Ratings after Sucrose, Aspartame and Water

Blundell and Hill[50] briefly reported comparisons of various hunger ratings among three occasions—after drinking 200 ml doses of a 0·73M solution of glucose (50 g), a 0·275 mM solution of aspartame (162 mg), or plain water. The report correctly terms the sweetened fluids 'loads', not 'drinks'. Plain sweetness is not a normal drink flavour in the UK. Furthermore, the intensity of the sweet taste of both the sucrose and the aspartame solutions would be far above the levels of palatable drinks in which sweetness is balanced with other tastes and with a food-flavouring principle such as a fruit.

The expectations created by the sensory characteristics of such experimental loads are known to be strong influences on motivation to eat.[29,30,51] The report states that the loads were 'of equivalent sweetness', but the basis for this claim is not given. Qualitatively, aspartame is the intense sweetener that tastes most similar to sucrose. Quantitative sweetness matches are generally carried out at much lower concentrations of sugars than 25%, and indeed to be practically useful they have to be done in the complex mixture constituting the real drink or food to be marketed. The relative sweetness values widely quoted in the industrial literature are sometimes based on quite irrelevant and bias-ridden comparisons, such as dilution to threshold.

Whether or not the sweetness levels differed, a highly sweet, non-caloric solution is liable to evoke expectations of the taste-satiating and longer-term appetite-suppressant effects of sugary drinks. Yet the report mentions no measures of the expectations generated by the tastes of the loads, let alone effects of such expectations on eating motivation.

The validity and sensitivity of the ratings and also of food-intake tests as predictors of variations in appetite in everyday life are no less crucial. Raw rating values are coherently analysed only within individuals (one person's 'extremely hungry' may not be another's, by any objective comparison). Furthermore, changes in each individual's ratings have to be translated into an objective measure for any empirical generalisations to be possible. In this case, a response criterion like change in food intake is of interest (although stimulus criteria like load energy, sweetener concentration, gastric volume or sugar absorption rate are crucial in other studies[52,53]). The authors do not state what particular rating descriptors they used but the generally phased appetite ratings they report elsewhere are less sensitive measures of energy-induced and cognitive changes in intake of normal foods than are sets of food-specific ratings of momentary acceptability.[30] Moreover, staple foods must be rated. The sugar-water pleasantness rating advocated by Cabanac,[19] and emulated by Blundell and Hill[50] and many others, is a very poor measure of general appetite and its post-ingestional suppression, let alone of the pleasures of eating: sweet water is not a food that most people would want to eat, and nobody can get pleasure out of the repeated tasting and spitting of plain sugar solution.

Finally, the size and the timing of any statistically reliable differences in ratings or food-test intakes have to be related to individuals' behaviour towards food in normal life. A significant difference is uninterpretable objectively unless scaled into, for example, a difference in food intake at the time of a habitual meal or snack on a familiar menu.[30]

Blundell and Hill[50] report that a mean difference in raw rating values between the aspartame solution and water (by implication, the stronger hunger ratings after aspartame) reached a statistically significant level in a *t*-test yielding a *P*-value below 0·05 only. The size of this difference in ratings is not reported. In any case, the ratings difference value would not predict the strength of any effect on eating, because no scaling of rating differences onto predicted food intake differences is mentioned. It is stated that increased food intake after ingestion of calorie-reduced foods has been seen in other studies, but again the size of the effect or its relation to normal eating habit is not given.

The timing of the loads in relation to the subjects' meal pattern(s) is not stated. It may well have been early mid-afternoon, since that significant result was not found in subjects fasted for 4 h—which most conveniently would involve omission of lunch. If the loads, the statistical rating effect and the increased intake in a food test took place between meals, the observations are potentially relevant to the common use of soft drinks, for they are consumed more frequently on drink occasions than with meals or substantial snacks. However, an appetite-stimulating effect within an hour of a drink between meals (or indeed with a meal) is unlikely to coincide with an eating occasion or even with the ready availability of food items to most people. Yet these results have been presented in other media as evidence that 'diet drinks make you hungry'. The paper itself goes much further and concludes with the claim that aspartame may contribute to the eating disorder of bulimia nervosa. The responsibility of a clinical research journal publishing such a statement on the basis of these observations can be questioned. Since the apparently slight difference in effect of aspartame solution and water is absent after a mild restriction of the diet, the extrapolation to ineffective dieting, let alone to clinical eating disorder, is all the more extraordinary.

The paper's less immoderate conclusion is that there was 'residual hunger' after the aspartame load, i.e. that aspartame has appetite-stimulating properties in some circumstances. However, there is no reason to expect observations in this design of experiment to relate to normal uses of diet drinks. The observed pattern of ratings and intakes is what should be generated by cognitive mechanisms known to be activated by the experimental conditions reported. Many subjects would expect a strongly sweet load to produce persistent suppression of interest in eating. The failure of the aspartame solution to do so could well produce sufficient surprise to raise hunger ratings and food intake slightly above the level they reach half an hour or so after a drink of water, which subjects would expect to have at most only a transient filling effect.

This report fails to cite the evidence for such cognitive effects of one-off experimental comparisons between loads or foods differing in energy content, reported and reviewed by Booth *et al.*[30] and apparent also in the patterns of ratings reported by Rolls.[54] Instead, the introduction mentions only the possible physiological effects of the amino acids in aspartame (at higher doses). This is to suggest that the observations should be considered to be specific to aspartame, and not to be general to diet drinks. The authors do not report or cite tests of intense sweeteners not containing amino acids, such as saccharin and acesulfame-K. The results of animal tests would be even less relevant than such human experiments to eating habits in everyday life.

Furthermore, the report fails to cite the evidence which exists that a day's use of genuine soft drinks sweetened with aspartame in fact has no effect on human food intake, compared with presumably more satiating sucrose-sweetened drinks. This evidence comes from the first day of substitution of aspartame for sucrose in normal food and soft drinks in the trio of studies by Porikos and Van Itallie:[55] in each study, whether of an obese group or a normal-weight group, average daily energy intake initially dropped by almost exactly the energy substituted.

Indeed, the results that Porikos and colleagues obtained after many days of replacement of sucrose by aspartame are consistent with other evidence cited earlier in this chapter, that diet drinks could reduce long-term daily energy intake. All the evidence is that aspartame does not stimulate eating.

If the use of intense sweeteners could stimulate daily food intake in everyday life by any known mechanism, it would be via difficulties created for normal slimming by the usually trivial achievements of sweetener substitution in reducing the energy contents of meals, and by mistaken reliance on 'slimming' effects of substitutes excusing the consumption of extra calories—perhaps even more than substituted! There is no reason to think that, by itself, the elimination of drink energy between meals by the use of intense sweeteners would create a net stimulation of daily energy intake.

3.4. Longer-Term Effects of Substituting Aspartame for Sucrose

Porikos and colleagues[55] found that, when they replaced all the sucrose in the diet by aspartame in either obese and non-obese subjects, the initial 25% reduction in average daily energy intake was incompletely 'compensated' within the duration of the experiment. After several days and even after over two weeks, energy intake rose from about 75% to only about 85% of the intake of sucrose-containing diet on the 'baseline' days preceding substitution. This increase from 75% to 85% is a mere 40% 'recovery' of energy intake.

Such results might be thought to be evidence that the sucrose had been stimulating intake, and that its replacement by aspartame removed much of its potential contribution to obesity. If so, then sucrose itself could be blamed as a cause of obesity and its indiscriminate replacement or removal from the diet could be recommended as a method of reducing weight or preventing obesity.

Such interpretations must be rejected as scientifically entirely unfounded. Porikos and Van Itallie[55] themselves repudiate such conclusions and there are several additional reasons for an entirely different interpretation. This focuses on any energy taken between meals—which includes sugars, but also any other macronutrient, and most especially fats and alcohol. Furthermore, the culprit is the timing of the habit, not 'empty calories'.

One consideration is that an incomplete return to 'baseline' energy intake could arise because '100% compensation' would involve a large increase in volume or weight of food intake. A rise in energy intake from the aspartame-sweetened diet from 75% to 100% of any 'baseline' value would have required one-third as much again as initially consumed of the foods and drinks in the aspartame diet. That means that half the subjects would have had to have taken more than one-third extra to attain that group average in a normal distribution. Yet it is notoriously difficult to persuade most people to take substantial extra amounts of food for long periods. What is hard to do deliberately is unlikely to occur unintentionally. Indeed, any large increases in intake or in the desire to eat might well have been noticed, which would have breached the 'blind'.

Two other considerations follow from the fact that the 'baseline' intake for these studies was under circumstances designed to give a high proportion of sucrose in the diet, in order to produce a reduction in energy intake on substituting aspartame that was large enough to bring about measurable changes in a moderate-sized group. As already implied, the average proportion of dietary energy as sucrose was 25%. The average proportion of sucrose in the US diet is thought to be about 11% and so the experimenters may have succeeded in more than doubling their subjects' normal sucrose intake.

The first point that follows is that 11% is 44% of 25%. That is, the eventual 40% compensation on the sucrose-free diet may in fact be complete compensation, back to habitual energy intake. In other words, the experimenters' successful manoeuvres to get a high sucrose intake also succeeded in raising total energy intake above habitual and, when aspartame-substitution enabled the subject to escape the experimenters' inveiglement into raised sucrose-energy intake, they compensated well-nigh perfectly on average (both the obese and the normal-weight groups). The groups were indeed observed to be gaining weight on average during the 'baseline' period, even though this was relatively short. Such an average rate of weight gain is most unlikely outside the experiment, even for the obese. So the results can be accounted for by a mechanism through which the experimental conditions stimulated energy intake, rather than sucrose stimulating intake.

The second, rather stronger, point that follows is that the 15% of energy intake 'uncompensated' after substitution is an estimate of the overeating induced by the experimental design. It may be an overestimate, if there was any difficulty in increasing the volume of food consumed by more than about 13% (i.e. the observed 10% on top of 75%). Much of this experimenter-induced intake is accounted for by the sucrose in the minimum of two soft drinks daily that each subject was required to consume on the pretext of washing down the fake medicine that the experiment purported to be about. Furthermore, sweetened snackfoods were available between served meals, to encourage still higher baseline sucrose intake. The point is that these extra drinks and foods could equally well have contained caloric or non-caloric sweetener or creamer, starch and bulking agent or alcohol and some substitute. The experimenters designed the energy difference to be in sweeteners—sweetness did not induce the energy difference. It would indeed be invaluable to repeat these experiments with fats and non-caloric fat substitutes (or with alcoholic and alcohol-free drinks, if that difference could be disguised in the long term), and with between-meal (and dessert) energy differences properly distinguished from within-meal (especially entree) energy differences.

The required soft drinks were taken by themselves between the served meals. The additional snackfoods, when used, were probably eaten after the evening meal and before sleep (K. Porikos, personal communication). This is consistent with the alternative hypothesis that the subjects were inveigled into higher intake of sucrose (and of energy) with less resistance than otherwise might have been met, because the satiating effect of the absorbed sucrose energy had declined before the next meal was presented.[12]

Most or all of the observed increase in energy intake after replacement of sucrose by aspartame could have come from the sucrose during breakfast and lunch.

3.4.1. Avoiding Slipping in Energy between Meals
As Booth and Mather[35] calculated, a large supper (or late-evening snack) that is partly absorbed during sleep will to that extent by-pass the intake-suppressant effect of rapid absorption.[26,28,31,32] Similarly, as reviewed earlier, a modest dose of starch or sugar is absorbed within an hour or so, and so gastrointestinal and systemic effects, and even after-effects of rapid absorption,[31] will have faded if a sugar-containing drink, cookie or piece of confectionery is eaten between meals or better-balanced snacks. Thus, the energy-based satiety signal will fail to suppress later food intake to produce any degree of compensation if readily assimilable energy is taken in modest amount more than an hour or so before a normal eating occasion. Hence, non-caloric drinks and non-caloric solid items to nibble with them could enable a relatively easy reduction in energy intake by those who habitually take more energy at such times than they expend over the day, whether persuaded by experimenters or by availability and convention in their culture.

Analysis of the effects on 'baseline' and 'compensation' of different individual patterns of food choice in the experiment of Porikos and colleagues could well provide some tests of this energy-timing hypothesis of intake-control efficiency. However, as is unfortunately almost universal practice both in experiments and in surveys, only group data and univariate analyses have so far been presented, although individualised analysis is now being undertaken. The numbers of subjects may be too small to use the power of multivariate analysis, but specific hypotheses might be testable by a correlational approach.

4. HOW TO RESOLVE THE ISSUE AND WHY WE MUST

4.1. How to Evaluate Low-Calorie Sweeteners

Ultimately the question is what happens in everyday life. As we have seen, even quite naturalistic experiments cannot be extrapolated validly. So we must turn to methods that have been little used as yet outside psychosocial research but have been statistically practicable for some time.[56] These methods apply in areas where there is some causal theorising around which to design questionnaires and analyse individuals' reports. Yet it is crucial that the data be elicited with minimal distortion of respondents' records of their everyday habits.

Relevant theory, such as the usefulness of energy-free drink occasions, can increase the power even of simple correlational analysis and path modelling of interview data.[44] The proposal that intense sweeteners are useful in weight reduction only when they replace most of the energy habitually consumed on drink occasions, can be evaluated conclusively using suitably analysed data on weight history, changes in the amount of energy habitually consumed in and with drinks, and other habit-changes likely to affect weight change.

4.2. Current Marketing Practices

The manufacturers of intense sweeteners, and of beverage and food products containing them, regard the provision of such products as a service to those who wish to use them in their effort to control body weight. They meet a demand for a sweet taste while worrying less about putting on weight.

Nevertheless, major manufacturers and distributors often make no claim that the 'artificially sweetened' product will itself cause weight loss or prevent weight gain. If they do claim that a product may be of help in weight reduction, they are legally required in the UK, for example, to add the proviso that it be used with a calorie-controlled diet. This is scientifically legitimate but reduces the claim to a tautology. A properly informative statement would say that there is no evidence whether or not the product can be of help in moderating intake and contributing to weight control. This chapter has shown that it is unlikely that users of intense-sweetened products know how to use them to help reduce energy intake. It would be irresponsible to provide nutrition information and cite the facts of thermodynamics while ignoring the influences such food labelling may have on dietary behaviour.

This labelling regulation has the odd consequence that the caveat can suggest a claim without any claim being made. Furthermore, permitted labels such as 'low fat' or 'reduced energy' are taken to imply usefulness in weight control that has not been demonstrated. Brand names, health-oriented tags and advertising slogans like 'Diet...' or 'light' convey similar implications to many consumers.

The legitimacy of allowing the consumer to act on such assumptions would be affected if there were clear evidence one way or the other on the effects of common usage of such products on reducing of overweight or preventing of unhealthy weight gain. The methods are now available, and testable hypotheses have been formulated, so it is in the public interest for the issues now to be resolved by appropriate research.

ACKNOWLEDGEMENTS

I am indebted to the Nutrasweet Company (USA), Slim Dietetics Products Ltd (Brazil), Hoechst AG (FRG), Hermes Sweeteners (Switzerland) and Boots plc (UK) for their various most helpful responses to my request for the information available to date concerning the role in weight control of aspartame, acesulfame-K and saccharin, respectively. Of course, I bear the sole responsibility for the use to which I have put their responses in writing this chapter.

REFERENCES

1. DESOR, J. A., MALLER, O. and TURNER, R. (1973). *J. Comp. Physiol. Psychol.* **84**, 496.
2. CROOK, C. K. (1976). *J. Exp. Child Psychol.*, **21**, 539.
3. GRILL, H. J. and NORGREN, R. (1978). *Brain Res.*, **143**, 281.
4. STEINER, J. E. (1973). In: *Taste and Development: The Genesis of Sweet Preference*', J. M. Weiffenbach (Ed.), US Govt Printing Office, Washington, DC.
5. PLINER, P. (1982). *Appetite*, **3**, 283.
6. BIRCH, L. L. (1980). *Child Dev.*, **51**, 489.
7. BOOTH, D. A. (1985). *Ann. NY Acad. Sci.*, **443**, 22.
8. BIRCH, L. L., BILLMAN, J. and RICHARDS, S. S. (1984). *Appetite*, **5**, 109.
9. BOOTH, D. A. (1977). *Psychosom. Med.*, **39**, 76.
10. BOOTH, D. A. (1980). In: *Obesity*, A. J. Stunkard (Ed.), W. B. Saunders, Philadelphia.
11. BOOTH, D. A. (1986). *Appetite*, **7**, 236.
12. BOOTH, D. A., CONNER, M. T. and MARIE, S. (1987). In: *Sweetness*, J. Dobbing (Ed.), Springer Verlag, Berlin.
13. MOSKOWITZ, H. R. (1972). *J. Appl. Psychol.*, **56**, 60.
14. BECK, C. I. (1978). In: *Low Calorie and Special Dietary Foods*, B. K. Divivedi (Ed.), CRC Press, Boca Raton, FL.
15. BOOTH, D. A., THOMPSON, A. L. and SHAHEDIAN, B. (1983). *Appetite*, **4**, 301.
16. BOOTH, D. A., CONNER, M. T., MARIE, S., GRIFFITHS, R. P., HADDON, A. V. and LAND, D. G. (1986). In: *Measurement and Determinants of Food Habits and Food Preferences*, J. M. Diehl and C. Leitzmann (Eds), University Department of Human Nutrition, Wageningen.
17. THOMPSON, D. A., MOSKOWITZ, H. R. and CAMPBELL, R. G. (1976). *J. Appl. Physiol.*, **41**, 77.
18. WITHERLEY, S. A., PANGBORN, R. M. and STERN, J. S. (1980). *Appetite*, **1**, 53.
19. CABANAC, M. (1971). *Science*, **173**, 1103.
20. CONNER, M. T., HADDON, A. V. and BOOTH, D. A. (1986). *Lebens.-Wiss. -Technol.*, **19**, 486.
21. WITTGENSTEIN, L. (1953). *Philosophical Investigations*, Blackwells, Oxford.
22. TORGERSON, W. S. (1958). *Theory and Methods of Scaling*, Wiley, New York.

23. Booth, D. A. (1987). In: *Eating Habits*, R. A. Boakes, M. J. Burton and D. Popplewell (Eds), Wiley, Chichester.
24. Stunkard, A. J., Sorensen, T. I. A., Hanis, C., Teasdale, T. W., Chakraborty, R., Schull, W. J. and Schulsinger, F. (1986). *New Eng. J. Med.*, **314**, 193.
25. Booth, D. A. and Toase, A.-M. (1983). *Appetite*, **4**, 235.
26. Booth, D. A., Campbell, A. T. and Chase, A. (1970). *Nature* (*London*), **228**, 1104.
27. Cabanac, M. and Fantino, M. (1977). *Physiol. Behav.*, **18**, 1039.
28. Booth, D. A. (1981). *Br. Med. Bull.*, **37**, 135.
29. Booth, D. A., Lee, M. and McAleavey, C. (1976). *Br. J. Psychol.*, **67**, 137.
30. Booth, D. A., Mather, P. and Fuller, J. (1982). *Appetite*, **3**, 163.
31. Booth, D. A. and Jarman, S. P. (1976). *J. Physiol., London*, **259**, 501.
32. Booth, D. A. (1972). *J. Comp. Physiol. Psychol.*, **78**, 412.
33. Geliebter, A. A. (1979). *Physiol. Behav.*, **22**, 267.
34. Hunt, J. N. and Stubbs, D. F. (1975). *J. Physiol., London*, **245**, 209.
35. Booth, D. A. and Mather, P. (1978). In: *Hunger Models: Computable Theory of Feeding Control*, D. A. Booth (Ed.), Academic Press, London.
36. Booth, D. A. and Davis, J. D. (1973). *Physiol. Behav.*, **11**, 23.
37. Booth, D. A. (1979). In: *Chemical Influences on Behaviour*, K. Brown and S. J. Cooper (Eds), Academic Press, London.
38. Smith, G. P. and Gibbs, J. (1979). *Prog. Psychobiol. Physiol. Psychol.*, **8**, 179.
39. Richardson, J. F. (1972). *Br. J. Nutr.*, **27**, 449.
40. Walker, A. R. P. (1974). *S. Afr. Med. J.*, **48**, 1650.
41. Garn, S. M., Solomon, M. A. and Cole, P. E. (1980). *Ecol. Food Nutr.*, **9**, 219.
42. Walker, A. R. P., Walker, B. F., Jones, J., Walker, C. and Ncongwane, J. (1982). *Am. J. Clin. Nutr.*, **36**, 643.
43. Werner, D., Emmett, P. M. and Heaton, K. W. (1984). *Gut*, **25**, 269.
44. Lewis, V. J. and Booth, D. A. (1986). In: *Measurement and Determinants of Food Habits and Food Preferences*, J. M. Diehl and C. Leitzmann (Eds), University Department of Human Nutrition, Wageningen.
45. McCann, M. B., Trulson, M. F. and Stulb, S. C. (1956). *J. Am. Dietet. Assoc.*, **32**, 327.
46. Adams, P. H. S. O., Grady, K. E., Lund, A. K., Mukaida, C. and Wolk, C. H. (1983). *J. Am. Dietet. Assoc.*, **83**, 306.
47. Parham, E. S. and Parham, A. R. (1980). *J. Am. Dietet. Assoc.*, **76**, 560.
48. Stellman, S. D. and Garfinkel, L. (1986). *Prev. Med.*, **15**, 195.
49. Royal College of Physicians Working Party on Obesity. (1983). *J. Roy. Coll. Physns*, **18**(2).
50. Blundell, J. E. and Hill, A. J. (1986). *Lancet*, **i**, 1092.
51. Wooley, O. W., Wooley, S. C. and Dunham, R. B. (1972). *Physiol. Behav.*, **9**, 765.
52. Booth, D. A. (1981). *Appetite*, **2**, 237.
53. Kissileff, H. R. (1984). *Neurosci. Biobehav. Rev.*, **8**, 129.
54. Rolls, B. J. (1987). In: *Sweetness*, J. Dobbing (Ed.), Springer Verlag, Berlin.
55. Porikos, K. P. and Van Itallie, T. B. (1984). In: *Aspartame: Physiology and Biochemistry*, L. D. Stegink and L. J. Filer (Eds), Marcel Dekker, New York.
56. Jöreskog, K. G. and Sörbom, D. (1981). *LISREL VI: Analysis of Linear Structural Relationships*, Scientific Software, Mooresville, IN.

INDEX